KEY TECHNOLOGIES OF METRO CONSTRUCTION IN HARD ROCK STRATUM

Other Related Titles from World Scientific

Modern Railway Engineering Consultation: Methods and Practices
by Ying Zhu and Lie Chen
ISBN: 978-981-3238-87-9

Practical Railway Engineering
Second Edition
by Clifford F. Bonnett
ISBN: 978-1-86094-515-1

KEY TECHNOLOGIES OF METRO CONSTRUCTION IN HARD ROCK STRATUM

Quanwei Liu

Qingdao West Coast Rail Transit Co., Ltd, China

World Scientific

NEW JERSEY • LONDON • SINGAPORE • BEIJING • SHANGHAI • HONG KONG • TAIPEI • CHENNAI • TOKYO

Published by

World Scientific Publishing Co. Pte. Ltd.
5 Toh Tuck Link, Singapore 596224
USA office: 27 Warren Street, Suite 401-402, Hackensack, NJ 07601
UK office: 57 Shelton Street, Covent Garden, London WC2H 9HE

Library of Congress Cataloging-in-Publication Data
Names: Liu, Quanwei, 1977– author.
Title: Key technologies of metro construction in hard rock stratum /
Quanwei Liu, Qingdao West Coast Rail Transit Co., Ltd, China.
Other titles: Ying yan di ceng di tie xiu jian guan jian ji shu. English |
880-02 Ying yan di ceng di tie xiu jian guan jian ji shu.
Description: Hackensack, NJ : World Scientific, [2020] | Translation of: Ying yan di ceng di tie xiu jian guan jian ji shu | Includes bibliographical references and index.
Identifiers: LCCN 2019058094 | ISBN 9789813278080 (hardcover) | ISBN 9789813278097 (ebook)
Subjects: LCSH: Subways--China--Qingdao--Design and construction. |
Underground construction--Technological innovations--China--Qingdao. |
Tunneling--China--Qingdao. | Rock excavation--China--Qingdao.
Classification: LCC TF847.Q56 L58 2020 | DDC 625.4/2095114--dc23
LC record available at https://lccn.loc.gov/2019058094

British Library Cataloguing-in-Publication Data
A catalogue record for this book is available from the British Library.

Originally published by China Communications Press Co., Ltd.

For any available supplementary material, please visit
https://www.worldscientific.com/worldscibooks/10.1142/11206#t=suppl

Typeset by Stallion Press
Email: enquiries@stallionpress.com

Editorial Committee

About the Author

Liu Quanwei is a researcher majoring in planning construction and safety quality management of urban metro. Under the tutelage of Academician of Chinese Academy of Engineering Wang Mengshu at the Beijing Jiaotong University, he obtained his Ph.D. in tunnel and underground engineering. He is currently a postdoctoral student in Shandong University. He was formerly an engineer at China Railway 14th Bureau Group Co., Ltd. engaged in construction technology management in important construction projects such as Qinghai-Tibet Railway and Nanjing Yangtze Tunnel. In 2010, he was transferred to Qingdao Urban and Rural Construction Municipal Engineering Management Office as director of the Metro Construction Administration Division and established the safety supervision system of metro in hard rock stratum. In 2014, he joined Qingdao Metro and worked as vice-general manager of Qingdao West Coast Rail Transit Co., Ltd., mainly engaged in production, safety and quality administration of metro construction.

Thanks to his long-term commitment to theoretical research, technological innovation and standardization construction in Qingdao Metro characterized by dual-structure strata and complex geological conditions, he has summarized a whole set of mature construction technologies and methods, promoted the wide application of some technological achievements such as arch-cover method of subsurface excavated station, established a standardization system of safe and civilized metro construction, and presided over two local standard specifications of Qingdao, namely, *Code for Safe and Civilized Construction Management of Qingdao Urban Rail Transit Engineering* and *Standardized Atlas for Safe and Civilized Construction of Qingdao Urban Rail Transit Engineering*.

Foreword I

Since China completed the construction of the first urban metro or metro — Beijing Subway Line 1 — in 1965, there have been 43 cities with metros up to 2016, and the total operating mileage exceeded 4,000 km. Especially in recent years, China's metro construction has aroused the world's attention with its extraordinary fast speed and large scale. In the past 50 over years, many metro constructors endured great hardships in pioneer work, summed up their experience into a relatively integrated standard system of design and construction to provide technological support for the large-scale metro construction in China. However, these standards and codes used to be stipulated on the empirical basis of metro construction in large cities such as Beijing and Shanghai with Quaternary soft soil as the main stratum. With the increasing scale of metro construction, for cities with hard rock geology such as Qingdao, Chongqing and Dalian, previous standards, codes and methods are no longer suitable for the geological environment and require updating urgently to provide new technologies, schemes and standards to guide metro construction in these cities.

Based on the technologies and engineering management experience of Qingdao Metro construction, the book systematically introduces the key technologies of urban metro construction in hard rock stratum. For every construction method, its technological development history, construction principle and key points of technical program are briefly introduced and followed by typical cases and summaries. With great practicality and operability, it can serve as a good technological assistance for those who are engaged in construction, supervision, design and management of metro construction. The book is concise and clear in writing by omitting the derivation and theoretical calculation of the complicated formulas and referring to necessary theories and principles. With detailed and accurate data, figures and tables obtained from engineering practice, the conclusions are quite meaningful, reliable and operable as a summary of the author's considerable experience in construction management. For example, dual structure stratum should be avoided and the metro deep burial scheme should be adopted; subsurface excavated station construction technology should be chosen when the surrounding environment of the metro station is complicated to ensure a minor impact on the surrounding traffic, rapid construction speed and good engineering benefit. These concepts are all right and should be recognized, promoted and applied in the construction. Observed from the construction of Qingdao Metro Line 3, Line 2 and Line 1, the construction security and speed were significantly improved with the increasing of burial depth of stations.

The author is my Ph. D. graduate with high theoretical level and long-term practice in construction management. He has participated in some important projects such as Nanjing Yangtze Tunnel, and gained solid foundation in construction technology and rich experience in project management. Other members of editorial committee involved in compiling the book are mostly young, experienced, innovative and enterprising technical workers of construction or research frontline. It is believed that this practical book will be truly helpful to enhance the development of metro construction technology of China, and I am earnest to recommend it to more engineers and technical staff.

Wang Mengshu
Academician of Chinese Academy of Engineering
Summer, 2017

Key Technologies of Metro Construction
in Hard Rock Stratum

Foreword II

As an important central city along the eastern coast of China, Qingdao is boosting metro construction intensely to solve the urban traffic congestion problem. Qingdao has a long-term plan of constructing 18 metro lines with a total mileage of 814.5 km, among which 6 lines with a total mileage of 305.8 km currently under construction. Such a construction scale is at the foremost in China. Construction of the test section of Qingdao Metro Line 3 was initiated in 2009, and its northern part was put to trial operation since December 16, 2015, which marks Qingdao, a city with a history of over one hundred years, truly and formally stepping into "Metro Era". By the year 2021, Qingdao will boast a fundamental metro network covering the whole city with a total mileage of over 300 km.

Qingdao is a beautiful coastal city as well as a national famous historico-cultural city well known for its blue sea, azure sky and undulant mountains dotted with red tiles and green trees. During large-scale metro construction, it is equally important to protect our environment and reduce the impact on city operations. Qingdao is also a city with very distinct geological features of dual structure, i.e., soft soil and weathered rock in the upper stratum underlain by hard rock (granite). There must be corresponding design and construction schemes based on such special geological environment. During the course of construction, there will be a lot of blasting operations; meanwhile, the abundant ground water in coastal areas poses a lot of threats to construction and future operations. Therefore, it is necessary to summarize the core concepts of metro construction as soon as possible to guide design and construction of metro in hard rock areas.

The author has experienced the construction of the first line of Qingdao Metro. In this book, with seven years' construction experience, he has provided a summary of key technologies and construction concepts for metro construction in hard rock stratum and answered applicable work methods in different geological conditions and environments. Each method has its key points in design and construction with the aid of cases. His viewpoints are very exact and convincing. Most of these ideas have truly originated from practice, therefore, the publication of this book is of great reference value and will provide significant guidance for massive metro builders, especially those working in hard rock geological areas. Hopefully, it will help to promote the healthy and orderly development of China's metro engineering construction.

Jia Funing
Executive deputy general commander of Qingdao
Metro Engineering Construction Headquarters
CCP secretary and board chairman of Qingdao Metro Group
July, 2017

Preface

With the acceleration of urban modernization in China, urban metro construction is in full swing. By the year 2016, 30 cities were involved in rail traffic which came to 133 lines with a total mileage of 4,125.8 km. It is estimated that by the year 2020, the total mileage of China's urban rail transit will exceed 8,000 km. Geological conditions in large and medium-sized cities in China vary significantly, but they generally fall into two categories. The first category is mainly soft soil stratum of the quaternary system, e.g., Beijing, Shanghai and Chengdu. Metro construction in these cities has been going on for a long time. Beijing metro was initiated in 1965. After more than 50 years of construction endeavors, Beijing metro has accumulated quite rich technical and management experience and developed relatively mature technical standards and construction techniques, offering favorable guidance to the large-scale metro construction in cities with similar geological conditions. The second category is mainly hard rock stratum, e.g., Qingdao, Chongqing and Dalian. In recent years, it has sprung up and developed rapidly. However, its construction technologies still refer much to the metro construction experience in soft stratum, its design and construction processes and methods are all under exploration, and no uniform design standards and mature construction techniques have been stipulated. It is urgent to summarize technical standards and construction methods to guide the metro construction in hard rock stratum. The author is engaged in this field for seven years, and has encountered a lot of difficult problems such as disturbance to residents, blasting, damage in structures, upper-soft and lower-hard dual stratum, treatment of hazardous geological phenomena. On the basis of systematic summarization of the research results in recent years, he has proposed a set of key technologies for urban metro construction in hard rock stratum, which is hopefully to provide a reference for the metro construction in cities with similar geological conditions.

The book is divided into six chapters. In Chapter 1, history and current situations of metro development in China were illustrated, and stratal characteristics of China's representative metro cities were summarized. Development of drilling-blasting tunnel construction technology in hard rock stratum was analyzed, and characteristics of typical hard rock stratum in Qingdao area were introduced. In Chapter 2, drilling-blasting tunnel construction technology in hard rock stratum was illustrated in details. The following aspects were analyzed: design and construction of vertical and inclined shafts, excavation, support, waterproofing and secondary lining construction technologies of tunnel. Key technologies of drilling-blasting method were mainly summarized for urban metro construction in hard rock stratum. In Chapter 3, open-cut station construction technology was illustrated. Based on the hard rock geological conditions of Qingdao, corresponding supporting types and waterproofing measures in deep foundation pit were studied. Soilnail wall and pile-anchor support were specially prosed, their application scope, key points of construction and engineering applications were analyzed, and key construction technologies of open-cut station in hard rock stratum were summarized. In Chapter 4, construction technology of subsurface excavated station was illustrated. According to construction features of subsurface excavated stations in Qingdao, corresponding construction method and waterproofing-drainage system were explicated for hard rock stratum, and key construction technologies of subsurface excavated station in hard rock stratum were summarized. In Chapter 5, reinforcement measures of poor geological section were illustrated. Reinforcement methods of strengthening ground surface and tunnel in Qingdao hard rock stratum were summarized, their applicability and feasibility were studied, and key reinforcement technologies of poor geological section in hard rock stratum were proposed. Chapter 6 mainly dealt with green construction and standardized site construction. In this chapter, development history of green construction and key management points of "Four savings, one environmental protection" were illustrated. Standardized design should cover construction preparation from layout of site, main facilities, office and living area to temporary electricity. The construction standardization of open-cut station and subsurface excavated station were clearly put forward.

Main compilers of different chapters are listed as follows: Chapter 1 was compiled by Liu Quanwei and Wang Yunlong, Chapter 2 by Liu Quanwei and Liu Songwang, Chapter 3 by Yang Zhongnian and Huang Xixi, Chapter 4 by Liu Quanwei and Liu Songwang, Chapter 5 by Wang Hongbo and Liu Yankai, and Chapter 6 by Liu Quanwei and Liu Linsheng, respectively. The whole book was overally arranged, checked, revised, reviewed and approved by Liu Quanwei.

Gratitude should be addressed to academician Wang Mengshu for his careful guidance on the application of a great number of research findings involved in this book. He has been to Qingdao Metro for many times to survey and help solve various difficult problems on construction site. He guided the important theories and schemes of metro construction in hard rock cities, suggested deep burial scheme in hard rock stratum, and advocated learning from Moscow metro. Hereby, I would like to extend my heartfelt thanks to academician Wang for his caring and help. The compilation of this book was also supported by many peers from Qingdao Metro Group, Shandong University, Qingdao University of Technology, First and Third Harbor Engineering Co., Ltd. of CCCC, who deserve special thanks from the author.

Although we've made the greatest efforts in compiling this book, there must be some inevitable errors, omissions or inappropriate expressions in it. We're earnest to receive any criticism or correction advice from massive experts and readers.

Liu Quanwei

July, 2017

Contents

Chapter 1

Brief Introduction

With the rapid development of China's economy and society, comes higher requirements for traffic and transportation by its people. Urban metro is the ideal traffic mode to implement our national policy of energy conservation and emission reduction, develop low-carbon economy and save energy, and improve the living environment. The recent two decades have been a period of great development in China's urban rail transit (URT or metro) construction. By 2016, there had been 30 cities with URT in China. It is estimated that the number of metro cities will exceed 45 with a total mileage of 8,000 kilometers (km) by the year 2020. Metro has a tremendous impact in supporting urban development. Not only does it solve urban traffic problems, but also stimulates economic growth from its source and accelerates the formation of new economic growth belts. In this way, URT will assemble the population and industries along metro lines, guide the expansion of urban space and improve the overall urban economic layout. Therefore, the development and construction of URT is an undeniable accelerator in transforming urban development patterns, adjusting urban economic structures and enhancing urban construction levels.

Urban metro construction is developing rapidly, which poses higher requirements for metro construction technology. China's metro construction was initiated in such first-tier cities as Beijing and Shanghai, where a set of relatively mature metro construction technologies have been developed. However, metro construction technologies have prominent regional characteristics, closely related to geological conditions of different regions. The current metro construction technologies mainly revolve around geology of soft rock stratum. Though metro construction in hard rock stratum is seemingly the same as mountain tunnel, there is still a lack of engineering experience reference due to its particularity in construction environment. Based on metro construction in hard rock stratum of Qingdao, this book generalized various construction technologies such as drilling and blasting construction technology, open-cut station construction technology, subsurface excavated station construction technology, grouting reinforcement technology in adverse geological section, standardized metro construction technology and so on.

1.1 Overview of metro construction

1.1.1 *Metro system development of the world*

In 1843, an Englishman named C. Pirsson put forward a suggestion of constructing an underground railway in England. In 1860, construction of the first metro line in the world started in London, adopting open-cut station method and single arch brick structure. The metro line was completed and opened for service on January 10, 1863. It had a total mileage of 6.4 km and was driven by steam locomotive.

The birth of the world's first metro line provided valuable experience for developing public transportation in densely populated metropolises. Particularly, with the successful research of electric locomotives in 1879, the underground passenger transport environment and service conditions were improved at an unprecedented level. As a result, metro construction flourished vigorously, and some metropolises constructed underground railways or metro lines successively. From 1863 to 1899, there were seven cities in five countries with metro lines, e.g., London and Glasgow in UK, New York and Boston in USA, Budapest in Hungary, Vienna in Austria and Paris in France.

London has improved its metro construction technologies since it constructed the world's first underground railway or metro line in 1863 (Fig. 1.1), and now its metro system has become one of the most advanced networks in the world. London metro has been making new progress almost every year, in particular since electrification was realized. At present, London metro boasts 12 lines with a total mileage of 410 km, comprising

Fig. 1.1. First metro line in London.

Fig. 1.2. New York metro.

171 km of underground tunnels, 275 stations and 419 metro vehicles. Its annual total passenger transport volume has exceeded 850 million.

The first metro of New York was completed in 1867 (Fig. 1.2). With the expansion of New York City, urban population keeps increasing and metro construction is also on the rise. Currently there are 31 metro lines in New York with a total mileage of about 443.2 km, comprising 258 km of underground tunnels, 504 stations and 6,561 metro vehicles. Its annual total passenger transport volume has exceeded 1 billion.

Fig. 1.3. Paris metro station

In 1900, in order to hold the Versailles Exhibition, Paris constructed its first metro line from Bastille to Maillot. Its total length was merely 10 km; however, it laid the foundation for the continuous development and improvement of the Paris metro network. Nowadays there are 14 main lines and two branch lines in the downtown area of Paris, including two loop lines and four lines of VAL vehicles with rubber wheel system. The total mileage of Paris metro lines is about 221.6 km, and underground tunnels account for 175 km. Altogether 308 stations (Fig. 1.3) and 347 metro vehicles are kept running routinely. Its annual total passenger traffic volume has already exceeded 1.2 billion. Reginal express railway (RER) in Paris is highly developed, with a total mileage of 363 km in operation, including 14 km collinear with the metro and 249 km of urban express railway of SNCF (abbr. for Société Nationale des Chemins de Fer: the French national railway system). The annual total passenger transport volume of RER is about 400 million.

From 1900 to 1924, another nine cities in Europe and America began their metro construction, like Berlin and Hamburg of Germany, Philadelphia of USA, and Madrid of Spain.

The first metro of Berlin was opened in 1902. Its urban metro has been extended in all directions until now (Fig. 1.4), and automated running technology has been adopted in some lines. At present, Berlin has 10 metro

Fig. 1.4. Berlin metro.

Fig. 1.5. Madrid metro.

lines, with a total mileage of about 146 km, including 90% underground tunnels, 173 stations, 2,410 metro vehicles. Its annual total passenger transport volume has exceeded 660 million.

Spain is also one of the earliest countries in Europe with metro (Fig. 1.5). The first metro of Madrid was opened in 1919, and it has now developed to 12 metro lines and one branch line, with a total mileage of

(a) (b)

Fig. 1.6. Tokyo metro.

about 281.58 km, comprising 281 stations and 1012 metro vehicles. Its annual total passenger transport volume has exceeded 400 million.

From 1925 to 1949, with the explosion of World War II, security became a priority and metro construction receded, but there were still a small number of cities constructing metro, such as Tokyo and Osaka of Japan and Moscow of the former Soviet Union. The first metro of Tokyo was completed and opened to traffic in 1927 (Fig. 1.6). Though Japan's metro was modeled on European technology, the focus during metro construction was on exploiting its main stations and adjacent public gathering places, which can promote the construction of underground business centers. Since the business centers were connected to the underground stations, the public welfare infrastructure of metro gained new energy and hence achieved good economic and social benefits.

In December 1927, the first metro of Tokyo as well as Asia was opened to traffic. Until now, Tokyo metro boasts 13 lines with a total mileage of abou 312.6 km, which contains 230 stations and 2,450 metro vehicles. Its annual total passenger transport volume has exceeded 2.5 billion, which makes Tokyo one of the cities with the highest metro passenger volume in the world.

Construction of the first metro in Moscow was started in 1932. Its total mileage was about 11.6 km, which contains 13 stations. It was completed and opened to traffic in May 1935. Its construction speed was unprecedented at that time. Even during the World War II, Moscow's metro

(a)

(b)

Fig. 1.7. Moscow metro.

construction was not interrupted. Until now, Moscow boasts 12 metro lines with a total mileage of about 277.9 km containing 171 stations. Its metro has always been recognized as one of the most beautiful metros in the world with the title of "Underground Art Palace". The gorgeous and elegant buildings of Moscow's metro stations differ from each other in pattern and shape (Fig. 1.7). All ending stations are connected to bus lines, trolleybus lines and light rail transit systems, and some of them are connected to railway stations, which provide convenient transfer for passengers. Currently Moscow's metro system boasts about 4,000 metro vehicles, and its annual total passenger transport volume has exceeded 2.6 billion.

After World War II, the world's metro construction was in a period of boom for 24 years from 1950 to 1974. During this period, there were about 30 cities constructing metro, such as Toronto and Montreal of Canada, Rome of Italy, San Francisco and Philadelphia of USA, Leningrad and Kiev of the former Soviet Union, Nagoya and Yokohama of Japan, Seoul of South Korea and Beijing of China.

The development history of the world's metro is also the advancement history of underground construction technologies. For instance, the first London metro line was constructed entirely using the open-cut excavation method, while subsurface excavation method is widely adopted at present. Traditional surrounding rock relaxation theory used to be guidance in strong support construction method, while theory of joint action between surrounding rock and support is applied in new Austrian tunneling method.

Average burial depth of Seoul metro is 100 meters (m), maximum burial depth of Mosco metro is 90 m, while minimum burial depth of Tianjin metro and Beijing metro is less than 6 m. Mechanical excavation has gradually substituted blasting excavation. All these achievements indicate that metro construction technologies have also undergone a great number of revolutionary changes.

1.1.2 *Development of China's metro*

Beijing began to construct the first metro line in China with dual functions of traffic and civil air defense in 1960s and it was open to traffic in January, 1969. Thus the construction of China's metro was initiated. However, metro construction at that time was in the beginning stage was relatively simple. Directed by the guiding thought of "Revolving around combat readiness and meanwhile giving consideration to traffic", its purpose was to construct a metro line mainly based on civil air defense facilities. In the 1980s, Beijing constructed its second metro line with a total length of 54 km, while Tianjin constructed its Metro Line 1 with a total length of 7.4 km and opened in August 1980. This alleviated the urban traffic congestion to some extent, but still could not form an urban rail transit network.

During the rapid development period since the implementation of the Reform and Opening-up Policy, the development of urban rail transit in China has undergone three successive stages: (1) Initiation, (2) Metamorphosis and (3) Booming.

(1) Initiation stage — from the end of 1980s to the middle of 1990s

Marked by the construction of Shanghai Metro Line 1 (21 km), Beijing Fuxingmen-Bawangfen Line (13.6 km), Transformation Project of Beijing Subway Line 1, Guangzhou Metro Line 1 (18.5 km), truly traffic-oriented metro projects were started in China. Inspired by the metro construction in Shanghai and Guangzhou, a large number of cities, including Shenyang, Tianjin, Chongqing, Wuhan, Shenzhen, Chengdu and Qingdao, began to submit their applications for constructing rail transit projects to the government authorities for approval.

(2) Metamorphosis stage — from 1995 to 1998

Though the construction of China's urban rail transit was started in 1960s, it was delayed for a long time for various reasons. The technology in professional fields such as vehicle, signal as well as automatic ticket vending and inspection service was advancing very slowly in China. In the early 1990s, when Shanghai and Guangzhou began metro construction, their technical equipment and instruments were basically dependent on imports, which resulted in many negative effects such as high cost (preliminary estimated cost 800 million RMB/km and actual cost 670 million RMB/km), long construction period and high maintenance cost.

Many regions opted for rapid development of metro construction, but did not consider the economical capability and practical needs of social development. Due to such problems as high cost as well as dependence on imports of all vechicles and most of the equipment and instruments, the General Office of the State Council issued an official notification informing that except for the metro projects under construction in Beijing, Shanghai and Guangzhou, all other metro projects should be suspended. The notification also proposed the development planning of metro networks and the localization of high-tech equipment. During this period of nearly three years, the country suspended the approval of urban metro transit projects.

(3) Booming stage — since 1999

At the end of 1997, the National Development and Reform Commission studied the implementation plan for localization of urban rail equipment. A batch of engineering projects were proposed as the experimental projects of localization, such as the First-stage Project of Shenzhen Metro Line 1 and 4 (19.5 km), Shanghai Metro Pearl Line (current Line 3, 24.5 km), Guangzhou Metro Line 2 (23 km), Nanjing Metro Line 1 (17 km). These four projects were approved in 1998 and hereforth the urban rail transit construction was restarted.

During this period, with the implementation of the active fiscal policy, the country provided the urban rail transit construction with a strong capital support. Meanwhile, with the technology imports and cooperation between international advanced manufacturing enterprises and domestic

Table 1.1. Development survey of urban rail transit in China (by the end of 2016).

City	Operation (line)	Operation (km)	Planning for 2020 (line)	Planning for 2020 (km)
Beijing	19	574	30	830
Tianjin	5	166	14	513
Shanghai	14	617	18	800
Guangzhou	10	308	17	677
Shenzhen	8	285	16	596
Chongqing	4	213	10	500
Nanjing	7	258	13	540
Wuhan	5	185	14	400
Qingdao	1	25.2	11	470

enterprises, localization of the urban metro transit vehicles and equipment was realized. Therefore, the cost of urban rail transit projects was greatly reduced, and thrived in China (Table 1.1).

1.2 Introduction to metro construction technology

1.2.1 *Overview of construction methods of underground stations*

1.2.1.1 *Open-cut station method*

Open-cut station method is a preferred method of metro construction where the ground transportation and environment permit. It has some advantages such as multiple operation surfaces, rapid speed, short duration, easy feasibility to ensure construction quality and low project cost. However, because it poses a significant disturbance to city life, its application is limited by various factors, especially when the local traffic and environment do not permit; therefore, it is usually substituted by subsurface excavated station method or new Austrian tunneling method. Open-cut station method is suitable for the conditions of shallow buried station

Fig. 1.8. Open-cut station method.

with wide construction site and large construction space, e.g., underground complex stations with transfer station, underground store, rest and entertainment venues, parking garage.

The construction with open-cut station method mainly includes four parts: construction of envelope structure, soil excavation in the station, construction of main station structure and refilling of overlying soil, as well as restoration of pipelines (Fig. 1.8). According to different geological conditions, dimension of station structure and foundation depth, envelope structure of the open-cut station method can adopt some structures such as underground diaphragm wall, anchor line (cable), drilled pile sealed up with chemical churning pile to sealing up, SMW (Soil Mixing Wall) cement soil with structural steel. At present, the open-cut station method still occupies the largest proportion in the construction of the underground station.

1.2.1.2 *Subsurface excavated station method*

Subsurface excavated station is a method for constructing the metro station after the construction of temporary (or perpetual) pavement system to maintain ground traffic. According to the composition and construction order of the temporary road system, there are generally two kinds of methods for subsurface excavated station: (1) Subsurface excavated station and (2) Cover and cut-top down.

Fig. 1.9. Subsurface excavated station in metro station.

(1) Subsurface excavated station (Fig. 1.9)
This construction method excavates earth from top to bottom designed elevation, then builds the structure from the bottom to top after building the temporary road or support to maintain the ground traffic. The temporary road system contains girder, pavement plate and envelop enclosure. It will be demolished after the station construction is completed. When the cover plate is only partially laid out, it is known as half subsurface excavated station.

The job program and construction plan of the subsurface excavated station procedure is completely consistent with the cut-cover method except for the temporary pavement system. Its characteristics are:

(a) Closed road time is relatively short and can be implemented in phases. Subsequent construction will have little effect on ground transportation once the road surface is restored or the excavation system is completed.

(b) The project cost is relatively high for the excavation system. The construction period is longer than the excavation method for the excavation is in a closed (or semi-closed) top, large mechanical applications are limited.

(2) Cover and cut-top down method

The order of operation is the opposite of open-cut station method.

This method excavates the earth from top to bottom under the roof to layered construction after excavating the ground to build the roof and vertical bracing structure.

The road system contains station roof, intermediate support, envelop enclosure, which are usually permanent.

Its characteristics are:

(a) The disturbance to the surrounding environment is short. It is effective for preventing the ground from sinking drop, protecting the neighboring buildings and underground utilities.
(b) The main force component of the structure often has dual functions: temporary structure and permanence.
(c) There is need to set up middle vertical temporary support system to bear the vertical load in front of the structure with the side wall.
(d) The connection node between the underground diaphragm wall, middle support column and the baseboard, floor system needs to be processed.

The excavation method solves the problems of cut-cover method affecting road traffic, concealed-digging construction having high risk and long construction period, and is widely applied in urban metro construction. At present, major cities all have construction examples of the subsurface excavated station method in metro construction, with more than 20 of them in construction in 2014 alone.

1.2.1.3 *Undercutting method*

The undercutting method is the construction method of excavation and lining in the ground without affecting the ground transportation.

The main construction methods are center cross-diagram (CRD), double side drift method, hole pile method (PBA) and arch cover method.

The corresponding construction methods should be selected according to the comprehensive comparison of multi-factors such as the site engineering geology and hydrogeology conditions, urban planning requirements, situation of neighboring buildings and roads, site condition, buried depth of the structure, structural style, time limit and civil work cost for the project.

Side drifts-support method is a common way to build a large cross tunnel, but it has low safely for the large disturbance to the earth's surface.

The core of the middle hole method is the CRD method which has high construction safety in accordance with the principle of small block, short steps, quick closing, consolidate step by step. The ground subsidence and the influence range are smaller than the side hole method. However, there are a lot of changes in the middle hole method. It is difficult to restrict the additional displacements generated in the process conversion. In addition, compared with the side hole method, it has more temporary support and waste engineering.

The principle of PBA method is to combine the open-cut station method and subsurface excavated station, in other words, when the foundation pit envelope structure construction is not available on the ground, construct some structures such as retaining pile side and pile bolccarling in the tunnel dug in the ground, to make the enclosure pile, vertical pile head beam, roof arch together constitute pile, beam, arch support frame system, to bear the external load of the construction process. Then, under the protection of the top arch and the side pile, excavate layer by layer. If necessary, set the pre-loading lateral support, construct internal structure. Finally, a permanent host system is formed with the outer edge of pile, initial support of crown arch and the inner secondary lining. Construction methods are in Table 1.2.

1.2.2 *Overview of metro tunnels construction methods*

The earliest recorded underground artificial buildings in China appeared in the early eastern Zhou dynasty (about 700 BC). *Zuo Zhuan* described a very famous historical story about Zheng Zhuang meeting with his mother to dig the earth and spring.

Other ancient civilizations also have many famous ancient tunnels, for example, in 2189 BC to 2160 BC, the pedestrian tunnel under the

Table 1.2. Undercutting construction method.

Construction method	**Diagram form**	**Application condition**
CRD method	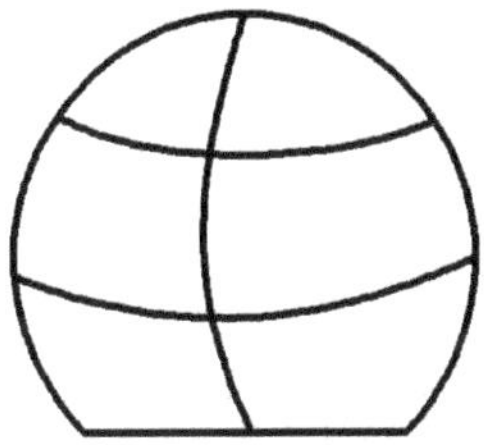	Suitable for large-span, poor formation tunnels, but requires high construction investment for support and demolition work. Underground station construction is mainly excavation of pilot tunnel.
Double side drift method	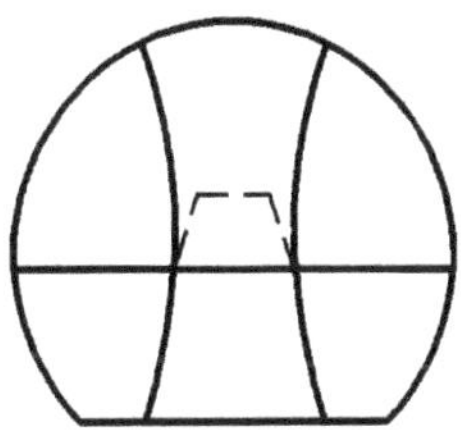	Suitable for large-span, poor formation tunnels. Construction investment is as large as CRD. There are still more applications in poor formation underground station.
Central excavation method	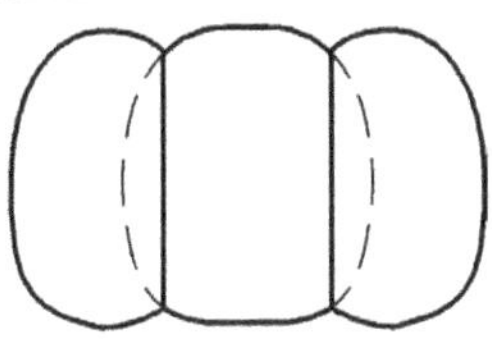	Construct two side guides first, then the middle part. Suitable for large-span, poor formation tunnels. Good to control the surface subsidence.
Side drifts-support method	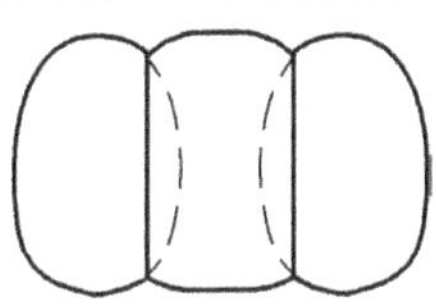	Construct the middle drift first, then, the both sides. Suitable for large-span, poor formation underground station, has less good control effect on the surface subsidence control.
Cavern-pile method	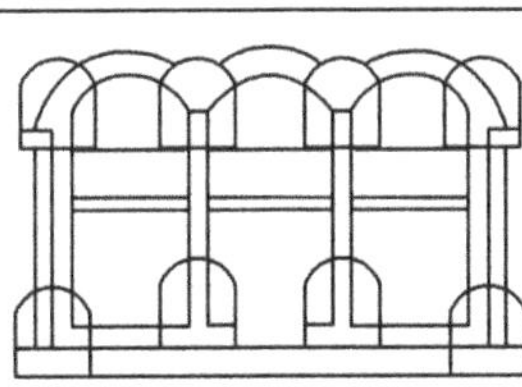	Construct each pilot tunnel first, then expand to the whole station. Force is clear, not limited by the span and the number of layers. Economic input is big.
Arch cover method	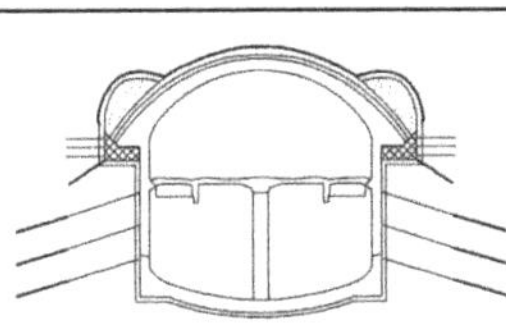	Suitable for the long-span and multi-layer underground station construction in upper-soft lower-hard ground.

Fig. 1.10. Quench method used to build this tunnel in Shimen, Hanzhong.

Babylonia and the Euphrates river is the earliest known traffic tunnel. The largest tunnel in ancient times could be a tunnel between Naples and Priscilla, completed in 36 BC, which is still in use now.

To avoid lining, most ancient tunnels are built in hard rock formations. The main instruments used for tunneling were original tools, such as hammer and drill before the appearance of gunpowder. The later discovery about the physical properties of rocks, for example, the quench method (Fig. 1.10), makes the rock easier to dig by burning to high heat, then jetting it with cold water or vinegar suddenly.

The Swede Nobel invented dynamite in 1866, which provided conditions for hard rock excavation. Dynamite was first used in the construction of the Gotthard tunnel (Fig. 1.11), the world famous one in South-central Switzerland. Compared to the first use of shield construction in the Thames tunnel in 1825, the tunnel used for blasting excavation was nearly 50 years late.

With the development of highway tunnel, railway tunnel and metro project, the construction of metro tunnels gradually diversified.

Fig. 1.11. The Gotthard tunnel used dynamite.

1.2.2.1 *Drilling-blasting method*

Drilling-blasting is a method of excavation by drilling, charging and blasting. Drilling- blasting is a traditional construction method. The law is developed in the long-term construction practice. Wood or steel are used as temporary support, after the tunnel is dug into type, the temporary support is replaced as monolithic lining instead of permanent supporting structure. Drilling- blasting method is the most common construction method of mountain tunnels. Most of China's railway, waterway, highway and other underground tunnels and part of the metro tunnel project all use this method. The basic construction procedure of drilling and blasting is shown in the Fig. 1.12.

The basic principle of modern drilling and explosion is that after the excavation of the tunnel, the blasting effect causes the rock mass fracture to form a relaxed state, which may slump any time. Based on the theory of relaxation load, the method of construction is to excavate one piece of land according to the sequence of parts, and support while digging to ensure the

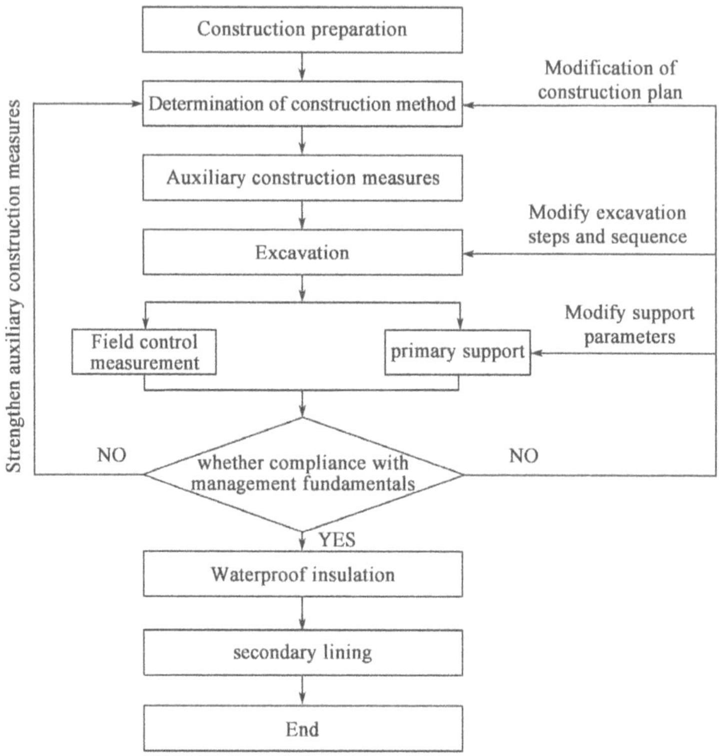

Fig. 1.12. Drilling and blasting construction procedures.

firmness. So the support is complicated, wood consumption is high. With the emergence of combined bolting and shotcrete, number of branches is reduced, which became the New Austrian Tunnelling Method.

New Austrian Tunnelling Method is a tunnel road construction method that gives full play to the surrounding rock self-bearing capacity and space constraint effect of excavation surface, using bolt and jet coagulation as the main support, strengthening the surrounding rock in time to constrain it to relax and deform, and guiding construction through monitoring measurement.

Using the principle of New Austrian Tunnelling Method, after the successful test on loess section in Jundushan tunnel in 1984, academician Wang summarized the basic principle of shallow buried excavation in the pioneering Beijing Fuxingmen turn line project which is full of risk and complexity. In shallow tunneling method, the primary support bears all basic loads as the safety reserve, and the primary support and the secondary lining bear special loads together.

When using shallow buried excavation to design and construct, various auxiliary methods are used to improve the reinforced surrounding rock at the same time. This helps to mobilize some of the surrounding rock force. In addition, different excavation methods are used to support the sealing ring, forming the joint support system with the surrounding rock, to achieve non-collapse, less settlement, safety construction and so on through monitoring measurement, information feedback and optimization design in the process of construction. At present, the drilling-blasting method of domestic metro is guided by the principle of shallow embedding.

The whole section division is excavated to the design contour and the lining is constructed when using borehole-blasting method. When the local layer is soft, it can use simple mining machinery, and it should be excavated and supported when needed according to the stability of the surrounding rock. When using partial excavation, drift heading is firstly excavated in the section, then expanded to the design contour. Partial excavation is mainly to reduce the disturbance of surrounding rock. The size of the branch and the number depends on geological conditions, tunnel section size, support type. It can use full section excavation in solid and monolithic rock layer, middle or small section tunnel. It needs partial excavation, and the temporary support should be set up in time to prevent the rock collapse when in soft and broken formation. Basic operations are drilling and blasting, transportation and slag, support and lining. Auxiliary operation are construction ventilation and dust removal, construction drainage and water supply, construction power supply and lighting, compressed air supply, etc.

1.2.2.1.1 Technical principles to be followed in the construction of drilling and blasting

(1) The surrounding rock must be fully protected during construction for the surrounding rock is the main bearing unit of the tunnel. It should adopt control blasting technology based on the geological section, section size and construction method.
(2) In order to give full play to the structure of the surrounding rock, the surrounding rock should be allowed to have control deformation.

(3) The following must be decided reasonably in construction: support structure type, support structure participation time, interplay of various means of support, bottom closing time, one tunneling length.
(4) During the construction, field monitoring must be carried out, providing reliable quantity measurement information in time to guide construction and design.
(5) Establish the industrial integration management system: design-construction inspection-geological forecast-measurement feedback-revise during the tunnel construction to improve tunnel construction technology.
(6) Generally, when choosing support means, it is necessary to choose a large area to be closely connected with surrounding rock, so that it has strong strain ability, can be set up and strained in time. Therefore, shotcrete and bolt metal mesh are usually used in combination, sometimes with steel support or grille to complement. Temporary invert is also important and cannot be neglected.
(7) Special attention needs to be paid since tunnel construction is a process of changing mechanics state, reducing the division, or reducing the influence of the stress change in surrounding rock caused by too many divisions and the relaxation of surrounding rock. Therefore, full section or large section division excavation method should be adopted as much as possible.
(8) In order to make the tunnel section close in a short time, an inverted arch (or temporary invert) should be constructed or a paved construction method should be used.
(9) The secondary lining shall adopt the construction sequence of wall first, arch second.

The drilling-blasting method can be divided into general drilling and shallow drilling. General drilling-blasting method is used in mountain tunnel, highway tunnel, hydraulic tunnel, etc. In deep geological environment, the vault buried depth is less than twice the hole diameter in shallow drilling method, mostly used in cities' metro tunnel, pipeline common ditch project, it is characterized by thin soil, poor geological conditions, small carrying capacity and fast deformation. There are often important buildings, underground pipes, urban roads and so on near the tunnel, with strict requirements on noise and settlement of construction.

1.2.2.1.2 Construction method used in drilling and blasting

(1) Full section method

The full section method is the construction method of pushing forward through the whole design. Using the explosive method, perforate all vertical surfaces in the work surface plane, and then blast simultaneously to push the entire working face into a ruler. According to the using development trend of drilling and explosion in all kinds of underground engineering, the full section method will be the preferred construction method.

This method is mainly used in tunnels that have stable, hard and integrity surrounding rock, and type I and II surrounding rock without temporary support after excavation. Middle and small section roadways whose height are no more than 5 m, and fracture surface no more than 30 m^2 are also suitable for this method. However in recent years, with the emergence of large construction equipment, the construction mechanization degree and the construction technology continuously improve. There are more and more tunnels under full section construction, use rail transport, with construction methods of rail transport and mixed transport also increasing.

The advantage of full-section method is to make maximum use of the operating space in the hole, its spacious working face, and use of large and efficient equipment to increase construction speed. The section is dug up once, construction organization and management are relatively simple, making good use of the advantage of deep hole blasting. Ventilattion, transportation, drainage and other auxiliary work and various pipelines laying work are more convenient. Its disadvantage is that heavy and expensive drilling rigs should be used in the large section tunnel construction, as a primary investment. This requires corresponding construction of road, assembly area, check equipment and energy for the use of large machines. It will take more time to change to other construction methods for the long tunnel and under changeable geological conditions. Some bench drill machines also make a lot of noise when working at the same time.

(2) Bench cut method

The bench cut method is to divide the tunnel section into several layers (usually two or three). Each layer carries forward as a step at a certain distance. The characteristic of this method is to reduce the section height, without needing heavy drilling equipment, and there are two neighboring empty surfaces during the construction of the next bench, make the blasting more efficient. According to the bench length, bench cut method can divided into long bench method (generally more than five times the width of the tunnel) and short bench method (less than two times the tunnel width). According to the bench arrangement form, bench cut method can divided into positive (under) bench and negative (up) bench.

The advantage of bench cut method is the relatively straightforward process with little interference, which allows upper drilling to be done with rock loading at the same time without support and scaffolding, using shotcrete or mortar as temporary support when necessary, and using bolt for permanent support. This method is not only suitable for surrounding rocks which have good stability and no need or require only temporary support tunnel after excavation, it is also available for soil tunnel.

(3) Drift heading construction method

Drift heading method is to dig first for some distance according to one or some small section drift heading, then gradually expand excavation to the design section and masonry. This method is mainly used in tunnels with complicated geological conditions or large sections engineering.

Drift heading method is divided into central lower drift heading, central upper drift heading, up and down drift heading, side wall drift heading. The side wall drift heading method is divided into single side wall drift heading (median wall method, CD method, CRD method), double side drift heading (glasses method, double-glasses method) and double side drift heading arch after wall first method.

The surface subsidence caused by the double side drift heading is only half of the short bench method. Although the excavation section is more partitioned and has more perturbation, the duration of the initial support full section closure is long. The intermediate deformation of construction is hardly developed for each block is closed immediately after excavation. The double side drift heading construction is safe, but the speed is slower and the cost is higher.

The following conclusions can be obtained by comparing the applicable conditions of the above three methods:

- Full-section method is mainly used for drilling and blasting. Therefore, it is only suitable for the excavation of hard rock soil layer, its tunneling is fast and easy to support.
- Bench cut method and drift heading method are suitable for the relatively soft clay layer earth's tunnel excavation, therefore, it is widely used in practical engineering.
- When using the drift heading method, the control of surface subsidence rate and final settlement value is better than the bench method. But it has more construction steps, the construction project is more difficult than the bench cut method. In various branch excavation methods, drift heading method has better control on ground subsidence than branch cut method and septum method. Among them, the two-side drift heading method is the most stable one. But the settlement for the arch excavation is much larger than the other two methods when using these two kinds of drift heading methods.

1.2.2.1.3 Advantages and disadvantages of drilling and blasting

In conclusion, the main advantages of drilling and blasting construction are as follows:

(1) Applicable to all kinds of geological conditions and ground water conditions.
(2) Suitable for various sections (single line, double lines, multi-line, station) and high flexibility of multi-section (transitional section, transition section).
(3) The surface subsidence and collapse can be effectively controlled by division excavation and auxiliary method.
(4) Comparing with shield method, it is more economical than short excavation.
(5) Comparing with the tunneling machine, there is no requirement for the homogeneity of surrounding rocks.
(6) Comparing with open cut method, it can greatly reduce the impact on the ground transportation and commercial activities to avoid a lot of demolition.

The main disadvantages of drilling and blasting are as follows:

(1) As a labor intensive work, it has higher labor costs.
(2) Construction speed is slow.
(3) It has poor working environment, high security risk, and its construction will likely influence the surrounding environment.

1.2.2.2 *Shallow tunneling method*

The technology core of shallow tunneling method is based on the basic principle of new Austrian tunneling method. This uses a variety of auxiliary measures to reinforce the surrounding rock in the construction, fully mobilize the surrounding rock's self-bearing capacity, provide timely support after excavation, closed ring, to combine with surrounding rock to form a joint support body department. It is a comprehensive construction technology that inhibits the large deformation of surrounding rock.

The basic principle of shallow tunneling method is as follows: through proper reinforcement and treatment of strata, reasonable mobilizing of surrounding rock capacity, using mining short footage to dig, timely implementation of the initial support structure and closure of the ring, making the surrounding rock and initial support structure form a joint support system to jointly undertake the construction stage load. After the deformation is basically stable, the secondary mold lining is made to complete tunnel construction. Primary support bear all the basic design load, while the secondary lining consist of safety reserves. The special load is assumed by the initial support and secondary lining. This new type of supporting structure system which is both rigid and flexible helps to overcome the unpredictability in tunnel engineering. Due to the formation conditions and the complexity of the surrounding environment, it is necessary to monitor the stratum, supporting material and construction objects during the construction process, revise the design and construction method according to the monitoring information to ensure the safety of construction.

According to the construction technology of shallow buried excavation technology, the construction technology has several advantages such as being flexible and changeable, no demolition, no effect on traffic, does not damage the environment, low comprehensive cost, high strength of tunnel support structure, making it especially suitable for China's national conditions. The 20 years of practice of shallow tunnel method is a process of continuous development. In particular, the long-span stealth technology has made great progress, allowing the shallow tunnel method to become a mature technology widely used.

1.2.2.2.1 Shallow tunnel construction principle

Shallow tunnel construction should adhere to the policy of fourteen words: catheter in advance, strict slip casting, short footage, strong support, close early, frequent measurement. This set of construction points accurately summarize the shallow tunnel construction technology.

- Catheter in advance — Before the excavation of the working face, infiltrate the leading small catheter as the advanced support around the arch of the tunnel according to the design. The surrounding rock between tube and tube has the arch effect after the excavation, the pipe shed itself forms a simple support to support and suppress deformation effect of the surrounding rock, in order to complete initial support.
- Strict slip casting — Cement grouting for reinforcing soil after infiltrating the leading small catheter, cementing the loose gravel together, in order to prevent collapse after excavation. After the catheter advance support, press cement or cement water glass slurry immediately, filling the pores of the sand. After solidification, gravel consolidation becomes a stone body with a certain strength, forms a shell around it to enhance the self-stabilizing energy of surrounding rock. In addition, the strict slip casting also includes the backside grouting and secondary lining backing grouting.
- Short footage — Each excavation cycle should be short, so as to shorten the excavation and support time as much as possible, and because inlaid use and vertical surrounding rock are less exposed, in order to ensure the construction is complete.
- Strong support — Use steel frame and quick concrete as support for the initial stage to limit formation deformation, The net shotcrete

support safety factor of shallow tunnel method is large, secondary support is generally not considered.

- Closed early — Early support should be closed as early as possible in order to change the stress condition.
- Frequent measurement — Construction measure refers to the measurement of surface subsidence and the subsidence and convergence of the arch roof. The measurement is the main means of dynamic tracking of surrounding rock and structural change in the construction process. It is the deformation monitoring of surrounding rock and supporting structure. The information is timely and accurately transmitted to the design and construction department to modify the design or take special construction measures.

While adhering to the policy, in the design of shallow tunnel method, special attention should be paid to site engineering geology and the surrounding conditions, in order to ensure the compatibility with the construction method and construction process.

1.2.2.2.2 General specifications of shallow tunnel

(1) Control the deformation of surrounding rock to the ground

The deformation amount not only includes the settlement deformation of surrounding rock caused by excavation, but also branches as the flexible deformation of the protection system caused by the surrounding rock, and the structural displacement of the foundation caused by the foundation subsidence in each stage of construction. To avoid damage the ground and the network of lines embedded in the stratum, protecting the natural landscape, get over the impact on traffic on the ground, better adapt to the environmental needs, the surface subsidence deformation must be strictly controlled.

(2) Rigid support or formation improvement is required

In order to suppress the ground and the ground deformation subsidence, when using the shallow tunnel method, the support time must be as early as possible, the supporting stiffness should also be appropriately increased, proper excavation method, support mode and construction technology must be selected. In addition, some basic measures such as improving the surrounding rock condition and advancing support should often be used.

The auxiliary method used in shallow tunnel method include grouting technology, stir and spray and reinforce technology, small catheter support technology, lock foot anchor pipe technology and freezing technology, etc.

(3) Guide the design and construction through test section

If the surrounding environment of the engineering site and the location of the tunnel are very complex, after finish the construction design, construction plan, test and the measurement plan, it is often necessary to select the geological conditions and the structure of a representative section of the project as a test section to open work first. In the process of construction, to observe, measure, analyze and research some conditions such as the subsidence deformation, support structure and surrounding rock stress state, affected degree of the ground environment. According to the data obtained in the construction of the test section, we can also use the inverse analysis method to get more and more realistic surrounding rock mechanics parameters, on this basis, the mechanical analysis and calculation are carried out. Through the research and analysis of test section construction, optimize the design of the overall construction plan, verify the measurement data management standard.

Shallow tunnel excavation method, such as the preceding drilling-blasting method, has become the main theory and practice of underground tunnel excavation for mountain tunnel construction in our country.

1.2.3 *Shield method*

Shield method is a fully mechanized construction method in the construction of underground excavation. Push the shield machine in the ground, to prevent the collapse of the tunnel through supporting the surrounding rock by the shield machine and the tube plate. Cutting device is used to carry out soil in front of excavation body, carry out the soil out of the hole by unearthed machine, jack up the back, and assemble precast concrete pipe to construct tunnel structure.

Shield construction technology was first pioneered by British engineer Bruno Mars in 1806, it is also used in the Thames underwater tunnel in

London, England, more than 200 years ago. The technology does not require a large number of construction sites along the tunnel. The influence on traffic, residence and so on was very small, and it was soon admired by all countries. Through the unremitting efforts of several generations of technicians, the technology of shield tunnel construction was originally applied in only a few developed countries in Europe and the United States, but gradually spread to urban municipal construction of current developing countries.

Since the 1950s and 1960s, the shield method was gradually applied in China's coastal (Shanghai, Guangzhou, Shenzhen, Tianjin) and inland cities (Beijing, Nanjing, Wuhan), and also has been applied in energy and other fields of rail transit. In recent years, with the development of municipal infrastructure in China, the effective utilization and development of underground space resources become the consensus of local governments. Therefore, the construction technology of shield tunneling in China has developed rapidly in all kinds of engineering practice.

There are a many shield types, which can be classified by method of excavation, structural type, section shape of shield structure, shield front construction, drain and stabilize digging face and so on.

According to the shape of shield, the shape can be divided into round, arch, rectangle and horseshoe, the circle kind application is more extensive for it is resistant to earth pressure in the stratum, the water pressure is good, the lining assembly is simple, common components can be used and are easy to replace. Shield excavation is divided into three types according to the different shield excavation: manual mining, semi-mechanical mining and all-mechanical mining. According to the front construction of shield, there are two different types: open chest type and closed chest type. According to the method of removing groundwater and stable excavation, it is divided into mud and water pressure, and the earth pressure balance type has no pressure shield, local pressure or total pressure shield.

Shield method has the following advantages:

(1) Safe excavation and lining, quick driving speed.
(2) The whole process of shield construction, excavating and assembling lining can be automated, and the construction labor intensity is low.

(3) Does not affect ground transportation, facilities and the underground pipeline and other facilities.
(4) When crossing the river, it does not affect the shipping. The construction is not affected by the weather conditions such as season and wind and rain. There is no noise or disturbance in the construction.
(5) The construction of a long tunnel in soft aquifer usually has technical and economic advantages.

Shield method has been widely used in the construction of soft soil, sand pebbly stratum metro tunnel. With technical improvements, it has been widely used in Guangzhou and Shenzhen for upper-soft lower-hard stratum and other short distance rock tunnel (rock uniaxial compressive strength is up to 30–60 MPa).

1.2.3.1 *Tunnel Boring Machine*

Tunnel Boring Machine (TBM) — the whole section tunnel boring machine, tunneling, supporting, slag and other construction process parallel connection, follow-up work. It is the factory assembly line tunnel construction equipment integrated with the machine, electricity, liquid, light and gas, that has some advantages such as fast speed of tunneling, beneficial for environmental protection, higher comprehensive benefit. It can realize the construction of complex geomorphic deep buried long tunnel which is difficult to realize by traditional drilling and blasting. Its use has been growing rapidly in China's water conservancy, hydropower, transportation, mining, municipal and other tunnel projects.

The hard rock TBM squeezes and shears rock through the hob on the rotary knife disc. It is then transported back to the main machine belt, and can be carried through the slag car or the tunnel to the outside of the tunnel.

In Europe and the United States, the total cross-section tunneling machine is referred to as TBM, while in Japan it is generally referred to as the shield machine. A subdivision can be called hard rock tunnel machine and soft formation tunneling machine. In China, it is customary to call the hard rock tunneling machine TBM, and the soft formation tunneling machine a shield machine.

Hard rock TBM is suitable for hard rock tunneling in mountain tunnel. Compared with the traditional drilling and blasting method, there are the following advantages and disadvantages:

Advantages

(1) Fast: the driving speed is about four to six times that of the drilling explosion method.
(2) Quality: after excavation, the wall is smooth, and the over excavation is small.
(3) High efficiency: saves lining, manual labor, and shortens construction period.
(4) Safety: no blasting operation, more security.
(5) Environmental protection: dust, gas, noise pollution, impact on the surface is very small.

Disadvantages

(1) Poor adaptability of the strata, and the tunnel construction is not suitable for short distance.
(2) The cross section has poor adaptability, for the variable section. The application of construction is restricted.
(3) The structure of TBM is complex, and the durability of the material parts is high and the manufacturing price is high.

Due to the unique advantages of TBM and its applicability to hard rock tunneling, TBM operation is widely used in metro tunnel construction in Chongqing, Qingdao and other cities in China.

At present, China is in a leading position in terms of the construction technology of underground engineering internationally.

1.2.3.1.1 The broad application of shallow tunneling method

The shallow buried excavation method originated from the Beijing metro redevelopment gate in 1986. It is a kind of tunnel construction method created by the Chinese for China's national conditions. This method is based on some theories of NATM and on the basis of the new Austrian law. On the basis of China's specific engineering conditions, a complete set of construction theory and operation methods of metro tunnel were developed. The difference from the new Austrian law is that it is suitable for the surrounding rock of loose soil in urban areas, and the depth of the tunnel is less than or equal to the tunnel diameter, the technical method of constructing tunnel with small surface subsidence.

Its outstanding advantage is that it does not affect city traffic, pollution-free, is noiseless and suitable for all sizes and sections of tunnel chamber. In Beijing and Tianjin, the construction of the underground tunneling of 3 m was realized by the concept of shallow buried excavation.

1.2.3.1.2 Research and utilization of shield excavation

With the deepening of the study of shield method, the shield construction technology increased. The development of the shield machine repair technology was also improved. Shanghai metro tunnel was built with the shield method. In addition to the range of single circular shield, double shield machine currently used to construct two parallel intersections, a square section shield machine was also used to construct underground passage. The road tunnel of Dalian Road was built by a mud shield with a diameter of 11.2 m. It is also the largest diameter shield machine in China. Guangzhou Metro successfully used a new composite shield machine with earth pressure balance, air pressure balance and semi-earth pressure balance to construct the underground tunnel of the complex strata with both soft soil and hard rock and fractured fracture zone. This greatly expanded the application of the shield method. Shenzhen, Nanjing, Beijing, Tianjin and other cities have different geological and hydrological conditions, but the shield tunneling method was successfully applied.

China's shield tunneling speed has reached more than 400 meters per month. The average progress is usually 160 to 200 m per month, although the maximum average progress can be up to 240 m per month. The surface subsidence can be controlled within –30–10 mm, tunneling work can be done with 1 m safe distance, and the deformation of the structure can be controlled under 2–5 m, with tunnel axis error controllable within 30–50 mm.

1.2.3.1.3 Exploration of new excavation methods and techniques

In addition of the above technological progress, with the change of geological conditions and engineering needs, China's engineers are also gradually exploring new construction techniques and methods, such as the curved tube curtain of the Gongbei tunnel of the Hongkong–Zhuhai–Macao cross-sea project. Others include the construction of freezing method and immersed tube tunnel, the application and improvement of the arch cover method in the upper soft substratum of Qingdao Metro, and so on.

1.3 Geological features of major metro cities of China

1.3.1 *The geological characteristics of major cities in China*

The metro construction in each city has its unique characteristics due to its own geological conditions, the following is a brief introduction to the representative cities.

1.3.1.1 *Beijing*

Beijing quaternary stratum has the greatest influence on tunnel construction. The stratum is mainly alluvial and diluvium in plain area. Due to the frequent diversion of rivers, the geological conditions are complicated. The general trend is such that the thickness of sedimentary strata increases gradually from west to east. Lithofacies distribution is characterized by the obvious transition from mountain to plain. The top of the alluvial fan in the western rivers is thick sandy soil and pebbles and gravel; most of the transitional strata in the east are clayey soil, silt and sandy soil, pebbles and gravel interlayer; then to the east the thick layer viscous soil and silt. In short, the strata in Beijing consists of three typical layers: gravel, pebble, sand and silt, clay–. Metro projects are built in these three typical formations or the resulting mixed strata (Table 1.3). Therefore, shallow tunneling method and shield method were widely used in metro construction in Beijing.

Table 1.3. Representative formation mechanics index of a line in Beijing metro.

Layer no.	Rock name	Natural density (kPa)	Cohesion (kPa)	Internal friction angle (°)	Modulus of compressibility (MPa)	Uniaxial compressive strength. (MPa)	Depth (m)
①	Filling	1.82	19.75	14.01	6.80	—	0.0–2.3
②	Silt	1.97	18.27	18.20	6.47	—	2.3–4.5
$②_3$	Silty-fine sand	1.94	—	25.53	17.60	—	4.5–7.0
③	Silt, silty clay	1.99	25.14	23.49	9.01	—	7.0–11.2
$③_3$	Silty-fine sand	2.02	40	27.81	20.41	—	11.2–15.3
④	Ball clay, silty clay	1.99	32.66	16.97	8.58	—	15.3–18.6
④	Silt	2.03	22.98	26.78	14.41	—	18.6–21.5
④	Silty-fine sand	2.04	18.5	29.24	24.60	—	21.5–23.8
④	Medium-coarse sand	2.08	—	33.43	32.88	—	23.8–26.9
⑤	Pebbles, round gravel	2.15	—	42.08	56.89	—	26.9–29.8

1.3.1.2 *Tianjin*

Tianjin is located in the intercalation zone of northeast new Huaxia second settlement belt and northeast of the north China plain. The basal fault structure is relatively developed, with multiple periods of activity. The late pleistocene has tended to be stable. There are some signs of activity in the local area, overlying a thick loose accumulation layer of tertiary system and quaternary system.

Affected by the Haihe river and the Bohai sea, the formation is primarily alluvial and diluvial strata and marine sediments. Therefore, the shield method is better suited for the metro tunnel construction which mainly involves silty clay, silt, silt soil, silty fine sand (Table 1.4).

Table 1.4. Representative formation mechanics index of a line in Tianjin metro (average value).

Layer no.	Name of rock	Natural density (kPa)	Cohesion (kPa)	Angle of internal friction (°)	Compression modulus (MPa)	Uniaxial compressive strength (MPa)	Depth (m)
①	Filling	1.81	18.23	15.11	6.90	—	0–2.9
②	Silt	1.98	19.99	17.23	6.60	—	2.9–4.3
$④_1$	Silty clay	1.97	18.00	20.90	4.87	—	4.3–5.3
$④_2$	Silt	1.95	10.20	24.70	9.11	—	5.3–8.8
$⑥_1$	Silty clay	1.93	15.00	24.80	7.12	—	8.8–12.5
$⑥_3$	Silt	1.92	9.00	26.90	12.32	—	12.5–18.6
⑦	Silty clay	1.99	12.60	22.80	4.76	—	18.6–22.1
$⑧_1$	Silty clay	2.02	13.50	19.80	5.87	—	22.1–25.2
$⑧_{11}$	Clay	1.95	20.00	18.90	5.11	—	25.2–28.3
$⑨_1$	Silty clay	2.04	12.60	23.20	5.9	—	28.3–33.1

1.3.1.3 *Yangtze River Delta area*

The Yangtze River Delta region is characterized by the impact of alluvium in the Yangtze River Basin. In addition, a well developed drainage network, more lakes, parts of the region affected by marine facies have led to the geological conditions in this area, such as Shanghai, Hangzhou, Suzhou, Nanjing, Ningbo and so on, being dominated by soft soil strata. Sand silt, silt and silt clay are the main soil layer affecting metro construction (Table 1.5). Shield construction is the most beneficial for the metro tunnel construction in soft soil area.

1.3.1.4 *Guangzhou*

The geological conditions of Guangzhou metro are complicated, and the sections through which the metro lines pass include the Pearl River Delta alluvial plain area, the platform hilly area (granite or mixed rock) and the limestone

Table 1.5. Representative formation mechanics index of a line in Shanghai metro (average value).

Layer no.	Rock Name	Natural density (kPa)	Cohesion (kPa)	Angle of internal friction (°)	Compression modulus (MPa)	Uniaxial compressive strength (MPa)	Depth (m)
①	Filling	1.86	19.22	15.86	5.50	—	0–0.8
②$_1$	Silty clay	1.97	24.30	16.50	5.48	—	0.8–2.9
②$_3$	Sandy silt	1.98	10.42	20.22	7.14	—	2.9–7.5
③$_1$	Muddy silty clay	1.92	15.00	14.00	3.98	—	7.5–8.6
③$_2$	Sandy silt	1.96	11.52	22.81	6.55	—	8.6–9.9
③$_3$	Muddy silty clay	1.93	14.00	10.50	2.87	—	9.9–13.2
④	Muddy clay	1.92	13.35	14.22	3.12	—	13.2–22.8
⑤$_{11}$	Clay	1.92	12.00	10.00	2.57	—	22.8–26.3
⑤$_{12}$	Silty clay clamp silt	1.98	6.00	26.00	2.95	—	26.3–30.5

karst basin. The lithology shows that the distribution of the quaternary over burden is uneven, including some poor engineering properties such as thick soft soil, muddy sand layer, loose sand, residual silt, etc. There are red sandstone, mudstone, permian and carboniferous limestone caverns with uneven weathering in cretaceous and tertiary and partial intercalation of weak layers in lower voluminous rocks, which are rich in fissure water, and locally intrusive granite bodies or mixed granites of varying degrees of metamorphism.

Because of the complex geology, the tunnel through the Guangzhou metro strata is diverse, has soft soil and hard rock strata, sand and gravel strata, and part of the line is also affected by karst, boulder and so on (Table 1.6). Hence the shield machine (polyshield majority), drilling and blasting method are often used in the Guangzhou metro tunnel construction.

1.3.1.5 *Chengdu*

Chengdu metro tunnel basically goes through quaternary strata. Most of them are in the quaternary system fourth series, the sedimentary deposits,

Table 1.6. Representative formation mechanics index of a line in Guangzhou metro (average value).

Layer no.	Rock name	Natural density (kPa)	Cohesion (kPa)	Angle of internal friction (°)	Compression modulus (MPa)	Uniaxial compressive strength (MPa)	Depth (m)
①	Filling	1.86	10	15	4.0	—	0–2.5
$③_1$	Silty-fine sand	1.90	—	30	—	—	2.5–3.6
$③_2$	Medium-coarse sand	1.95	—	32	—	—	3.6–6.9
$④_2$	Sludge	1.56	10	5	1.8	—	6.9–9.2
$⑤_1$	Silty clay	1.93	20	19	6.0	—	9.2–16.7
⑥	Total weathering detritus	1.99	28	22	8.0	—	16.7–19.5
$⑦_1$	Strongly weathered conglomerate	1.95	33	25	8.0	—	19.5–25.6
$⑧_1$	Intermediate weathered conglomerate	2.50	200	30	—	12	25.6–32.8
$⑨_1$	Breezy conglomerate	2.63	500	35	—	32	32.8–36.2

alluvial deposits, and the sedimentary deposits of the sedimentary layer and the alluvial pebble soil sand lenses. Therefore, the typical characteristics of Chengdu metro tunnel are sand pebbles and high-rich water stratum (Table 1.7). It is proved that most of the metro tunnels are constructed by shield method, and only some large sections are used for undercutting methods.

1.3.2 *General situation of metro construction in Qingdao area*

Qingdao's terrain is high in the east and low in the west, with a sag in the middle. Its physiognomy consists of low eroded mountains, hills, plains. At the main body for Jiaolai plain, the average elevation is about 50 meters. In

Table 1.7. Representative formation mechanics index of a line in Chengdu metro (average value).

Layer no.	**Rock name**	**Natural density (kPa)**	**Cohesion (kPa)**	**Angle of internal friction (°)**	**Compression modulus (MPa)**	**Uniaxial compressive strength (MPa)**	**Depth (m)**
①	Filling	1.9	20.2	15	5	—	0–1.8
$②_3$	Silty clay	2.0	30	20	7	—	1.8–2.7
$②_4$	Silt	1.95	20	10	4	—	2.7–5.9
$②_5$	Silt	1.75	—	15	5	—	5.9–7.8
$②_6$	Medium sand	1.85	—	25	15	—	7.8–12.6
$②_{81}$	Pebble soil (loose)	1.9	—	30	25	—	12.6–16.4
$②_{82}$	Pebbly soil (slightly)	2.0	—	35	30	—	16.4–21.8
$②_{83}$	Pebbly soil	2.1	—	38	36	—	21.8–25.4
$②_{84}$	Pebble soil (compacted)	2.2	—	40	49	—	25.4–29.5
$③_{84}$	Pebble soil (compacted)	2.25	—	45	55	—	29.5–38.1

the hilly north and south areas, the elevation is about 200–500 m. Jiaozhou bay is a semi-closed bay in southeast coast of Shandong peninsula. It is about 27.8 km wide from east to west, 33.3 km from north to south with nearly 210 km long coastline, and the average water depth is 7 m, with maximum depth 65 m. The bay mouth is toward the southeast, it joins the yellow sea through a 3 km wide, 30–40 m deep sink. Exterior drainage originates in northern and northeast low hilly land, flowing into the yellow sea from the south, divided into a three river system of Dagu river, north Jiaolai river and coastal river. The fault structure mainly consists of the near east-west direction and the north east trend, which constitutes the right figure structure of Yishu fault. Wulian-Qingdao fault and Ming-Jimo fault are located at the boundary between China and North Korea and the central orogenic belt, which is the main fracture of the tectonic framework in Qingdao.

There are many coastal and river systems in Qingdao area, and all river flows are obviously controlled by precipitation, as the seasonal variation is obvious. In terms of groundwater conditions, the main fissure water in Qingdao area are unconsolidated rock pore water, clastic rock pore fissure water, bedrock fissure water supply sources to atmospheric precipitation and river, littoral lateral surface water and underground water supply and water supplies in some coastal areas. The general flow of groundwater is consistent with the slope. The groundwater level fluctuates from 1.5 to 2.0 m. In the vicinity of the sea, the groundwater is affected by the tides, and the groundwater level changes regularly with the change of the tide and ebb tide.

According to the geological investigation, the typical soil structure in Qingdao is binary compound rock strata. In general, quaternary soil layer of different thickness coversthe strong, medium, micro rate of decay granite batholith. The fourth system consists of soil thickness of 0–20 m, the station and interval through various weathered rock and soil layer, with uneven longitudinaland upper-soft and lower-hard stratum.

Based on the geological conditions of Qingdao, Qingdao metro tunnel project has the following characteristics:

1.3.2.1 *Upper-soft lower-hard ground*

The upper-soft lower-hard ground is the representative characteristic of Qingdao metro tunnel project (Fig. 1.13). Taking Qingdao Metro Line 13 as an example, the single-channel tunnel in the whole line passes through the weathering section 68, and the access or out station tunnel goes through totally 82 the upper soft and lower hard stratum. It is difficult for tunnel drilling and blasting in the upper soft and lower hard stratum (Tables 1.8 and 1.9), which is mainly reflected in:

(1) Risk of tunnel vault rock instability: surrounding rock of the arch is weak, mainly with silty soil, strong weathering strata and sand stratum, can easily cause collapse under mismanagement.
(2) Influence of the tunnel bottom rock blasting construction: although the surrounding rock of the tunnel vault is weak, even only using mechanical or manual excavation, due to the lower part of tunnel rock mass strength being higher, with the drilling and blasting method, blasting vibration has a certain disturbance on the arch surrounding rock.

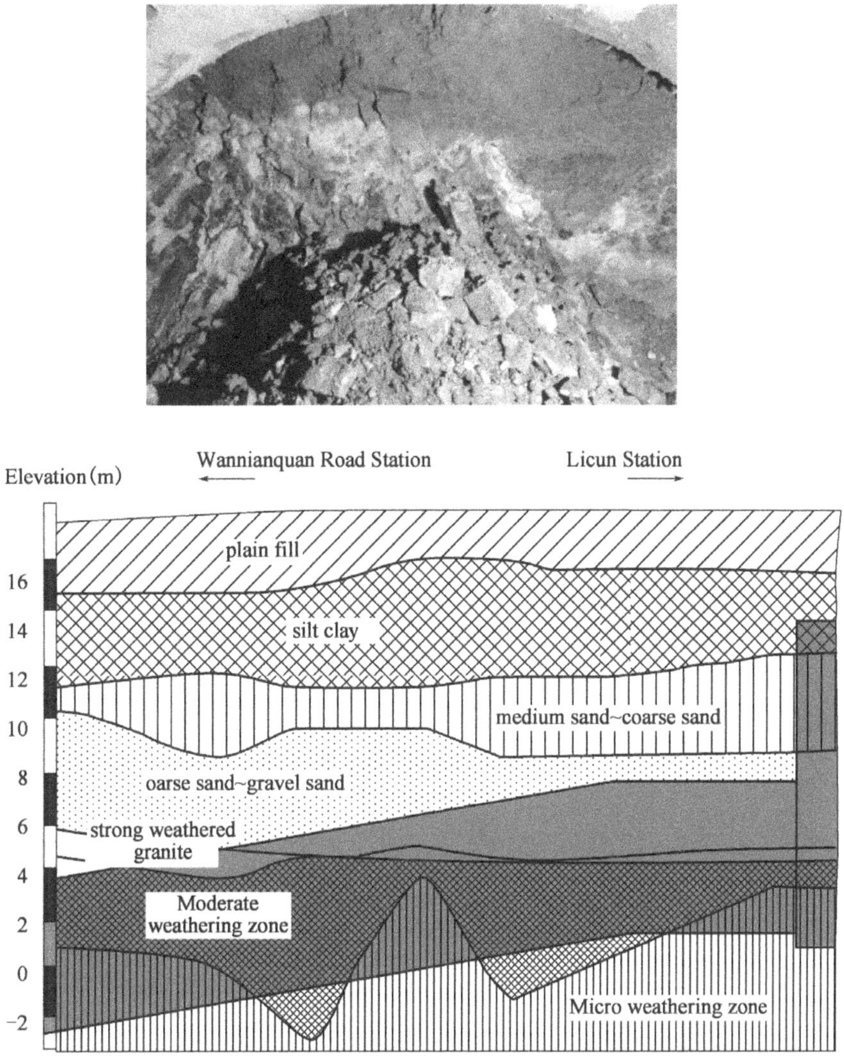

Fig. 1.13. Typical upper-soft lower-hard ground.

1.3.2.2 *Some marine strata*

Part of the metro line construction environment is characterized by marine conditions near the sea. For example, Qingdao Metro Line 13 has the most tunnel fracture, high water chloride ions content, with the highest amounting to 13,000 mg/l. Marine conditions pose a big challenge to the durability of tunnel second lining and mechanical maintenance, etc.

Table 1.8. Typical mechanical index of hard segment of Qingdao metro (average value).

Layer no.	Rock name	Natural density (kPa)	Cohesion (kPa)	Angle of internal friction (°)	Compression modulus (MPa)	Uniaxial compressive strength (MPa)	Depth (m)
①	Filling	1.75	20	15	5.4	—	0–2.5
④	Coarse sand in silt.	1.85	—	—	6.3	—	2.5–4.8
⑥	Silty clay.	1.91	14.4	5.3	4.3	—	4.8–8.2
$⑯_3$	Strongly weathered granite porphyry.	2.30	—	45	10.1	—	8.2–16.3
$⑰_3$	Weathered granite porphyry.	2.60	—	55	—	32.77	16.3–23.5
$⑱_3$	Light breeze granule porphyry.	2.65	—	65	—	68.32	23.5–35.9

Table 1.9. Typical mechanical index of hard segment of Qingdao metro (average value).

Layer No.	Rock name	Natural density (kPa)	Cohesion (kPa)	Angle of internal friction (°)	Compression modulus (MPa)	Uniaxial compressive strength. (MPa)	Depth (m)
①	Filling	1.75	20	15	5.4	—	0–3.1
④	Coarse sand in silt.	1.85	—	—	6.3	—	3.1–5.8
⑫	Silty clay silt.	1.93	5.8	12.6	6.5	—	5.8–9.6
$⑯_2$	Strong weathered granite	2.50	—	50	31.1	—	9.6–17.7
$⑰_2$	Medium-weathered granite	2.60	—	55	—	38.66–61.3	17.7–26.0
$⑱_2$	Breeze granite	2.65	—	65	—	82.54–189.5	26.0–36.9

1.3.2.3 *Conditions of some reclamation strata and surrounding rocks suddenly change*

In some offshore areas in Qingdao, reclamation expansion engineering are carried out where the geological condition is complex. For example, some offshore rocks are affected by ongoing geological processes such as seawater erosion. With crack development (Fig. 1.14), rock face ups and downs frequently and haphazard, geological exploration drilling, formation characteristics are hard to reveal, which causes difficulties for metro tunnel construction.

The Jinggangshan road station to Jialianhua station in Qingdao Metro Line 13 passes underneath a reclamation area. The area between the sections of wall rock changes suddenly, and sometimes even appears to change every variety. The surrounding rock changes suddenly from level III to level V, without any sign of transition. In the construction of such surrounding rock, it is advisable to adopt advanced drilling and other means to forecast and frequently transform construction and support methods, leading to low the construction efficiency.

Fig. 1.14. Crack water in the tunnel.

1.3.2.4 *Environmental impact control of the depth and construction of local strata become the subjects*

As mentioned earlier, the Qingdao region quaternary soil layer thickness is 0–20 m, and the large rock layer thickness is rather changeable. Parts of rock buried depth is shallow, which has great significance on tunnel stability. At the same time, due to the shallow rock especially breeze rock buried shallow, the control of blasting vibration in the tunnel construction is very important on the influence of the surrounding buildings and pipelines.

This book, which is based on typical geological conditions of the Qingdao metro, introduced typical Qingdao metro construction technology from some aspects such as hard rock tunnel, station construction, poor geological water plugging and reinforcement, the angle of green construction, and so on.

Chapter 2

Drilling–Blasting Tunnel Construction Technology in Hard Rock Stratum

2.1 Design and construction procedures of drilling–blasting method in hard rock stratum

2.1.1 *The main design principles of hard rock stratum tunnel*

Urban metro tunnel design in hard rock strata needs comprehensive analysis, combined with geological analysis, and should not only meet functional requirements but also construction efficiency, economic efficiency, social requirements and many other aspects. The main design principles are as follows:

(1) The design of the underground interval structure should be based on the assumption of the structure being functional and meeting the requirements of city planning, traffic operations, environmental protection, earthquake protection, water resistance, fire prevention and corrosion resistance, while at the same time building a safe structure with advanced technology at a reasonable cost.

(2) The size of the structure should meet the underground railroad clearance standards, functional requirements of various types of equipment and construction technology requirements, and take into consideration the necessary amount of construction errors, structure deformation and displacement factors.

(3) Structure design should be based on the construction method, structure and component types, environmental conditions and load

characteristics, matching the characteristics of structural design specifications and design method, combined with engineering monitoring design information.

(4) The structure shall have sufficient strength, rigidity, stability and durability during construction and use. The structure strength, stiffness and stability should be calculated according to the construction stage and normal use stage. For a reinforced concrete structure, the crack width should be checked.

(5) The drainage pattern and lining structure type of the interval should be determined according to the surrounding rock conditions, groundwater situation, runoff conditions and features of topography.

(6) The tunnel should be waterproof with adequate drainage measures, following the principles of row cutting and plugging, adjusting measures to local conditions. Comprehensive management should be undertaken to make the main body of the concrete structure waterproof, making three-seam waterproofing the key and building a complete drainage network system to achieve the goal of unobstructed drainage and waterproofing, with reliability and reasonable economy. When there are strict requirements for environmental groundwater protection, measures to prevent waterlogging and discharge water should be adopted.

2.1.2 *Selection of construction method for the interval tunnel*

The tunnel is the underground structure connecting the station, and the structure form is usually determined by the construction method and functionality. The construction method has a decisive influence on the determination of the structure type and the construction cost of the metro. The selection of construction method, on the one hand, is affected by some aspects such as the engineering, geological, hydrogeological and environmental conditions (the status quo of the ground buildings and underground structures, road width and traffic conditions, etc.), line plane position, tunnel excavation width and embedment depth; on the other hand, the method has direct influence on the normal functioning of urban residents and ground traffic during the construction period, the complexity of the project, the timescale, cost, development and utilization of the underground space, operation effects and so on. Therefore, the determination of the construction method must be based on local conditions and overall consideration. The method of construction should be reliable, advanced and economical. According to the experience gained from other projects conducted in China and abroad, for the Qingdao Metro Lines 1, 2, 3 and 13, the main construction methods used for the underground tunnel were the Ming excavation method and the concealed excavation method (TBM method, shield method, drilling and blasting method).

2.1.2.1 *Open-cut method*

This method is simple and reliable. The construction risk is small and easy to control. The progress of the project is quick as it can be divided into sections according to convenience. Construction cost and operation cost are low when the excavation depth is shallow. There is low requirement for specific geological conditions and waterproof treatment is easy. However, the construction has certain influences on the normal activity of urban ground transportation and residents, and does some damage to the surrounding environment during the construction period. Based on the scope of the excavation and the number of underground pipelines to be removed, a large construction site is required. Generally, the open-cut method is applicable in a broad area or open field, where the buildings and underground pipelines are fewer and the traffic volume is small or the flow of traffic is organized, or else a combination of municipal engineering construction methods are needed. For long-span construction, complex structures, deep excavations and poor geological conditions, environment permitting, the construction interval should be given priority to reduce the risks involved in construction and reduce project cost.

2.1.2.2 *Drilling–blasting method*

The drilling–blasting method is suitable for a structure that is deeply buried since it provides a steadying ability for rocks in tunnel excavation, especially for deeply buried or hard rock excavation. The drilling–blasting method is also called excavation blasting. The basic process involves drilling, charging, blasting smoke, slag, initial support and secondary lining.

A whole section of excavation can be used for a small section of tunnel, or it can be used for the upper and lower steps of large sections of tunnel. When there are complex geological conditions present in a large engineering section, one can adopt a pilot tunnel excavation method step by step, wherein the first advance is at a certain depth of a small cross-section of the roadway, which then helps to open the top section, and then again using smooth blasting at a cross-section to design that section.

The special geological conditions of Qingdao are as follows: The bedrock is buried at both deep and shallow depths and the rock strength is high, which creates good conditions for the underground engineering to adopt the drilling and blasting method. Compared with other cities, it is not difficult to use drilling and blasting in Qingdao. The cost is relatively low, and it does not interfere with traffic, which makes it an especially good construction method. The disadvantage is that the working face is relatively narrow, and the drilling explosion has a certain adverse effect on the surrounding environment.

The drilling–blasting method is a kind of construction method which has been developed in recent years to adapt to the needs of shallow buried tunnels. At present, it has been widely used in the construction of metro areas. The drilling and blasting method is simple and flexible, and can be used to verify or modify the design parameters and adjust the construction process according to the information feedback measured by the construction monitoring so as to achieve the goal of safety and economy.

In addition, due to the site's location, the drilling and blasting method causes less interference to ground transportation and pipelines. Different engineering measures and construction methods are adopted for different geological conditions and surrounding environments. Different excavation methods can be used to deal with soft and hard non-uniform strata. For some locations with good geological conditions, the project is cost-effective. The drilling and blasting method is applicable in various geological conditions and various kinds of sections of high flexibility. As compared with the shield method, it does not need large machinery and equipment, can cut reinforced concrete foundation piles directly in the tunnel excavation face and is also economical in a relatively short time frame. At present, this method has been widely used in metro area and railway station tunnel constructions in China.

The drilling and blasting method also has its own weakness: it can easily cause groundwater loss during construction, which causes ground subsidence, and it is necessary to take practical protective measures for important pipelines and houses. Improper handling during construction can easily cause the ground to collapse, thus causing damage to the surrounding environment and causing construction accidents. When the scope is large, the excavation needs to be divided into several steps due to which the interference between the processes is large, leading to certain risks in the construction.

The horseshoe shape (Figs. 2.1 and 2.2) is generally used in the drilling–blasting method. The tunnel lining is composed of initial support, secondary lining and a waterproof interlayer. The initial support is

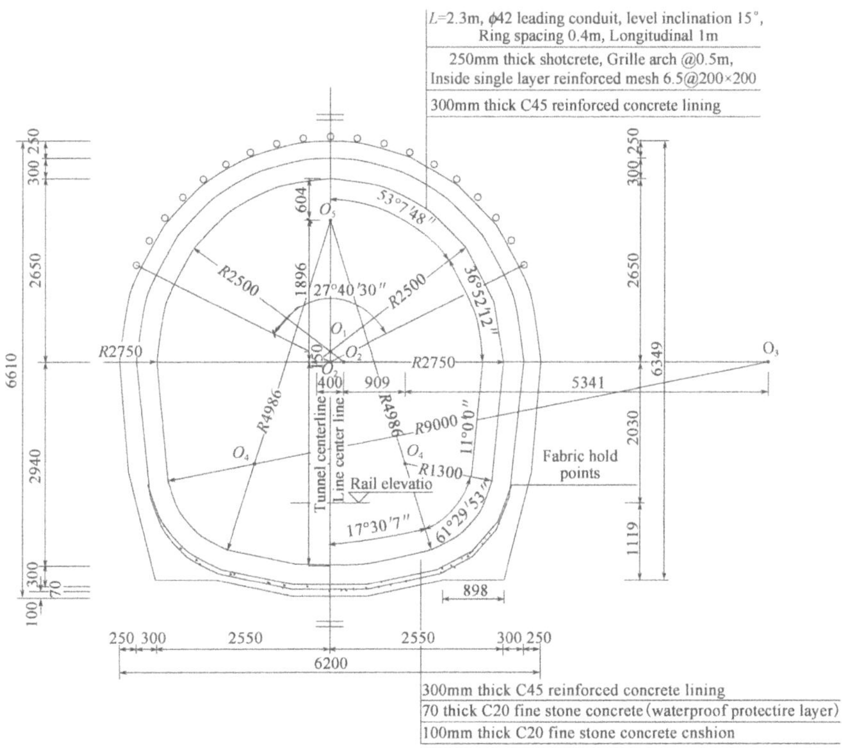

Fig. 2.1. Standard horseshoe section of tunnel during drilling and blasting process (train speed: 80 km/h) (unit: mm).

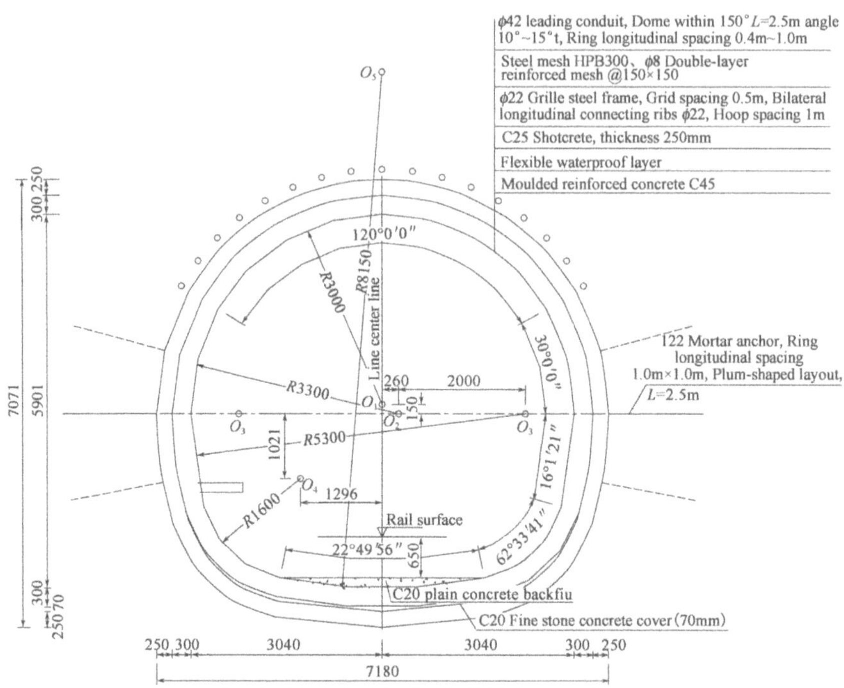

Fig. 2.2. Standard horseshoe section of tunnel during drilling and blasting process (train speed: 120 km/h) (unit: mm).

composed of a shotcrete and grill steel frame, and the secondary lining is waterproof reinforced concrete.

When the tunnel's depth is shallow, through loose unstable soil and broken rock, due to the poor conditions of the surrounding rock, one should emphasize an advance guard of formation and its reinforcement when using undercutting methods and reduce the excavation face as far as possible to ensure the stability of the cavern. At this point, the tunnel lining structure stiffness is greater and the early support allows smaller deformation. When one has to construct the tunnel through soil, one can use small duct grouting in advance to consolidate the soil around the arch in order to form a soil arch structure with certain supporting capacity, and provide security, stability and conducive conditions for the excavation. When the excavation is through a rock tunnel with shallow depth and rigid integrity, one can use small and medium prestressed bolts with leading oblique insertion, making the arch form a rocky bolt arch structure with considerable self-bearing capacity, providing safe and stable conditions for the second half of the step excavation support. When the interval tunnel is in strong weathering or coarse sand, it can be constructed by the drilling and blasting method in combination with precipitation (or the internal precipitation after the curtain at the outer stop).

2.1.2.3 *TBM method*

A whole-section tunnel boring machine (TBM) is commonly used in rock strata, under the protection of the shield shell, excavating the rock with a cutter head, driving forward using gripper shoes on the rock by the action of the jack, and generally not having maintenance constraints such as earth pressure and slurry pressure. The TBM is a large and efficient piece of tunnel construction machinery which integrates tunneling, dredging, guiding, supporting, ventilating and dustproofing.

2.1.2.3.1 The advantages of the TBM method

(1) The driving efficiency is high. Continuous operation can be carried out when the excavating machine is used, so that the broken rock, the chattering and the support train can be realized in a faster time frame.

(2) The construction quality of the excavator is good, and the amount excavated is low. The inner wall created by the tunneling machine is smooth, and there is no convexity; therefore, it can reduce the amount of supporting work required and reduce the project cost.

(3) The disturbance to the rock face is minor in comparison with other methods. The use of the excavator can greatly improve the construction conditions at the rock face, and the stability of surrounding rocks is improved, thus ensuring the health and safety of the construction workers.

(4) Construction conditions are safe, and the recent TBMs can be modified in a protective shed or a closed control room, and high performance of the machine has greatly improved the safety and operational environment.

2.1.2.3.2 The disadvantages of the TBM method

(1) The tunneling machine has poor adaptability to various geological conditions (faults, fracture zones, compression zones, water flow, hard rocks, etc.).
(2) Due to the complex structure of the driving machine, the durability requirements of the materials and parts are high, so the equipment price is higher. It is necessary to spend a lot of money to buy parts and make the machines before the construction, which makes the construction investment high and unsuitable for short tunnels.
(3) The excavation diameter and shape cannot be changed in the construction, and the application of the machine is restricted.
(4) The station has to fulfil the requirements of starting, receiving, adjusting and passing, thereby increasing the construction cost and restricting the construction duration.

As can be seen from the above description, the TBM method is characterized by a high mechanization degree and quick construction speed. Under the same conditions, the driving speed is 2.5–4 times that of conventional drilling. In addition, it produces low vibration and low noise, and offers safe and reliable operation. It has the advantages of low impact on nearby buildings, underground structures and the daily lives of the area's residents. According to general experience, the TBM method is more suitable when the length of the tunnel is greater than 6 km.

2.1.2.4 *Shield method*

The shield method is a mechanized construction method, whereby one shields the mechanical propulsion in the formation by providing a shield around the bearing shell and lining segments, thereby preventing a collapse of the tunnel's surrounding rock. At the same time in front of the excavation, the soil is excavated with a cutting device and carried out of the hole by an unearthing machine, a jacking machine is used in the back and the precast concrete pipe is assembled to form a mechanized construction method of tunnel structure.

2.1.2.4.1 The advantages of the shield method

(1) Both excavation and lining can be conducted safely, with fast driving speed.
(2) The whole process of shield construction, excavating and lining can be realized by means of automatic operation, and the labor intensity is low.
(3) It does not affect ground transportation, underground pipelines or other facilities.
(4) It does not affect shipping across the river. It is not affected by weather conditions such as seasons, wind and rain, and there is no noise or disturbance caused by the construction.
(5) It has technical and economic advantages in the construction of long tunnels buried deep in soft aquifers.

2.1.2.4.2 The disadvantages of shield construction

(1) The section size cannot be adjusted easily, and this method is not suitable for long durations of driving in hard rock formations.
(2) The new type of shield equipment is expensive, and it is not economical for the short constructions.
(3) To sum up, the various methods of construction are shown in Table 2.1.

Table 2.1. Comparison table of tunneling methods in metro interval.

Construction method	Advantage	Disadvantage
Open-cut method	The construction technology is mature, the method technology is simple and reliable, the construction risk is low, and the process is easy to control; the project is fast, and the cost is low. The requirements for geological conditions are not high.	The construction has great impact on urban traffic and the normal life of residents, and does a certain amount of damage to the surrounding environment with the removal of a large underground pipeline; the construction site occupies a large area.

(Continued)

Table 2.1. (*Continued*)

Construction method	Advantage	Disadvantage
TBM method	The pipe lining and tunneling operations are highly efficient with high construction quality, high construction safety, low disturbance to surrounding rock and moderate influence on process.	The method has poor adaptability to strata, along with the high price of the driving machine, long equipment procurement cycle, inability to change tunnel excavation shape, high requirements for and increased investment in the station and unsuitability for weak strata.
Drilling and blasting method	The method has good adaptability, high construction flexibility, low cost and ability to change tunnel excavation shape.	The tunnel construction has a great influence on the surrounding environment and affects the working procedure.
Shield method	This method offers safe excavation and lining, good driving speed, low labor intensity and strong formation adaptability; in the loose water-bearing strata, construction of long tunnels requires greater depth due to the technical and economic superiority.	The section size is changeable, and it is not suitable for hard rock above 30 mpa.

2.2 Design and construction of vertical and inclined shafts

2.2.1 *Vertical slope location*

When the metro tunnel adopts the drilling and blasting method, it is necessary to provide a reasonable setting according to the interval length, the total construction schedule and the design conditions. Because of the advantages and disadvantages of the tunnel approach, it is necessary to ensure a comprehensive understanding when selecting the method. The advantages and disadvantages of the vertical slope of Qingdao Metro Line 13, are analyzed in Table 2.2.

The analysis of the advantages and disadvantages proves that it is advantageous to set the inclined well for the positive line construction, but the site selection should be combined with several factors.

Table 2.2. Analysis of the advantages and disadvantages of vertical sloping well.

Parameters	Shaft	Slope	Notes
Area coverage	Covers less area	Covers more area	—
Cost	Low	High	It means the cost of the vertical shaft itself.
Construction time limit for project	Short	Long	—
Slag way	Crane vertical lift	Transportation vehicles	—
Normal operation efficiency	Low	High	—

2.2.1.1 *Design conditions*

When designing the permanent contact channel in the interval, the position of the vertical inclined well should be set first. When designing the permanent air well in the interval, the vertical shaft can be set first.

2.2.1.2 *Total construction time requirement*

In the construction of the inclined hole, under the same geological conditions, the construction period is longer than 2–3 months, so one needs to consider the interval length when setting the deviated well. Monthly single line footage statistics are shown in Table 2.3 for Qingdao Metro vertical slope efficiency.

Refer to Table 2.3 for overall consideration of the construction period.

2.2.1.3 *Total area covered*

Since the metro construction is generally close to the downtown area, the ground is more vulnerable for the duration of the construction. According to the vertical slope of Qingdao Metro, the shaft area is commonly 1,000–2,000 m^2 and the slope area is commonly 3,000–5,000 m^2, so one also needs to consider the possibility of land occupation when setting the

Table 2.3. Monthly single line footage.

Surrounding rock situation	Shaft footage per month (m)	Inclined shaft footage per month (m)	Notes
Surrounding rock of Levels II, III	60	90	—
Surrounding rock of Level IV	40	50	—
Surrounding rock of Level V	30	40	—
Surrounding rock of Level VI	25	30	—

vertical slope covers. While selecting an appropriate site, priority should be given to clearing the land, avoiding municipal public land, surveying the surrounding roads, buildings and pipelines, and only then indicating the corresponding position in the layout plan.

2.2.2 *Temporary facility and layout of vertical shaft*

The site layout of the vertical and inclined shaft should comply with the principles of reasonable compactness, safety and economical convenience, while at the same time guaranteeing the construction demands of minimizing the area used by following the below principles:

(1) The road should not be occupied to build anything, including the office and living facilities.
(2) The surrounding greenery should not move be moved.
(3) No gas or oil pipelines should be used to build production and living facilities.
(4) The construction unit shall rationally allocate the premises and production sites according to demands of the site, and meet the requirements of operation area, office area, and living area separately.
(5) The drainage ditch and water collecting well should be set in the field to meet the drainage requirements so that the groundwater is not affected.

2.2.2.1 *Key point of shaft layout*

In the design of the site, the shaft needs to be considered because it is responsible for the overall construction efficiency as it provides a guarantee for the main excavation work by enabling the setting up of the wind, power and water supplies and other necessary facilities such as a mixing station and temporary sludge field facilities. The main points of the temporary facility layout are as follows:

2.2.2.1.1 On-site mixing station

The shotcrete used in the hard rock stratum for the drilling and blasting method at the beginning of construction is generally mixed at the scene, so the site needs to set up a small mixing station. At the time of site planning, the proper mixing station area needs to be selected. According to relevant experience in the Qingdao area, the on-site mixing station is usually equipped with 50–75 Hz semi-automatic or fully automatic strong mixer and the mixer should be equipped with an automatic measuring device; the mixing station shall be closed to meet the needs of winter construction and prevent dust pollution.

2.2.2.1.2 Sand and stone storage area

Because of the particularity of urban metro construction, there is a requirement for one day's worth of raw materials to be stored on site for normal construction. Based on the principle of check before use, there might be an additional requirement for sandstone, needing the setting up of a sandstone temporary storage site, or if the site conditions are not conducive, the material can be set up outside the venue. According to the construction efficiency and calculations of four requirements a day in two loops for the main construction, 50 m^3 of spraying pulp, and 35 m^3 of gravel and 35 m^3 of sand are required every day; considering a variety of adverse factors, including timeliness of supply, sand storage should not be less than 70 m^3.

2.2.2.1.3 Collecting pool and sedimentation tank

The underground seepage drainage is the key link in the construction of the underground metro in the city. First of all, leakage in the tunnel and redundant construction water need to be discharged in time to ensure safety and sanitation; second, as water discharged in the tunnel flows into the municipal pipe network, there needs to be a carefully study conducted of regional geological conditions and groundwater conditions before setting a tank and collecting basin. According to Qingdao's coastal geological hard rock tunnel construction experience, because of rich groundwater supply, each shaft or slope at peak construction displaces on average 2,000–4,000 m^3 of muck a day; in the groundwater-rich location, the displacement can be up to 5,000–8,000 m^3 every day. In the Dalian area, shaft displacement can reach more than 20,000 m^3 daily during the tunnel construction. To prevent the municipal pipeline from silting, the water tunnel needs to be drained at multiple stages of precipitation; when the tank is set at least through the three levels of precipitation, the volume of the sedimentation tank will be manageable and in accordance with the water yield in the vertical slope layout plans.

2.2.2.1.4 Change and generator room

External power supply is guaranteed to ensure normal construction. Urban construction external power is easy to access at the site, and is stable and reliable, but there needs to be a temporarily structure at the site during the construction of overlay, along with an electricity machine with capacity of 600–1,000 kV·A. Table 2.4 shows the electrical equipment commonly used for the underground construction.

Table 2.4. List of main electrical equipment used in underground excavation.

No	Device name	Quantity	Power (kw)	Total power (kw)
1	Concrete mixer	1	55	55
2	Bridge crane	1	54	54
3	Injection twin screw compressor	2	132	264
4	Anchor grouting machine	1	5.5	5.5
5	Grouting pump	1	7.5	7.5
6	Mortar mixer	1	4	4

Table 2.4. *(Continued)*

No	Device name	Quantity	Power (kw)	Total power (kw)
7	Vibrator	4	2.2	8.8
8	Engineering drilling machine	2	18.5	37
9	Office lighting	20	0.02	0.4
10	Ground lighting	3	1	3
11	Fan	1	110	110
12	Concrete jet	3	7.5	22.5
13	Second lining trolley	2	30	60
14	Concrete transfer pump	1	75	75
15	Construction of the lighting	50	0.2	10
16	Water pump	15	10	150
17	Motor power			853.3
18	Lighting capacity			13.4
19	Combined total power			866.7

2.2.2.1.5 Temporary slag dump

There are generally no direct loading methods for the residue from the shaft to the ground, especially in urban construction, where large transport vehicles are required to move through traffic at peak hours with no dedicated thoroughfare. Therefore, basic outbound slag is mostly moved at night, so there needs to be a temporary sludge processing area at the scene. For the layout, one needs to plan ahead for the size of the temporary deposit slag field because of the sludge waste field is too big or too small it may cause a spillage of the waste residue and thereby influence the construction. According to the tunnel excavation efficiency in hard rock strata, the daily average speed of a single face is 1.5 m per day, the speed of a single shaft is $1.5 \text{ m} \times 4 = 6 \text{ m}$, the metro section tunnel is excavated at a rate of about 40 m^2 of area per day, and every morning at sunrise, the slag quantity is nearly $40 \times 6 \times 1.5 = 360 \text{ m}^3$ (virtual). Therefore, the temporary sludge field area should ideally not be less than 150 m^2.

2.2.2.1.6 Air compressor

The air compressor is mainly for wind supply for the tunnel construction equipment. The main requirement for the wind equipment in the tunnel is due to the drilling machine, concrete spraying machine and so on. While making preparations for the tunnel construction, the supply of wind equipment should be considered well in advance to avoid the occurrence of insufficient air supply. An example is given to illustrate the method of calculating the air volume required for the drilling–blasting method.

Calculation formula of air volume:

$$Q = (1 + K_{备}) \cdot (\Sigma q \cdot K + q_{漏}) \cdot K_m$$

where $K_{备}$ is the spare coefficient of the air compressor, usually 75–90%, mostly 75%, Σq is the airflow required of pneumatic tools (m^3/min), for example, an 8-typhoon drill, with 2 jets working at the same time, $q_{漏}$ is the leakage loss of pipe and accessories, the average leakage per kilometer is 1.5–2.0 m^3/min, mostly 2.0 m^3/min, K is the working coefficient is 0.75, K_m is the influence coefficient of altitude on the production capacity of air compressor is 1.0.

Construction needs of air volume calculation:

$$\begin{aligned} Q &= (1 + K_{备}) \cdot (\Sigma q \cdot K + q_{漏}) \cdot K_m \\ &= (1 + 0.75) \times [(8 \times 1.5 + 2 \times 3) \times 0.75 + 2 \times 0.75] \times 1 = 26.3\text{m}^3/\text{min}. \end{aligned}$$

According to an analysis of the calculation result, two 20 m^3 air compressors can be configured to meet the air supply needs of the pneumatic equipment.

2.2.2.1.7 Ventilation fan

In the drilling and blasting method in order to ensure the safety of workers, improve labor productivity and ensure the need for fresh air in the tunnel during blasting, along with proper discharge of harmful gas and reduction of dust concentration, the ventilation modes adopted in tunnel construction are mainly diffused ventilation, ejector ventilation, mechanical ventilation and auxiliary tunnel ventilation (Fig. 2.3).

Fig. 2.3. Ventilation equipment effect in the tunnel hole.

It is recommended to use mechanical a ventilation scheme in the tunnel construction. In order to ensure the ventilation effect, generally pressure ventilation is used.

In tunnel ventilation, pressure was applied, from the ground by the shaft fan and duct, into the ventilation chamber to carry out excavation in the four directions and the slurry was deviated into the wells through the horizontal channels. This configuration is appropriate when the tee is set as far away as possible. The construction ventilation air volume required is based on the number of holes. A one-time blasting hole allows minimum wind speed. The need to rule out fumes and combustion from the end result is taken into consideration while calculating the maximum air volume.

【Engineering project】

(1) Calculate the wind volume by the maximum number of people working in the tunnel at the same time,

$$Q = k \cdot m \cdot q,$$

where k is the wind volume reserve coefficient, $k = 1.1$–1.2, m is the maximum number of people working at the same time in the tunnel, on average 40 people/per working face, q in each hole, the required air volume per minute (3 m^3/min).

$$Q = k \cdot m \cdot q = 1.2 \times 40 \times 3 = 144 \text{ (m}^3\text{/min)}.$$

(2) Calculate the wind volume according to the minimum wind speed in the hole,

$$Q = 60 \times V \times S,$$

where V is the minimum wind speed allowed in the hole (0.15 m/s), S is the tunnel area (m^2),

$$Q = 60 \times V \times S = 60 \times 0.15 \times 37.3 = 336\ (m^3/min).$$

(3) Calculation of the air volume required for blasting of the surface,

$$Q = \frac{7.18}{t}\sqrt[3]{A(F \cdot L)^2},$$

where t is the ventilation time (30 min), A is the consumption of explosive charge (34 kg), F is the excavation section area (37.3 m^2), L is the ventilation length (460 m).

$$Q = \frac{7.18}{t}\sqrt[3]{A(F \cdot L)^2} = 516(m^3/min).$$

(4) Calculate the amount of air required for internal combustion machinery,

$$Q = (H_s \cdot \alpha_s + H_D \cdot \alpha_D + H_E \cdot \alpha_E) \cdot q,$$

Where H_s is the total power of slag loading machine (kW), H_D is the total power of transport vehicles (kW), α_D is the efficiency of transport vehicles, H_E is the total power of other machines (kW), α_E is the other mechanical work efficiency, q is the air volume of unit power of internal combustion machinery (3 m^3/(min·kW)).

A small excavator is used in the hole, of 116-kW power, along with 2 dump trucks of 162-kW power.

$$\begin{aligned} Q &= (H_S \cdot \alpha_S + H_D \cdot \alpha_D + H_E \cdot \alpha_E) \cdot q \\ &= [116 \times 0.6 + 162 \times 2 \times 0.6] \times 3 \\ &= 792\ (m^3/min). \end{aligned}$$

Therefore, assume the maximum of 792 m^3/min control volume and consider the duct's 100 m air leakage rate of 1% on average, leading to the total required air volume of 792/(1–0.01 × 713/100) = 853 m^3/min.

According to the calculation results, the air is supplied by two sets of 55-kW axial flow fans, selecting a total duct length of ~1,000–1,500 soft duct. The wind at the bottom of the shaft is constrained with the ~600 duct.

Mechanical ventilation is the main ventilation in the underground excavation, and pressure ventilation is also adopted. In order to ensure the ventilation effect and improve the working face environment, the tunnel can be equipped with a 4-kW fan.

2.2.2.1.8 Other facilities

In addition to the above facilities, the general shaft also needs to furnish various warehouses, offices and emergency personnel housing. In the site layout design, one also needs to consider the entrance and exit road, field road, gate, etc.

The layout plan for a vertical well of Qingdao Metro Line 13 is shown in Fig. 2.4. The list of temporary shaft facilities is shown in Table 2.5.

2.2.2.2 *Key points of inclined well layout*

The inclined well usually covers a large area, with ground wide enough and convenient for site layout and temporary facilities. The layout principles for shaft and well are basically the same. In addition, because there is a hole transporting the slag car Canon down the slope, the sediment can be directly shipped from the hole and can even be directly shipped to the abandoned slag field; so it can be determined whether there is a need to set up a temporarily abandoned slag field based on conditions in the field. An inclined plane layout for Qingdao Metro Line 13 is shown in Fig. 2.5, along with a list of temporary facilities in Table 2.6.

2.2.3 *Pipeline layout of wellhead and hole*

2.2.3.1 *Wellhead pipeline and layout points*

The main pipes in the shaft are the high-pressure duct, ventilation pipe, water supply pipe, drain pipe, power line and lighting power line. The concrete discharge pipe should be arranged at the mouth of the shaft.

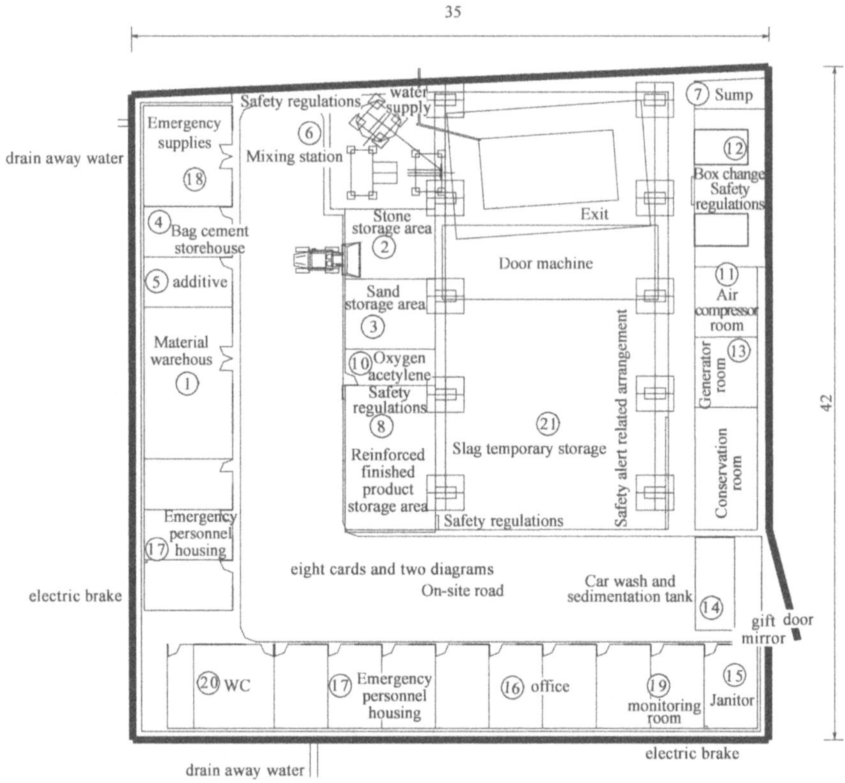

Fig. 2.4. Layout plan of vertical well in Qingdao Metro Line 13 (unit : m).

According to the requirements of wind volume, the high-pressure duct is required to use a steel pipe of more than $\phi 1{,}000$ mm. The installation requires smooth joint sealing to prevent air leakage and setting aside of the pipe when necessary.

In the wellhead area, the ventilator is usually used with the steel pipe or tube of $\phi 1{,}000$–$\phi 1{,}300$ mm. The hanging should be straight, and the hoop can be set up, and the connection of the well wall is strong.

The water supply and drainage pipe can be used for steel tubes of $\phi 75$–$\phi 100$ mm, the dressing should be smooth, straight, with few corners and connections, and the head should be tight and watertight.

The layout of the pipe in the hole is shown in Fig. 2.6, and the feng shui pipeline layout in the hole is shown in Fig. 2.7.

Table 2.5. List of temporary facilities for a vertical well in Qingdao Metro Line 13.

No.	Temporary works	Unit	Area
1	Material warehouse	m^2	64.8
2	Gravel storage area	m^2	30
3	Sand storage area	m^2	30
4	Bagged cement warehouse	m^2	21.6
5	Admixtures library	m^2	21.6
6	Mixing station	m^2	56
7	Set pool	m^2	24
8	Rebar finish stacking area	m^2	61.5
9	Turnover material stacking area	m^2	
10	Oxygen acetylene library	m^2	15.6
11	Generator room	m^2	18
12	Box change	m^2	56.6
13	Air compressor machine room	m^2	18
14	Sedimentation tank and car wash table	m^2	17.5
15	The gatehouse	m^2	21.6
16	The office	m^2	21.6×3
17	Emergency housing	m^2	21.6×7
18	Emergency storeroom	m^2	21.6×2
19	The control room	m^2	21.6
20	The toilet	m^2	43.2
21	Temporary slag dump	m^2	200
22	Total	m^2	980.6

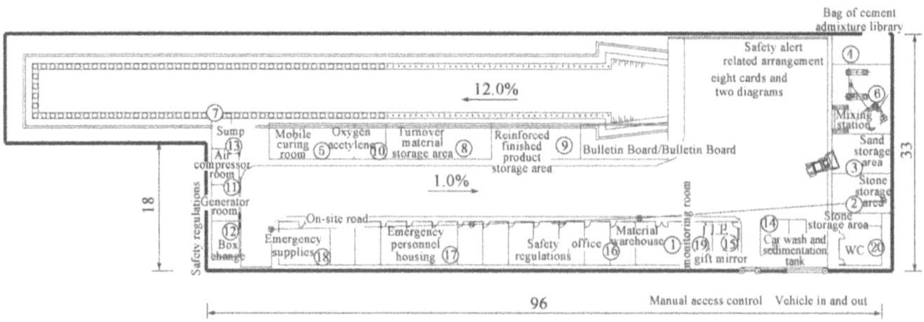

Fig. 2.5. Layout plan of an inclined well in Qingdao Metro Line 13 (unit: m).

2.2.3.2 *Interior piping and layout points*

The main pipes in the hole are the high-pressure duct, ventilation pipe, water supply pipe, drain pipe, power line and lighting wire. The pipeline layout is shown in Fig. 2.8 and the ventilation tube setting in Fig. 2.9.

The high-pressure duct is connected from the wellhead, and the wall is arranged along the hole's wall. The laying requirement is smooth, the joint is sealed and air leakage is prevented. The front end of the pipe is kept at about 30 m, and the air separator is connected with a high-pressure hose. The length of each working face hose should not exceed 50 m. The pipe shall be installed on the opposite side of the electrical circuit and on the same side as the water supply pipes.

The ventilation tube adopts a ϕ1,300–ϕ1,500-mm pull chain light wind pipe and a wellhead iron pipe or hose connection, which is hanging flat, with good stability, to avoid fold increase resistance, with a curved tunnel junction to avoid accidents. The ventilator should follow the plan.

There should be a distance of 30 m from the water supply pipeline front to the excavation face, with a 50-mm diameter high-pressure water segregator hose, at an intermediate oblique angle. The length of the hoses used on other working face water supplies should not be more than 50 m. The water supply pipe should be installed on away from the electric circuit as it should not interfere with transport and pedestrians.

The three-phase five-wire power supply line is adopted in the tunnel hole, and the lighting voltage in the operation lot is 36 V, and the lighting voltage in the hole and non-operating area is 220 V.

Table 2.6. List of temporary facilities for a slope of Qingdao Metro Line 13.

No.	Temporary works	Unit	Area
1	Material warehouse	m^2	64.8
2	Gravel storage area	m^2	35
3	Sand storage area	m^2	35
4	Bagged cement warehouse	m^2	21.6
5	Admixtures library	m^2	21.6
6	Mixing station	m^2	49
7	Set the pool	m^2	20
8	Rebar finish stacking area	m^2	60
9	Turnover material stacking area	m^2	100
10	Oxygen acetylene library	m^2	20
11	Generator room	m^2	20
12	Box change	m^2	56.6
13	Air compressor machine room	m^2	20
14	Sedimentation tank and car wash table	m^2	32.5
15	The gatehouse	m^2	21.6
16	The office	m^2	21.6×6
17	Emergency housing	m^2	21.6×8
18	Emergency store	m^2	21.6×2
19	The control room	m^2	21.6
20	The toilet	m^2	43.2
21	Field curing room	m^2	40
22	Total	m^2	1028.1

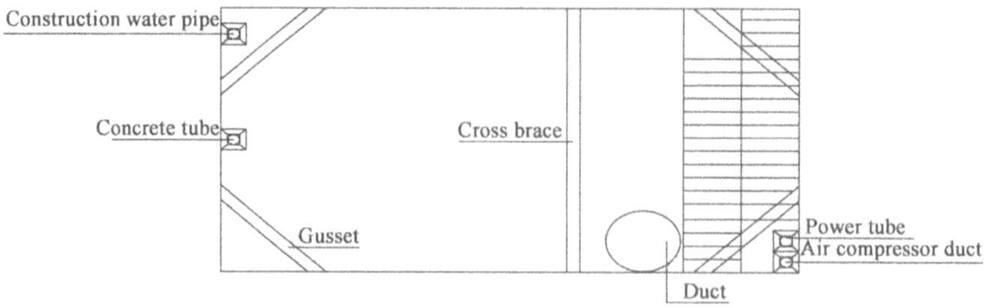

Fig. 2.6. Layout of vertical wellhead pipeline.

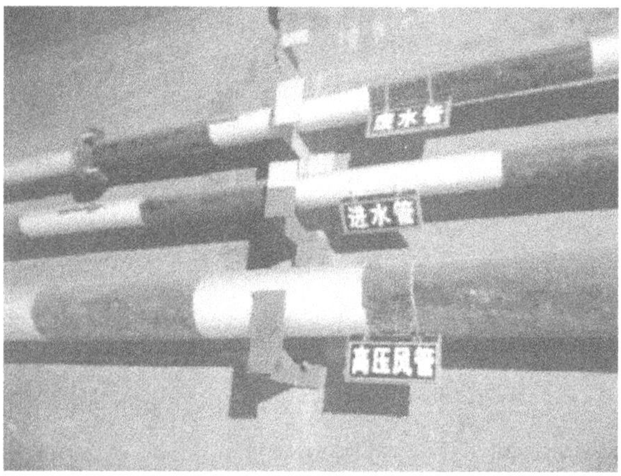

Fig. 2.7. Geomantic pipeline layout in the cave.

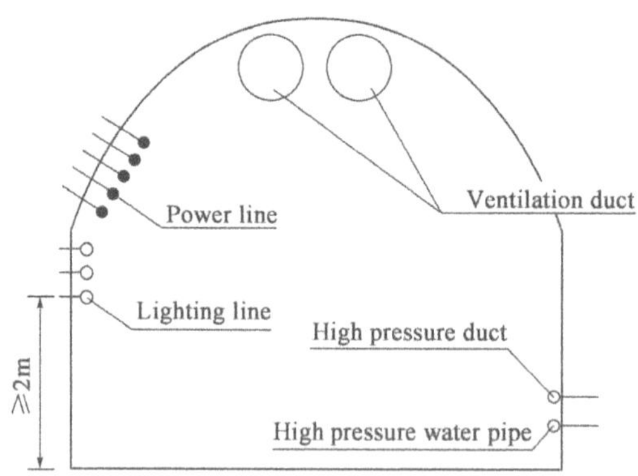

Fig. 2.8. Pipeline layout in the hole.

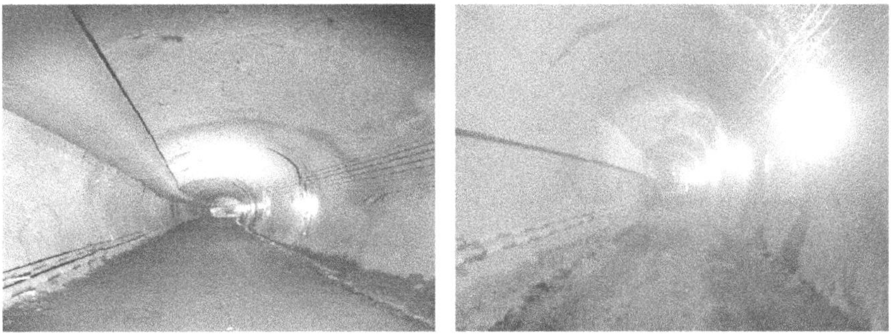

Fig. 2.9. Ventilation tube setting in the tunnel.

2.2.4 *Vertical shaft-hoisting equipment selection*

2.2.4.1 *Calculation improve equipment efficiency in vertical shaft*

Slag needs vertical shafts for transportation, so it is very important to select equipment carefully and check equipment performance status as the capability directly affects the hole construction. This book introduces the calculation example of hoisting equipment for hoisting of vertical shafts in the two No. 1 vertical shafts of the fifth work area of Qingdao Metro Line No. 13.

2.2.4.1.1 Calculation of single sunrise slag

The slag discharge will be maximum when the four sections of tunnel are excavated at the same time. The amount of excavation per meter for 4 sections is $42.8 \times 4 = 171.2$ (m^3), the interval starting range is YCK19 + 178.196 and the range of shaft center distance is YCK20 + 146.109; since the shaft to the interval starting point (the smaller range direction) tunneling length is 427.913 m, the interval end (the larger range direction) tunneling length is 353.891 m. According to the time limit for a project, the interval excavation time is about 12 months, and the daily driving single direction will approximately be 427.913/12/30 (working days) ≈ 1.2 (m); taking a loose coefficient of residue as 1.7, the slag quantity is $171.2 \times 1.5 \times 1.7 \approx 349$ (m^3/d).

2.2.4.1.2 Calculation of slag time for a single bucket, single cycle

The muck and lift are prepared for 2 min. A skip is lifted from the bottom hole to the wellhead in about 4 min (note: wire rope hoist speed is 8 m/min, hole depth is 29.451 m and the height from the mouth of the well is about 5 m). The time taken for the skip from the wellhead to finish discharging slag is about 2 min, with the skip spending nearly 4 min down at the bottom of the shaft. The total cycle time for another skip to de-slag is $2 + 4 + 2 + 4 = 12$ (min).

2.2.4.1.3 Hoisting equipment

Taking a single bucket's single slag time as 18 min (including muck time), the daily slag time of the improved system becomes 10h. The daily ascending number is $10 \times 60/18 = 34$ (time), and a single bucket's lifting capacity is at least $349/34 = 10.26$ (m^3).

One can select a 30-t hook crane as lifting equipment for slag and steel materials, and 30 t for slag, steel bars, grill material and dipping sono-buoys. The selected slag bucket volume is 10 m^3, and size is 2.5 m (length) × 2 m (width) × 2 m (high).

The maximum static tension for improved hooks is 300 kN (30 t), the 10 m^3 slag bucket's weight is about 3.5 t, the weight of the wire rope is 8.5 m × 4 × 5.05 kg/m = 0.17 t and the residue weight is 10 m^3 × 1.9 t/m^3 = 19 t. So, the largest load of the improved hook is $3.5 + 0.17 + 19 = 22.67$ (t) < 30 t, which satisfies the requirement of load.

Check out the slag ability: Taking the hoisting system slag time as 10 h/day, the boost time for one day becomes $10 \times 60/12 = 50$ times and hoisting system slag capacity per day is $10 \times 60/12 \times 10 = 500$ m^3 > 349 m^3, which can meet the requirement of slag.

The shaft is excavated and dredged through a small excavator, and the slag can be discharged through the lifting equipment.

2.2.4.2 *Selection of shaft-hoisting equipment*

The common shaft-hoisting equipment contains electric hoist, gantry crane, bridge crane, mobile crane, etc. The electric hoist is widely applied in underground shaft hoisting, but in recent years, due to the electric hoist's slightly lower safety and reliability, it has been replaced by the gantry crane and bridge crane which have higher safety performance and more enhanced ability; the mobile crane is commonly used in small areas of shaft hoisting in engineering. The lifting capacity and lifting speed of the lifting equipment are generally considered according to the amount of slag taken by the vertical shaft.

The selection of various shaft-hoisting equipment is shown in Table 2.7.

Table 2.7. Selection of various shaft-hoisting equipment.

Crane type	Advantages	Disadvantages	Scope of application	Notes
Electric hoist	Low cost and flexible, can work at the same time as other machines, improved speed, easy installation, small footprint.	Relatively small lifting capacity, high efficiency, high failure rate, low safety and reliability in performance.	Small sites, less slag, cross-work, short construction period.	
Gantry crane	High efficiency, high safety and reliability.	The cost is high, the foundation and other inputs are large, the installation is complicated, and the failure directly affects the construction.	Good site conditions, large slag quantity, long construction cycle and long shafts.	
Bridge crane	High efficiency, high safety and reliability.	The cost is high, the basic input is large, the installation is complicated, and it has a direct impact on the construction if it breaks down.	Good site conditions, large slag quantity, long construction cycle.	
Mobile crane	The ability to improve can be flexibly chosen according to the need, and the entrance is flexible and not restricted by the site.	High slag cost, high efficiency, low safety and reliability in performance.	Small site, less slag, short construction period, temporary slag shaft.	

2.2.5 *Setting of incline ramp and passing bay*

The incline slope design should consider the incline length and slope; the slope incline is generally 10–15% (Fig. 2.10). The slope is too small to increase the slope length and the slope is too large to affect the transportation capacity.

For the convenience of the passing bay over the incline well, a passing bay hole is set every 100 m, the passing bay fracture surface should be widened, the reserved hole width should be large than 7 m, the length should be about 15–20 m and the slope should be set up as a flat slope. Using gradient transition between a standard section and the reserved hole, the gradient can be used when unilaterally widened, or also with double-sided widening. Passing bay excavation is the same as the section excavation method. When the inclined shaft's construction reaches the passing bay mileage, the footage is shortened and the tunnel's excavation is strictly controlled. The survey team lofted the position of the side wall on the right and then proceeded to the passing bay by the step method.

When the passing bay excavation is near the design end mileage, the loop footage is strictly controlled to prevent over-excavation after entering the hole. The passing bay excavation is completed by the metering team,

Fig. 2.10. Site photos of slope 12 ramp.

after deciding the excavation contour points and strictly controlling the direction of the hole. Then, drilling–blasting is performed and the boundary stress concentration support can be appropriately increased. The dark hole of the passing bay plane has been shown in Figs. 2.11–2.13.

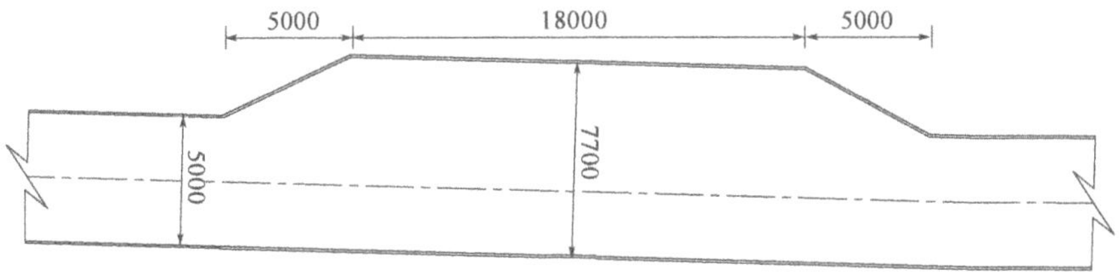

Fig. 2.11. Schematic diagram of single-side widening passing bay of dark hole (unit: mm).

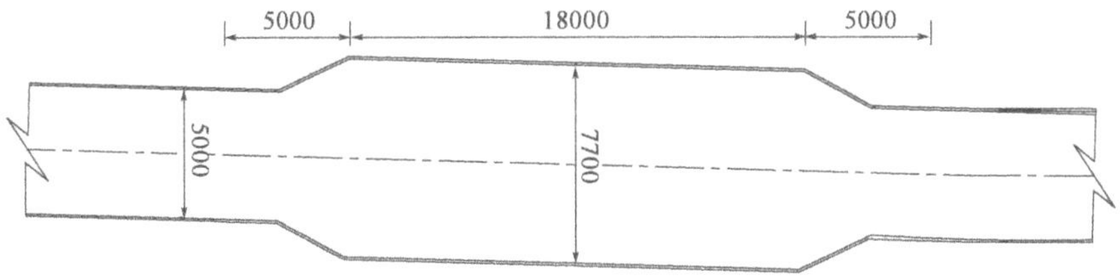

Fig. 2.12. Double-side widening plan of dark hole in the passing bay (unit: mm).

Fig. 2.13. Scene of passing bay in the dark hole.

2.2.6 *Horizontal transportation equipment selection in tunnel*

In drilling–blasting construction in hard rock underground stratum, efficiency is guaranteed based on the two key links that affect quick tunneling, excavation and transportation. Especially in urban metro construction, the blasting time has strict restrictions, such as from 7:00 to 19:00 in the Qingdao region, in which to do the two loops (explosion twice a day), thereby needing to continuously optimize the excavation and transport links.

Statistical analysis was carried out in each cycle of the operation time of each process, slag transportation, the transportation of sprayed concrete and steel transportation, all of which take up 40% to 60% of the whole cycle time. The process time of each cycle during the excavation of the drilling and blasting method is shown in Table 2.8.

Table 2.8. Statistical table of each cycle's time in drilling and blasting excavation.

No.	Cycle operation	Average process time (h)
1	Shot eye	2
2	Charge blasting	1
3	Digging and dredging	3.5
4	Establish the netting and advance support	3.5
5	Shotcrete	2
6	Total	12

2.2.6.1 *Option of shipment*

At present, there are three main types of transportation modes in domestic tunnel construction:

(1) Rail slag loading and rail transport mode are suitable for mechanical excavation of short-distance tunnels, such as shield tunnels.
(2) Trackless slag and rail transport mode are suitable for long-distance tunnels, such as ordinary railway single-line tunnels.

(3) Trackless slag and trackless transport mode are suitable for large-section tunnels or short-sized tunnels, such as highway tunnels.

In the drilling and blasting construction of the urban metro tunnel, the tunneling length of single hand surface is generally within 1 km, and the mode of transport is trackless slag and the non-rail transport mode.

2.2.6.2 *Selection principle of shipping equipment*

2.2.6.2.1 Size of the equipment should be matched with the tunnel clearance

Clearance dimensions for metro railway tunnels are much smaller than railway tunnels and highway tunnels. The standard section tunnel excavation contour line width and height are usually within 7 m. It should be ensured that the mechanical equipment can operate freely within the boundary dimension or hole size. Crossover operation machinery and equipment should be able to meet the requirements of a safe distance between each other, for example, one needs to consider whether the loader and slag dump truck can each be utilized satisfactorily in the limited space.

2.2.6.2.2 Performance and production capacity of mechanical equipment should be adjusted to the construction work

The production capacity of each kind of equipment should be matched with other equipment related to its work, for example, the slag equipment should be matched with the dregs equipment. Dynamic performance of mechanical equipment should satisfy the slope of the tunnel, especially when using a deviated well as the construction passageway, whereby the equipment must be rationally allocated according to the inclined shaft gradient and the equipment's production capacity (the number of configurations) should correspond to a workload per cycle, on the premise of meeting the time constraints of the project and trying to reduce the total investment on equipment.

2.2.6.2.3 Pay attention to equipment adaptability

In tunnel construction using the urban metro drilling and blasting method, the surrounding rock changes frequently and the change is generally big; during the selection of the loading equipment, one should fully consider the adaptability of the equipment to all kinds of surrounding rock types, to adapt to a variety of environmental requirements.

2.2.6.2.4 Universal equipment is preferred

If the equipment in the tunnel fails, due to the limited working space, it is easy for the passage to get blocked, and other processes cannot be carried out, thus affecting the entire tunnel's transportation. General equipment refers to the equipment manufacturers and models that have the features of sufficient spare parts and easy maintenance, so as to reduce the impact of equipment failure on the construction to the lowest.

2.2.6.2.5 Optimize environmental protection equipment

Tunnel construction has limited space, so ventilation is one big problem, requiring the selected equipment to be low-pollution or pollution-free equipment, in order to minimize air pollution in the tunnel.

2.2.6.3 *Typical case of shipping equipment type selection*

Case 1: In the construction of the main shaft and the interval of Qingdao Metro Line 13, the construction of four constraints is done at the same time, with driving at 6 m per day at the peak speed, average excavation section of 40 m^2 and tunnel clearance of about 6.6 m; a single constraint's tunneling is about 400 m.

(1) Equipment selection
 (a) **Grilled slag equipment:** Restricted by the space in the hole, as it has to be moved up the steps, the chosen grilled slag equipment is small excavators, model SY65C, with a bucket capacity of 0.22 m^2; the equipment covers a small area, does not need to be under the

new surrounding rock blasting operations, the operation is safe and flexible, and the slag can also be shipped out when the slagging work is ongoing.

(b) **Muck equipment:** A cartwheel loader is chosen as the muck equipment, model ZLC50C, with a bucket capacity of 3 m^3, and length × width × height dimensions of 8,100 × 2,990 × 3,430; the actual efficiency of muck is about 1.3 m^3/bucket.

(c) **Carrying slag equipment:** Shaft equipment selection should be based on the slag bucket capacity of the shaft-hoisting equipment. The slag bucket capacity is 12 m^3. Considering the muck needs to be loaded with a side loader in parallel, one must choose a dump truck which has a rated slag quantity of 6 m^3; two cars can fill up the lifting equipment dregs bucket.

(2) Equipment quantity selection

(a) **Grilled slag equipment:** Using small excavators, the time taken to find the top of the blasting face is about 0.5 h, and the grilled slag takes about 2 h. The four constraints are constructed at the same time, at least 6 cycles per day, with a theory operation time of about 15 h. Considering the four surfaces, the blasting interval is short. It takes a long time to configure two pieces equipment in the early stage of the construction, so the number is increased to three.

(b) **Slag loading equipment:** Compared with the slagging efficiency, the actual loading efficiency of the side flip loader is higher, and the two can meet the demands of on-site construction.

(c) **Slag transport equipment:** Shaft equipment is affected by the shaft-hoisting capacity. Considering the continuous operation of the shaft-hoisting equipment, four pieces of equipment can meet the requirements of transporting slag, and they can increase or decrease the capacity according to the length of the main heading.

Case 2: For a certain inclined shaft and construction of the positive line interval in Qingdao Metro Line 13, the four constraints are constructed at the same time, driving 9 m per day at the peak speed, with an average excavation section of about 40 m^2, clearance of the tunnel initial support of about 6.6 m and single constraint tunneling of about 500 m.

(1) Equipment selection
   (a) **Grilled slag equipment:** Restricted by the space in the hole, as it has to be moved up the steps, the chosen equipment is small grilled slag prophase of the model SY65C. But it found later in the construction that it took up too much time to look for the top and grilled slag, so then the medium EC210B backhoe excavator was selected; although the grilled slag reduced operating flexibility, the time it took to look for the top and grilled slag reduced from 2.5 h to 0.5 h, which greatly increased the efficiency.
   (b) **Muck equipment:** A cartwheel loader, model ZLC50C, is chosen as the muck equipment, with a bucket capacity of 3 m^3 and length × width × height dimensions of 8,100 mm × 2,990 mm × 3430 mm; the actual efficiency of muck is about 1.3 m^3/bucket.
   (c) **Slag transport equipment:** The selection of the deviated well slag transport equipment is without the limitation of the lift system; only parallel working with the muck loader is taken into consideration, and a machine whose rated slag carrying volume is 10 m^3 is chosen.

(2) Equipment quantity selection
   (a) **Scum equipment:** The EC210B backhoe excavator has high efficiency, and can be configured to meet the needs of the site; based on optimization, one can configure 1 or 2 small excavators to work at the top of the blasting face, which can further improve the working efficiency of the grilled slag equipment.
   (b) **Muck equipment:** Two previous configurations can meet the site construction requirements, but with a deepening of the hole; because of the high requirements of the slag in the inclined well, the hole's depth needs to be appropriately increased, and considering transportation of other materials, the slope height can be increased to 4 units.
   (c) **Slag conveyance equipment:** The amount of slagging equipment can be configured flexibly based on daily excavations, and

it is not restricted by other conditions. The slant well is equipped with six self-discharging vehicles, which can meet the requirements of slag transport.

2.2.7 *Brief summary*

In the construction in hard rock stratum, the setting scheme of the vertical inclined shaft is closely related to the construction efficiency of the positive line, so it is necessary to make a plan before the construction starts, and the following points should be considered:

(1) According to the construction efficiency of the vertical shaft, normal construction time can be estimated. After the completion of the setting of the vertical inclined shaft, the length of the interval should be made as consistent as possible.
(2) The wellhead should be as close as possible to the normal line, but it is not appropriate to set it directly above the positive line. When the line spacing between the two lines is narrow, the shaft should not be set between the two lines.
(3) In urban metro construction, the settings of the vertical slope should fully take into consideration the limitation that large vehicles place on city transportation. There should be ample ground space for the possible temporary storage of slag to avoid the slag causing a traffic jam; traffic control and relevant laws should be enforced.
(4) The lateral channel of the general vertical slope needs a large hole diameter. The geological report should be analyzed in advance and a good location should be chosen with the surrounding rock having a set vertical slope. This is conducive to a quick start of the hole construction and the main line construction.
(5) The general site area of the vertical slope is small. Therefore, the temporary facilities' layout should be according to the arrangement of the main job in order to avoid insufficient distribution of power, short supply of the air compressor, insufficient shaft-lifting capacity and so on.

2.3 Excavation and support technology of drilling–blasting method

2.3.1 *Support structure theory and evolution*

2.3.1.1 *Main points and differences of the traditional support mechanics theory and the modern support mechanics theory*

With the increasing demand for underground space, the practice of and theoretical research in underground engineering have made great progress. In large-scale underground engineering practice, it is generally realized that the core of the issue and underground engineering problems are due on the excavation, which supports two key processes. In other words, the manner of excavation has to be more conducive to the stability of cavern and ease of support and transportation. In tunnel engineering, the practice and research around these core problems have put forward different theories and gradually established different theoretical systems. One theory is the relaxation of load theory which was put forward by representative figures such as Karl von Terzaghi and M. M. Plotkiyakonov in the 1920s. Its core content is a stable rock mass has the ability to withstand load, while an unstable rock mass may collapse without a supporting structure. At the same time, another theory that was proposed is the theory of traditional drilling and blasting, which is similar to the concept of ground engineering. Another theory was put forward by representative figures such as Rabcewicz, Miller-Fecher, J. Talobre and H. Kastner in the 1950s. Its core content is the stability of the surrounding rock. It is obvious that a rock mass has self-stabilizing ability, and the loss of stability of the surrounding rock has a course. In this process, necessary help is provided and the surrounding rock is still maintained in a stable state. On the basis of engineering practices and reasoning, the Austrian rock mechanics engineering group led by Rabcewicz put forward the origin of the modern support theory and named it the new Austrian law.

The mechanics of the traditional support theory include soil mechanics, structural mechanics and so on. Soil mechanics considers the surrounding rock as a granule, calculates the load size and distribution of the supporting structure, calculates the internal force, and makes it reasonable in the structural mechanics visual support and lining as the bearing structure. Through the establishment of a "load structure" mechanical system, the most adverse load combination is used as the structural design load. In accordance with this theory, the following points should be carefully observed: (1) It is inevitable that the surrounding rock should be relaxed or steady after excavation. However, the occurrence of collapse is accidental, so it is necessary to accurately judge the possibility of the collapse of various surrounding rocks. (2) Even if the surrounding rock does not collapse, the relaxation could still impose load on the supporting structure. Therefore, it is necessary to accurately determine the size and distribution of the load. (3) In order to ensure the stability of the surrounding rock, temporary support and permanent lining should be designed according to the load size and distribution to carry the structure and make the structure force reasonable. (4) Although the bearing structure is designed according to the most unfavorable combination of loads, one should mitigate the development of slacking and collapse.

The modern supporting theory includes rock mass mechanics, elastic mechanics, elastic–plastic mechanics, etc.; it takes the surrounding rock stress as per the rock mass, analyzes and calculates the stress and strain states and the change process, and takes support from the stress rock boundary condition and holds the control of the surrounding rock's stress and strain. The effect of the analysis is optimized, a surrounding rock-supporting mechanics system is established and the actual stress and strain states are used as the design support.

In the implementation of this theory, the following points should be carefully observed: (1) The surrounding rock is the main stress-bearing structure, so it is necessary to protect the surrounding rock and reduce any disturbance during the construction. (2) The primary support and permanent lining are the only constraints on the surrounding rock, which allows finite deformation of the surrounding rock within its carrying capacity, and prevents excessive deformation which can result in instability. Therefore, this method uses a thin-walled flexible structure as the appropriate primary support. (3) The stress and strain of the surrounding rock's dynamics indicates whether it is in a stable state; therefore, in order to provide timely, appropriate support, that dynamics should be controlled through monitoring and correctional design. (4) The overall buckling is usually caused by local failure development, so the supporting structure should be closed as soon as possible; for comprehensive constraint on the surrounding rock, there should be a timely closed loop, especially if the surrounding rock is broken with weak bearing capacity.

It can be seen from the above section that relaxation theory pays attention to the results and the processing of the results, while rock-bearing theory pays more attention to process control. Thus, the two ideas produce very different supporting results.

2.3.1.2 *The development of the tunnel support system*

Based on the two kinds of support calculation concepts, and the development in construction materials and technology, the tunnel support system is also undergoing corresponding changes.

Due to the influence of construction technology and the relaxation theory, generally wood or steel are used as more strong support in the initial stages after tunnel excavation. The support method causes a lot of stress on the supporting system, wastes a lot of wood or steel and at the same time takes up a lot of construction space.

In recent decades, according to the modern rock mechanics principle design of the tunnel support system, the tunnel structure is composed of surrounding rock and the supporting structure, of which the surrounding rock is the main weight-bearing part and the supporting structure is the auxiliary but often indispensable part, while in some cases, the supporting structure is the main bearing unit.

Therefore, as soon as the tunnel is excavated and exposed, it is suitable for the laying of an initial shotcrete support to protect and strengthen the surrounding rock and gradually stabilize it. The secondary lining is adopted to ensure permanent stability and safety of the tunnel. The primary support and secondary lining form the compound support of the tunnel.

There is widespread use of some materials such as anchor and shotcrete. Spray anchor bracing is widely used in engineering practice as it is constantly developed and perfected, not only showing great superiority under complex geological conditions but also accumulating a large amount of engineering practice experience. It summarizes a set of perfect and advanced construction methods, and puts forward the design principle and points out the direction for scientific construction of tunnel projects.

Shotcrete required an even mix of a certain proportion of cement, sand and gravel to form a supporting layer through inject on the surface of surrounding rock with water at a high speed (30–100 m/s), after which it condenses, causing a sclerosis quickly with the concrete jet mix. Since concrete is under certain pressure during injection, the sand, stone aggregates and cement particles cause repeated impact, which is equivalent to continuous compaction and consolidation. Therefore, to ensure that the sprayed concrete has high physical and mechanical properties, a smaller water–cement ratio can be used during the injection process.

The bolt support is needed to improve the stress state of the surrounding rock, and the strength of the surrounding rock can be improved by inserting the bolt into the rock mass.

According to the principle of support, shotcrete support can give full play to the self-bearing capacity of the surrounding rock, thus reducing the pressure on it. In the construction process, a joint operation with a high degree of mechanization is carried out, which is beneficial in reducing labor intensity and improving work efficiency. Based on economics alone, spray anchor bracing produces a thicker tunnel lining and required less dosage than traditional methods, according to the statistics, as compared with pouring concrete support. The thickness of supporting shotcrete can be reduced by 1/3–1/2, saving rock excavated volume by 10–15%, accelerating the support speed by 3–4 times, saving labor by more than 50% and reducing support costs by more than 30%. Based on a vast amount of engineering practice at home and abroad, the construction quality in this method is reliable. In all kinds of supporting operations of underground engineering technology in China, spray anchor bracing solved a large number of technical problems which are difficult to solve by using traditional support, proving the method's obvious superiority.

Since the 1980s, in a large number of engineering practices, China has also innovated the concept of "shallow buried tunneling method", along with all kinds of advanced treatment measures, such as lead bolt, lead pipe shed, lead small pipe, advance grouting and so on, thereby enriched the tunnel support system.

2.3.1.3 *Principle of suspension roof support*

The bolt was first used in a coal mine in 1912, and the main types are resin bolt, mortar bolt and pipe joint bolt. The mechanism of the anchor rod is in discussion within the engineering community, while the mainstream view is mainly based on suspension, combination, extrusion and reduction. The principle of anchor support is shown in Table 2.9.

Table 2.9. Principle of bolt support.

No.	Bolting action	Abbreviated drawing	Principle
1	Suspension action	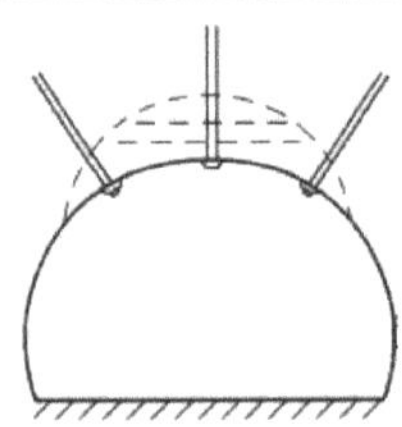	In the block structure or fractured rock mass, it can fix the loose block on the stable rock mass to prevent slips and falls, or it can join the pieces together. The bolt itself is pulled by the loose block; this effect is called the suspension action.
2	Span reduction effect	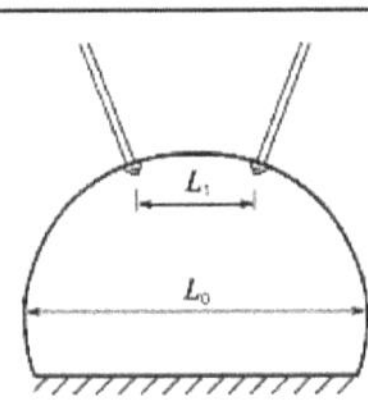	The bolt is inserted into the tunnel roof strata, equivalent to increasing the fulcrum in the roof; the tunnel span is shortened from L_0 to L_1 to reduce the surrounding rock stress, so as to maintain the stability of the surrounding rock.
3	Combination action	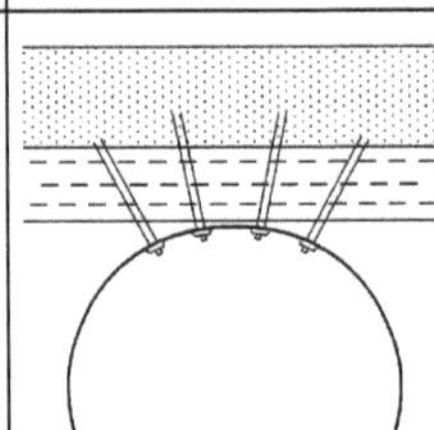	In a layered structure, the bolt is inserted into thin layers of rock. The aim is to mix a number of thin layers of rock anchor together, combined into a thick plate or beam, thus improving the overall carrying capacity of the surrounding rock and reinforcing the stability of the surrounding rock.

Table 2.9. (*Continued*)

No.	Bolting action	Abbreviated drawing	Principle
4	Squeezing action	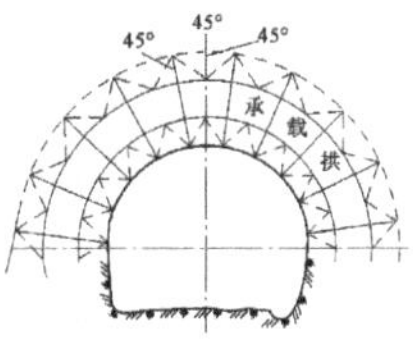	In the prestressed system, the anchoring bolt reinforces the surrounding rock and rock mass from conical extrusion near both ends. According to the certain spacing arrangement under the effect of a prestressed anchor, a uniform extrusion zone is formed, along with a bearing arch. Although the loading method is different for the unprestressed bonding bolt, it has a similar squeezing effect. The surrounding rock's displacement deformation causes the anchor rod to be pulled. At the same time, the bolt is binding to the surrounding rock, and the surrounding rock between the two adjacent bolts is arched and stable under the pressure of binding force.

2.3.1.4 *The history of the development of shotcrete*

Shotcrete is developed from gunite slurry, which was first used in mining and civil engineering in the United States in 1914. In 1942, a rotary concrete jet was developed by the Swiss Aliva company. In 1947, the federal German BSM company developed a dual-tank concrete jet. Mill tunnels, built between 1948 and 1953, were the first to use shotcrete support. Later, Switzerland, West Germany, France, Sweden, the United States, the United Kingdom, Canada, the Soviet Union, Japan and other countries adopted shotcrete technology in civil engineering. The metallurgy, water and electricity sectors in China studied injection machinery and concrete shotcrete technology in the early 1960s. In November 1965, in cooperation with the Third Metallurgical Construction Company, the Architectural Institute of Metallurgical Department successfully built a bow with shotcrete support in the mining roadway to the An Gang Chang Ling iron ore. Since the 70s, on both the domestic and overseas fronts, research and development work on sprayed concrete has increased, leading to many technical breakthroughs, and has since been extensively developed in fields such as underground engineering, thin-walled structure engineering, maintenance engineering, geotechnical engineering and protective engineering.

Shotcrete and pumping concrete are two different concepts. Shotcrete is mainly used in underground engineering, geotechnical engineering, municipal engineering, hydraulic engineering and tunnel engineering. Bolt spray rapid hardening concrete is often used early as a rapid and stable constraint after blasting excavation as it greatly reduces landslides and may speed up the construction progress. On the contrary, concrete is no more than an addition of early strength agent; there is not much room for development of the material, and the main development space is the improvement of the injection machine.

The dry type of concrete jet conveyor is a long, simple and durable equipment. But because it is difficult to make the dry mixing concrete flow easily out of the nozzle after it is mixed with water, construction dust gets sprayed back to a large area. The dry dust that is thus sprayed poses a health hazard to the workers who are working in tunnel engineering projects where the dust pollution is more serious.

From the 1960s, in western developed countries, wet shotcrete technology has been gradually utilized, and all types of wet shotcrete machines have also gradually been developed. The basic principle is to add water to the concrete in the finished concrete wet spraying machine, convey it to through the nozzle and spray the formed material after adding an accelerator.

In China, for a variety of reasons, the dry jet is still the main injection equipment. However, with the enhancement of environmental awareness and the improvement of the quality requirements of shotcrete, an increasing number of wet concrete jets have been used. In some large-scale water conservancy projects of international bidding, such as the Ertan hydropower station, Yellow River Xiao Lang Di project and Three Gorges project, wet concrete spraying operation was adopted. Many kinds of wet concrete or shotcrete used in China were imported from abroad. In recent years, some domestic units have developed several wet concrete jets, but the scale of production has yet to be expanded. The main advantages of wet shotcrete are as follows:

(1) The dust concentration outside the machine and the nozzle is greatly reduced, and any harm to workers' health is eliminated.
(2) Productivity is high. A dry shotcrete machine can generally work at a speed not more than 5 m^3/h, while with the use of a wet shotcrete

machine, manual operations can be at speeds up to 10 m^3/h and automated operations can be at speeds up to 20 m^3/h.

(3) The rebound degree is low. When using dry spray, the concrete rebound degree can reach 15–50%, and the rebound rate can be reduced to less than 10%.

(4) When using wet spray, because the water–cement ratio is easy to control, the concrete hydration degree is high; therefore, it can greatly improve the quality of sprayed concrete and improve the uniformity as well. As with the dry spray, the water–cement ratio of concrete is adjusted according to the experience of the jet worker and based on naked-eye observations. The quality of concrete depends largely on the tractor driver's ability to operate correctly.

2.3.2 *Forepoling*

Since the 1980s, in a large number of engineering operations, according to the different levels of the surrounding rock, bolts, advanced tube sheds and advanced ductule treatment are widely used, enriching the tunnel support system. During tunnel construction in the hard rock stratum using the drilling and blasting method, advanced support is also necessary to ensure construction safety.

The main advanced support measures are in many forms, such as leading bolts, leading pipes and small catheters. The types of support applications of the hard rock stratum tunnel in Qingdao area are shown in Table 2.10.

Table 2.10. Statistical table of application of main advance supporting types.

Advance support measures	Application environment	Specific parameter	Characteristics
Advanced anchor bolt	Levels III, IV of surrounding rock	Using normal hot-rolled ribbed bar or hollow grouting bolt, φ20–φ30 mm, length is 2–4 m.	It is suitable for the whole structure of surrounding rock and local crack development.

(Continued)

Table 2.10. (*Continued*)

Advance support measures	**Application environment**	**Specific parameter**	**Characteristics**
Leading small catheter	Levels IV, V, VI of surrounding rock	φ30–φ42-mm seamless steel tube with a length of 2–4 m.	It is widely used in the drilling and blasting method of tunnel construction.
Leading pipe shed	Levels V, VI of surrounding rock	For a large section, the tunnel with large load can adopt a big pipe shed of more than φ114. The length is generally 20–40 m.	It is applied to special sections such as large sections, poor geological sections, underground excavations, horse-head doors and underground buildings.
Advance grouting	Poor geologic sections in a broken belt or fault zone	The surface pre-grouting can be used for semi-section of the hole and the whole section.	Suitable for water or groundwater rich strata, use pre-grouting consolidation formation to prevent landslides.

2.3.2.1 *Main construction parameters of advanced anchor bolt*

2.3.2.1.1 Process parameters

The lead bolt can generally be applied in the whole structure of the surrounding rock in the hard rock formation and the local crack development area. One can use a normal hot-rolled ribbed bar or hollow grouting bolt, of φ20–φ30 mm, and a length of 2–4 m.

2.3.2.1.2 Construction technology

The bolt is arranged according to the design requirements and the YT-28 blower gun is used for drilling. Hole cleaning should be carried out after pore formation. The hollow grouting bolt with an anchor head will be installed into the bottom hole. The check plug, plate and nut are installed, then connected to the grouting pipe. Grouting through the tail hole is done with a grouting pump, using slurry, and the grouting pressure is controlled between 0.5 MPa and 1.0 MPa. The grouting sequence is from the bottom to top by the root. The stopper is plugged into the hole after grouting, sealing the hole with quick-setting cement.

(1) **Hole arrangement:** According to the design requirements based on the drawings, the feeding holes are painted red with an obvious mark on the rock face after excavation. The borehole should clear the pumice from the rock face and cliff stone. The bolt hole direction should be according to the design point of view; the local construction should be perpendicular to the main joint fracture surface of the rock strata according to the conditions of the rock mass.
(2) **Drilling:** Using a pneumatic drill for the borehole drilling, the drilling equipment, drilling depth and diameter should comply with the design and specification requirements, with a pitch error of less than 5 cm and oblique hole deviation of less than 5‰; the drilling should be straight, the aperture should be 30–40 mm larger than the bolt diameter and the bolt hole must be examined carefully before installing the anchor.
(3) **Clear hole:** Before grouting, clear water and high-pressure wind are used to clear the rock powder and water from the hole.
(4) **Hollow bolt installation:** According to the size of the design, the bolt is either purchased or custom-made. The hollow bolt is installed according to detailed instructions. The grouting rod of the hollow anchor is inserted with the anchor head to the bottom of the hole. The check plug, plate and nuts connecting the grouting pipe are installed. The length of the anchor rod should not be greater than 100 mm. Before the bolt is installed in the water section, the water inside the hole will be extracted and only then the bolt will be installed.
(5) **Grouting:** Taking the hollow grouting anchor rod body as a tube (Fig. 2.14), one can use electric pump grouting. The grouting pressure is not based on design requirements and can be controlled within 0.4 MPa. The grouting order is from the bottom to the top by the root of the grouting. The stopper is plugged into the hole after grouting, sealing the hole with quick-setting cement. Grouting must be dense, such as slurry, with timely recharge. Exhaust problems of the hollow

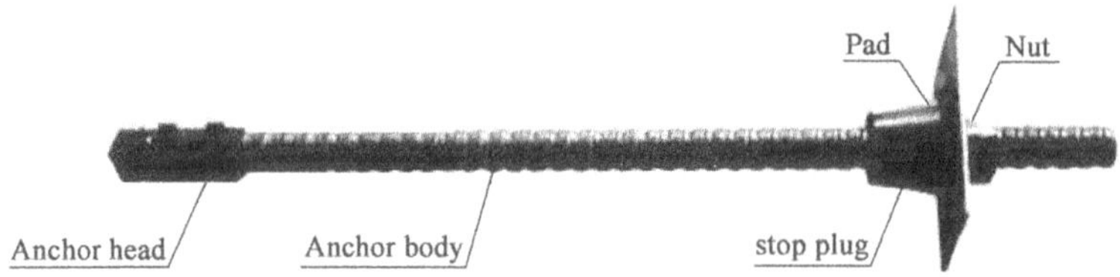

Fig. 2.14. Sample diagram of hollow grouting bolt.

grouting anchor should be attended to; one can stop grouting after being the pulp is out so as to vent it. Water is used to lubricate the filling tank and pipe after the grouting begins or 30 minutes after stop grouting midway.

2.3.2.1.3 Special case treatment

(1) **Grouting interruption:** One should find out the cause of the grouting interruption, resolve it as soon as possible and restore the grouting as soon as possible. If the grouting cannot be restored immediately, one should rinse the hole immediately and then restore the grouting.
(2) **String of pulp:** In order to prevent the stringing of slurry, one should construct odd holes first, then construct even holes. To grout at the same time for the string of grouting holes, one can use a plasma device or use a pump at the same time for multiple root bolt grouting. When a large number of slurry leaks occur, one should use low pressure, adjust the slurry concentration, limit the current, limit the quantity and use intermittent grouting for infusion or injection of other wall-filling channels first; then grout for the string of grouting holes can be continued, using a plasma device and a grouting pump at the same time for multiple root anchor groutings.
(3) **Surging water treatment:** In the grouting hole with gushing water from the orifice, the water pressure and water inflow are recorded before the grouting, and the following comprehensive measures are selected according to the water inflow condition: First, high pressure is applied during grouting from the top to the bottom. Second, the slurry is used for the screen, and then the slurry is closed after 1–24 h,

and it is coagulated. Third, pure pressure grouting is adopted. Fourth, treatment with rapid coagulation is utilized.

2.3.2.2 *Main construction parameters of advanced small ducts*

The advanced small duct method is a relatively commonly used method of forepoling in the construction of underground hard rock stratum, generally adopting a $\phi 42 \times 3.5$-mm seamless steel tube and a small pipe, with supporting parameters set by the designer according to the grade of the surrounding rock (Figs. 2.15 and 2.16).

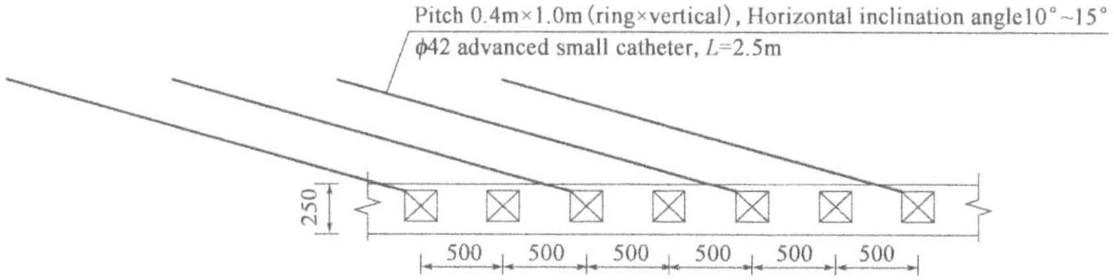

Fig. 2.15. Example of single-row small catheter design parameters (unit: mm).

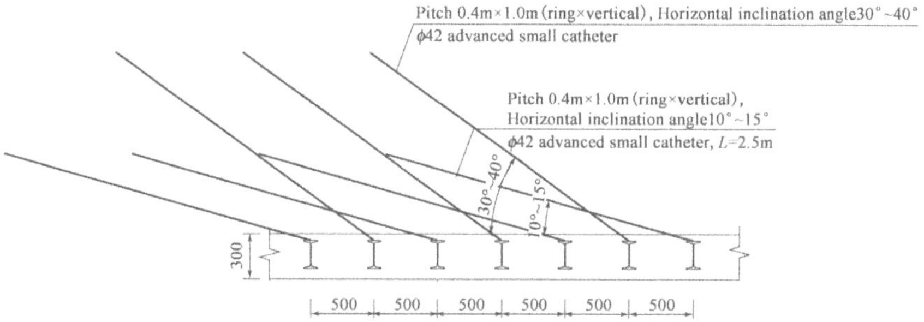

Fig. 2.16. Design parameters of double-row small catheters (unit: mm).

2.3.2.2.1 Ducting

For the convenience of small duct grouting, one needs to dill on the pipe shaft. The drilling and grouting holes can be round holes of $\phi 8$–$\phi 10$ mm, the tube end of 0.5 mm is not bored into, a spacing of 15 cm is used, a torx screwdriver is also used and the steel pipe's front end is placed into a cone, as shown in Fig. 2.17.

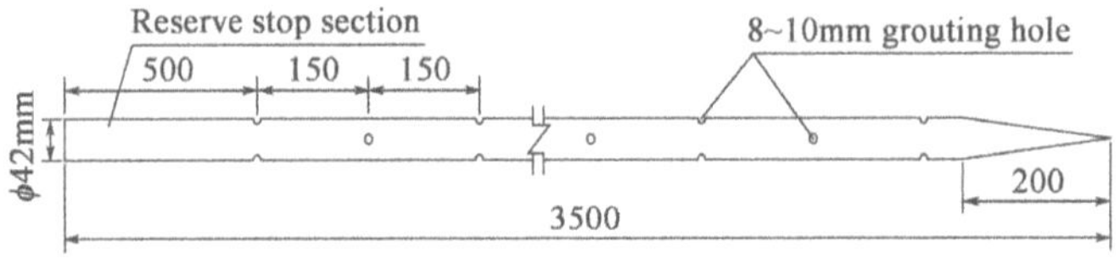

Fig. 2.17. Schematic diagram of small catheter (unit: mm).

2.3.2.2.2 Drilling

First, one should measure the lofting and mark the small tube hole with red paint. The extrapolation angle between the direction of the drilling and excavation surface is 15°. A pneumatic drill is used to make the holes. One should avoid the drill pipe's swinging during the borehole drilling and ensure that the hole is straight. After a deep hole is drilled, one should blow out the debris with a blow pipe and pay attention to avoid a hole collapse. Some of the drill holes are located above the bored piles; if drilling a hole is relatively difficult, the drilling position can be adjusted.

2.3.2.2.3 Ductule installing

According to the geological characteristics in the Qingdao region, small ductules are constructed using the drilling method. Drilling is first done using an air pressure gun, jacking the pipe to the designed hole, with an exposed orifice of 30 cm. The steel pipe tail and steel frame are welded firmly. The allowable deviation of the extrapolation angle of the advanced small pipe is 5‰, the jacking length is not less than 90% of the steel tube length, the hole diameter is larger than the steel pipe diameter of 30–40 mm, the drilling depth is greater than the length of the tube of 3–5 cm and the hole spacing of the borehole is plus or minus 50 mm.

2.3.2.2.4 Grouting

(1) The plain surface is sprayed with C25 shotcrete to prevent slurry.
(2) The sediment is washed before grouting, proceeding from the bottom up.
(3) The grouting pressure should meet the design requirements, and the slurry must be filled into the steel pipe and the surrounding space

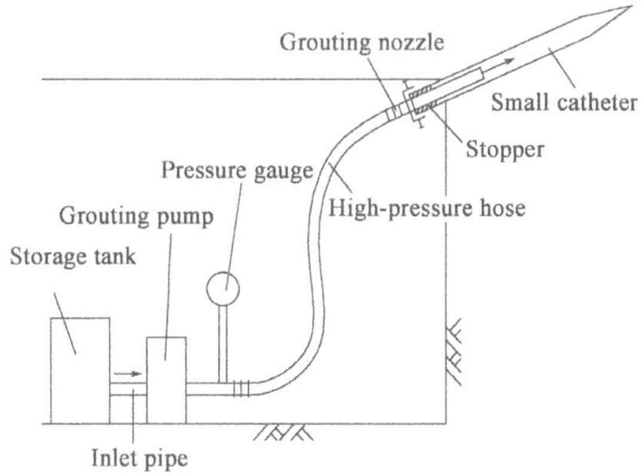

Fig. 2.18. Schematic diagram of small ductule grouting.

(Fig. 2.18). The grouting pressure is generally 0.5–1.0 MPa, and the grouting water–cement ratio is 1:1–1:1.5.

(4) Using a ball valve to check the slurry, the grouting standard is as follows: When the grouting pressure is stable for 10 min, the grouting quantity reaches more than 80% of the design allowance and one can finish the hole grouting. When finished, one should stop the pump first, then close the ball valve, washing the line in the end.

(5) There is a need to equip the slip casting machine with a flow meter to fill in the records carefully during grouting. In order to analyze and improve the operation at any time, one must pay attention to the construction and support the work. One must also adjust the grouting parameters according to the grouting test results and the field situation.

2.3.2.2.5 Treatment of abnormal phenomenon during grouting

(1) When encountering channel slurry, one can use a pulp machine to make multiple holes and use plugging string grouting to grout every hole, which will plug the channel hole in time.

(2) When the grouting pressure suddenly rises, one should stop grouting to find out the cause.

(3) When the volume of grouting is large, the pressure is constant, and the slurry concentration and mixture ratio should be adjusted to shorten the gelation time.

2.3.3 *Tunnel blasting*

During tunnel construction in the hard rock stratum, one should choose a suitable excavation and blasting method, design an initiating network form with a single period of initiating the maximum dosage and determine blasting footage cycles according to some requirements such as the section size of the tunnel and connecting passage, geological conditions and surrounding environment; also, safe distance should be maintained to avoid damage as a result of the blasting vibration. If there is a pipeline in the construction area and it cannot be moved, a construction plan shall be formulated according to the requirements of the property rights unit.

The blasting operation may affect the construction safety mainly due to blasting vibration, flying stones, air impact and the harmful effects of blasting dust. It is extremely important to ensure blasting safety in the working environment especially in during metro hard rock stratum tunnel construction in an urban area. Blasting is conducted in the city streets with numerous and complicated pipelines and buildings around. The blasting operation is bad for the overall environment, with negative effects mainly produced by the vibration, debris being thrown about in a slingshot manner, shock waves, noise hazards, etc. All this could possibly affect the surrounding security. Therefore, it is necessary to strictly control the maximum explosive charge of a single segment to ensure that it does not exceed the safe allowable value. Before construction, one should coordinate with the relevant departments and the nearby residents, initiate the pipeline migration and protection work, formulate a transportation guide plan and ensure safe construction.

In the initial blasting stage, the stability of the surrounding rock is poor due to the thin covering layer. It is necessary to understand the geological, hydrological situation of the hole, adopt reasonable excavation technology, reduce the blasting seismic intensity and reduce the disturbance to the retaining rock mass. One should prevent the occurrence of falling blocks or a cave-in accident, and ensure construction safety. The tunnel blasting operation should be according to the geological, hydrological situation and section size, which would determine the excavation process and the rational blasting parameters, thereby reducing the blasting seismic wave damage to the retain rock and any supporting structure disturbance.

Before the detonation, all personnel should be evacuated from the cave along with the mobile devices to avoid damage; the pipelines and non-mobile devices should be shielded. After blasting, the exhaust ventilation system will be started and the smoke from the hole will be discharged. Security personnel will be able to enter the hole, conduct a perimeter check, confirm security and continue to the next cycle.

2.3.3.1 *Calculation and verification of safety distance of blasting seismic wave*

According to the blasting safety procedures based on regulation GB6722-2014, there are certain criteria for blasting vibration to protect the buildings on the ground, with peak particle vibration velocity and vibration frequency being the main two indicators. The permissible blasting vibration safety standards are shown in Table 2.11.

In terms of urban metro construction, during construction blasting, the surrounding houses and pipelines need to be protected. In the construction blasting area, the distance between the source and the target is known, based on which the blasting safety distance can be determined according

Table 2.11. Permissible safety standards for blasting vibration.

No.	Class of protected objects	Safety allowance for the particles to vibrate V (cm/s) ≥10 Hz	10–50 Hz	>50 Hz
1	Earth cave, adobe house, rubble house	0.15–0.45	0.45–0.9	0.9–1.5
2	General civil buildings	1.5–2.0	2.0–2.5	2.5–3.0
3	Industrial and commercial buildings	2.5–3.5	3.5–4.5	4.5–5.0
4	General ancient buildings and monuments	0.1–0.2	0.2–0.3	0.3–0.5
5	The operation of hydropower station, power plant and central control room equipment	0.5–0.6	0.6–0.7	0.7–0.9
6	Hydraulic tunnel	7–8	8–10	10–15
7	Traffic tunnel	10–12	12–15	15–20

(Continued)

Table 2.11. (*Continued*)

		Safety allowance for the particles to vibrate V (cm/s)		
No.	**Class of protected objects**	**≥10 Hz**	**10–50 Hz**	**>50 Hz**
8	Mine openings	15–18	18–25	25–30
9	Permanent rock high slope	5–9	8–12	10–15
10	Large volume of new concrete (C20)			
	Age: initial set-3d	1.5–2.0	2.0–2.5	2.5–3.0
	Age: 3d–7d	3.0–4.0	4.0–5.0	5.0–7.0
	Age: 7d–28d	7.0–8.0	8.0–10.0	10.0–12.0

Notes: The monitoring of the blasting vibration determines the three components of the particle vibration. (1) The vibration velocity of the particle as given in the table is the maximum of three components, and the vibration frequency is the active frequency. (2) The frequency range is determined according to the field measured waveform or according to the following data: chamber blasting $f < 20$ Hz; open deep-hole blasting 10–60 Hz; open shallow hole blasting 40–100 Hz; underground deep-hole blasting 30–100 Hz; underground shallow hole blasting 60–300 Hz.

(a) To ensure the building's safety the allowed particle vibration velocity should be chosen based on the importance of the building, the building's quality, age of the building, the self-vibration frequency and the foundation conditions. (b) The safety of ancient buildings and monuments shall be ensured by expert demonstration at the provincial level and above. (c) While selecting the tunnel, the particle vibration velocity, the structure, the importance of the supporting conditions, the classification of the surrounding rock, the explosive source, the excavation span, the excavation depth, size, direction and the surrounding environment, and other factors should all be taken into account. (d) The safe allowable particle vibration velocity of the newly poured concrete with unblocked water can be selected according to the upper limit value given in this table. (e) For the permanent rock high slope, the importance of the slope, the initial stability of the slope, the support conditions and excavation height should be considered.

to Sadowski's formula, which calculates the maximum allowable blasting detonation dose or biggest delay blasting priming dose.

The formula is as follows:

$$Q_{\max} = R^3 (V/K)^{3/\alpha},$$

where $Q_{\max}$ is the blasting charge, the maximum loading capacity of delayed blasting (kg), R is the detonation point source to protect object distance (m), V is the location of the protected object is allowed based on the particle vibration velocity (cm/s), K, α are the coefficient and

Table 2.12. K and α values of different lithologies in the detonation zone.

The lithology	K	α
Hard rock	50–150	1.3–1.5
Medium-hard rock	150–250	1.5–1.8
Soft rock	250–350	1.8–2.0

attenuation index of the topography and geological conditions between the protected objects in the blasting calculation.

In the formula, K and α can be obtained from Table 2.12, and also can be determined by a field test.

In Table 2.12, K and α both have a larger scope, generally chosen according to project analogy, such engineering analogy, the terrain and rock properties at the jobsite, combined with years of practical experience; one can take the coefficient K, which is related to the topographical, geological conditions between the protection object and break point, as 200 and take the attenuation index α as 1.65.

2.3.3.2 *Design of tunnel blasting parameters for hard rock formation*

According to the present situation of the civil blasting market, the explosives used in the blasting operation in the hard rock formations of the urban metro are usually the emulsion explosives, non-electric millisecond detonators (1–20 sections) and detonation cords.

2.3.3.2.1 Drilling design

(1) **Hole diameter:** The blasting adopts emulsion explosives with a diameter of 32 mm; therefore, the diameter of drilling is $d = 42$ mm.

(2) **Hole depth:** According to the geological conditions and blasting construction procedures, the excavation of the tunnel is 0.5–1.5 m, so the depth of the tunneling hole is 0.6–1.8 m, and the cutting hole is deepened by 20 mm, and $L = 0.8$–2.0 m.

(3) **The total amount of explosives used for each driving cycle**

$$Q = qV = qSL_b\eta,$$

Where q is the unit powder consumption (kg/m^3), η is the utilization ratio of gun hole (generally 0.8–0.95, in this design 0.85 is taken), S is the tunneling section area (m^2), L_b is the hole depth, calculated by caving hole depth (m).

(4) **Hole parameter design:**
- Minimum resistance line, $W : W = (0.4 \sim 1.0)L$
- Pitch row, $a : a = 500$–700 mm.
- Row spacing, $b : b = 500$–700 mm.

(5) **A drill charge weight**

$$Q = qabH,$$

Where q is the unit explosive consumption (0.8–1.2 kg/m^3), a is the pitch row (m), b is the row spacing (m), H is the tunneling footage (m). For the specific amount of each shot hole, see the blasting parameter list.

(6) **The maximum dose of a single segment:** According to the "blasting safety regulations", and regulation and construction experience, combined with the geological conditions at the jobsite, the sheet section of the biggest dosage can be calculated by taking k as $k = 200$ and α as $\alpha = 1.65$ in the Qingdao area lithology, using Sadowski's formula $V_{max} = k(Q^{1/3}/R)^{\alpha}$. The allowed single period of maximum charge can be calculated as $Q_{\max} = R^3(V/K)^{3/\alpha}$, by adjusting k, α according to advanced geological forecast and adjusting the single charge according to the monitoring results in the construction and the blasting effect.

2.3.3.2.2 The choice of cut form

Cutting is the key to the success or failure of the blasting, and is usually the main source of the maximum vibration velocity. In order to achieve the purpose of shock absorption, the cutting according to the excavation method and the section size should be the v-cut with a straight hole cut to take full advantage of the easy casting characteristics of the v-cut, which allows maximum reduction of vibration. The blasting surface details of an international airport are shown in Figs. 2.19 and 2.20.

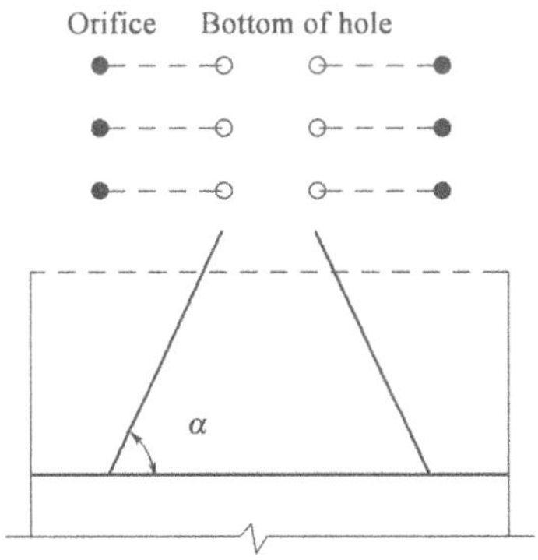

Fig. 2.19. Wedge cut.

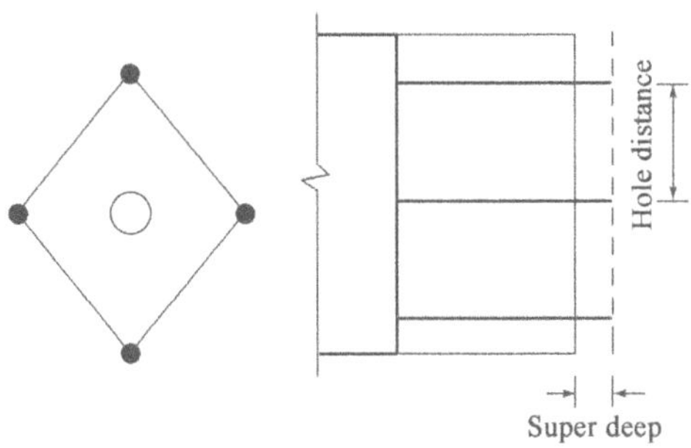

Fig. 2.20. Large diameter hollow straight hole cut.

2.3.3.2.3 Charge structure and filling

Charge structure and single-hole loading: The main hole and the auxiliary hole are filled with the continuous charge, and the residual space after each charge is filled with gun mud. The mortar can be made of a 1:3 clay and sand mixture. The gun clay can be a 1:3 mixture of clay and sand! The standard is to hold it in a ball by hand, to hold it loose by hand, or to pack the sand in a paper cylinder to make it into a medicine roll. See Fig. 2.21 for the loading structure.

The charge is usually 32 mm in diameter, 30 cm long with, roll quality 300 g of emulsion explosive. The peripheral hole can be used rather than a coupling interval charge structure. A charge of 20 mm in diameter, 20 cm long, with roll quality 100 g of explosives is used, and the orifice plug is strengthened.

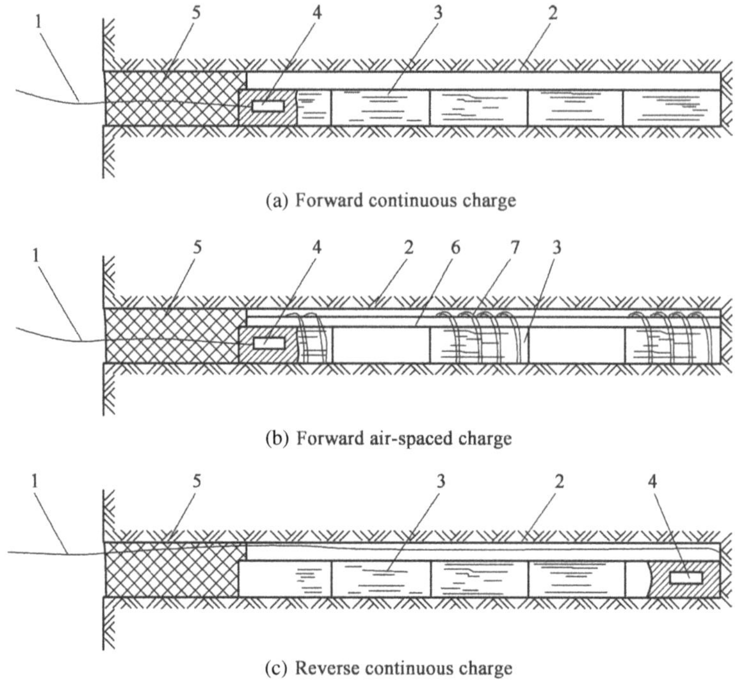

Fig. 2.21. Loading structure diagram. (a) Forward charging, (b) positive air separation charge, (c) reverse continuous charge. (1) Nonel tube, (2) hole wall, (3) cartridge, (4) cap, (5) gun mud, (6) bamboo strips and detonating cords, (7) bands.

2.3.3.2.4 Drill hole layout

First of all, one should choose the slot and slot hole, then mark the peripheral hole and then mark the auxiliary hole according to the section size. Slot holes are generally arranged in the center of the excavation face and are about 20 cm deeper than other gun holes. The peripheral hole is generally arranged on the contour line of the driving section and meets the requirement of smooth blasting. The bottom of the hole should be about 10 cm beyond the design contour line. The bottom hole should be about 15 cm higher than the design level of the bottom plate. The bottom of the hole should reach the base level of 10–20 cm, and the hole spacing and resistance line should be the same as the auxiliary hole. When using a straight hole cut, the hollow hole should be at a 10-m depth at a time.

According to the grade of the surrounding rock and the size of the cross-section, different forms of blasting excavation are performed: For level II surrounding rock, blasting construction method is used on the whole section; for levels III, IV, V, the up and down steps method is utilized, with the v-cut, tube detonator delay in 1–20 smooth blasting. Based on the different surrounding rocks, the supporting structure designs, the Qingdao area geological characteristics and the relevant requirements, excavation of surrounding rock at all levels is done as follows: II, III footage of the surrounding rock control in 1.5 m, IV level footage of the surrounding rock control in 1 m or 0.75 m, V class surrounding rock penetration of 0.5 m. It is difficult to locate the points in accordance with the requirements of geological and vibration velocity adjustment cycle. Specific blasting parameters are shown in Tables 2.13–2.17. The corresponding perforation layout is shown in Figs. 2.22–2.26.

The holes in the underground tunnel are inclined holes and horizontal holes. Except for the slot holes, which are inclined holes, the other holes are horizontal holes.

Table 2.13. The whole section of surrounding rock blasting parameters for tunnel of level II.

Blasting parameters of level II surrounding rock									
No.	**Paragraph**	**Name**	**Quantity (a)**	**Pitch of holes in pattern (minimum resistance line) (m)**	**Pitch of holes (m)**	**Borehole angle (°)**	**Hole depth (m)**	**Single-hole charge (kg)**	**Maximum amount of explosive charge (kg)**
1	No. 1	Grade 1 cutting hole	4	1	0.6	70	1.0	0.6	2.4
2	No. 3	Grade 2 cutting hole	6	0.6	0.6	73	2.1	0.8	4.8
3	No. 5	Reliever	7	0.8	0.8	90	1.8	0.7	4.9
4	No. 7	Reliever	7	0.8	0.8	90	1.8	0.7	4.9
5	No. 9	Reliever	5	0.8	0.8	90	1.8	0.7	3.5

(*Continued*)

Table 2.13. (*Continued*)

Blasting parameters of level II surrounding rock										
No.	**Paragraph**	**Name**	**Quantity (a)**	**Pitch of holes in pattern (minimum resistance line) (m)**	**Pitch of holes (m)**	**Borehole angle (°)**	**Hole depth (m)**	**Single-hole charge (kg)**	**Maximum amount of explosive charge (kg)**
6	No. 10	Reliever	5	0.8	0.8	90	1.8	0.7	3.5
7	No. 11	Reliever	5	0.8	0.8	90	1.8	0.7	3.5
8	No. 12	Inner ring eye	8	0.6	0.6	90	1.8	0.7	5.6
9	No. 13	Inner ring eye	8	0.6	0.6	90	1.8	0.7	5.6
10	No. 14	Inner ring eye	8	0.6	0.6	90	1.8	0.7	5.6
11	No. 15	Inner ring eye	7	0.6	0.6	90	1.8	0.7	4.9
12	No. 16	Periphery hole	11	0.6	0.4	92	1.8	0.45	5.0
13	No. 17	Periphery hole	11	0.6	0.4	92	1.8	0.45	5.0
14	No. 18	Periphery hole	11	0.6	0.4	92	1.8	0.45	5.0
15	No. 19	Periphery hole	10	0.6	0.4	92	1.8	0.45	4.5
16	No. 20	Periphery hole	14	0.6	0.4	92	1.8	0.45	5.0

Notes: Main technical indicators:

Whole section area: 40 m^2, penetration: 1.5 m.
Number of holes: 124, total dose: 73.7 kg.
The average consumption: 1.2 kg/m^3, volume: 60 m^3.
Number of detonators: 135.

Table 2.14. On the bench blasting parameter of tunnel of levels III, IV surrounding rock.

On the bench blasting parameter of levels III, IV surrounding rock										
No.	**Paragraph**	**Name**	**Quantity (a)**	**Pitch of holes in pattern (minimum resistance line) (m)**	**Pitch of holes (m)**	**Borehole angle (°)**	**Hole depth (m)**	**Single-hole charge (kg)**	**Maximum amount of explosive charge (kg)**
1	No. 1	Grade 1 cutting hole	4	1	0.5	75	1.55	0.5	2
2	No. 3	Grade 2 cutting hole	6	0.7	0.5	80	1.52	0.4	2.4
3	No. 5	Reliever	6	0.6	0.7	90	1.3	0.3	2
4	No. 7	Reliever	4	0.6	0.7	90	1.3	0.3	1.2
5	No. 9	Reliever	7	0.6	0.75	90	1.3	0.4	2.8
6	No. 10	Reliever	7	0.6	0.75	90	1.3	0.4	2.8
7	No. 11	Periphery hole	13	0.5	0.4	93	1.3	0.3	3.9
8	No. 12	Periphery hole	13	0.5	0.4	93	1.3	0.3	3.9
9	No. 13	Bottom hole	5	0.6	0.7	93	1.3	0.3	1.5
10	No. 14	Bottom hole	5	0.6	0.7	93	1.3	0.3	1.5

Notes: Main technical indicators:
Upper step area: 22 m^2, penetration: 1 m.
Number of holes: 70, total dose: 23.8 kg.
Consumption: 1.08 kg/m^3, volume: 22 m^3.
Number of detonators: 77.

Table 2.15. Under bench blasting parameters of levels III, IV surrounding rock.

Under bench blasting parameters of levels III, IV surrounding rock									
No.	**Paragraph**	**Name**	**Quantity (a)**	**Pitch of holes in pattern (minimum resistance line) (m)**	**Pitch of holes (m)**	**Borehole angle (°)**	**Hole depth (m)**	**Single-hole charge (kg)**	**Maximum amount of explosive charge (kg)**
1	No. 1	Heading eye	5	0.8	1	90	1.3	0.6	3.0
2	No. 3	Heading eye	5	0.8	1	90	1.3	0.6	3.0
3	No. 5	Heading eye	4	0.6	1	90	1.3	0.6	2.4
4	No. 7	Heading eye	5	0.7	0.75	90	1.3	0.4	2
5	No. 9	Heading eye	6	0.7	0.75	90	1.3	0.4	2.4
6	No. 11	Periphery hole	11	0.6	0.5	93	1.3	0.3	3.3
7	No. 13	Periphery hole	11	0.5	0.5	93	1.3	0.3	3.3

Notes: Main technical indicators:
Lower step area: 20 m^2, penetration: 1 m.
Number of holes: 47, total dose: 19 kg.
Consumption: 0.8 kg/m^3, volume: 20 m^3.
Number of detonators: 51.

Table 2.16. On the bench blasting parameter of level V surrounding rock.

On the bench blasting parameter of level V surrounding rock									
No.	**Paragraph**	**Name**	**Quantity (a)**	**Pitch of holes in pattern (minimum resistance line) (m)**	**Pitch of holes (m)**	**Borehole angle (°)**	**Hole depth (m)**	**Single-hole charge (kg)**	**Maximum amount of explosive charge (kg)**
1	No. 1	Grade 1 cutting hole	6	1	0.5	70	1.2	0.5	3
2	No. 3	Reliever	6	0.5	0.5	80	1.2	0.2	1.2

Table 2.16. (*Continued*)

On the bench blasting parameter of level V surrounding rock										
No.	**Paragraph**	**Name**	**Quantity (a)**	**Pitch of holes in pattern (minimum resistance line) (m)**	**Pitch of holes (m)**	**Borehole angle (°)**	**Hole depth (m)**	**Single-hole charge (kg)**	**Maximum amount of explosive charge (kg)**
3	No. 5	Reliever	6	0.5	0.6	90	1	0.2	1.2
4	No. 7	Reliever	6	0.5	0.6	90	1	0.2	1.2
5	No. 9	Reliever	6	0.5	0.6	90	1	0.2	1.2
6	No. 11	Reliever	4	0.5	0.6	90	1	0.2	0.8
7	No. 13	Inner ring eye	7	0.5	0.4	90	1	0.2	1.4
8	No. 14	Inner ring eye	8	0.5	0.4	90	1	0.2	1.6
9	No. 15	Periphery hole	18	0.4	0.3	93	1	0.15	2.7
10	No. 16	Periphery hole	19	0.4	0.3	93	1	0.15	2.85
11	No. 17	Bottom hole	11	0.5	0.6	93	1	0.2	2.2

Notes: Main technical indicators:
Upper step area: 22 m^3, penetration: 1 m.
Number of holes: 97, total dose: 17.55 kg.
Consumption: 0.79 kg/m^3, volume: 22 m^3.
Number of detonators: 106.

Table 2.17. Under bench blasting parameters of level V surrounding rock.

Under bench blasting parameters of level V surrounding rock									
No.	**Paragraph**	**Name**	**Quantity (a)**	**Pitch of holes in pattern (minimum resistance line) (m)**	**Pitch of holes (m)**	**Borehole angle (°)**	**Hole depth (m)**	**Single-hole charge (kg)**	**Maximum amount of explosive charge (kg)**
1	No. 1	Heading eye	8	1	0.6	90	1	0.25	2
2	No. 3	Heading eye	7	0.5	0.6	90	1	0.25	1.8

(*Continued*)

Table 2.17. (*Continued*)

Under bench blasting parameters of level V surrounding rock										
No.	**Paragraph**	**Name**	**Quantity (a)**	**Pitch of holes in pattern (minimum resistance line) (m)**	**Pitch of holes (m)**	**Borehole angle (°)**	**Hole depth (m)**	**Single-hole charge (kg)**	**Maximum amount of explosive charge (kg)**
3	No. 5	Heading eye	6	0.5	0.6	90	1	0.25	1.5
4	No. 7	Heading eye	3	0.5	0.6	90	1	0.25	0.8
5	No. 9	Inner ring eye	7	0.5	0.5	90	1	0.2	1.4
6	No. 11	Inner ring eye	8	0.5	0.5	90	1	0.2	1.6
7	No. 13	Periphery hole	13	0.5	0.5	93	1.1	0.2	2.6
8	No. 15	Periphery hole	14	0.5	0.5	93	1.1	0.2	2.8

Notes: Main technical indicators:
Lower step area: 20 m^3, penetration: 0.75 m.
Number of holes: 66, total dose: 14.5 kg.
Consumption: 0.7 kg/m^3, volume: 20 m^3.
Number of detonators: 71.

2.3.3.2.5 Design of detonation network

The surface blasting technology should be used in the excavation of hard rock tunnel, and the order of the detonation of each gun hole is as follows: cut hole and auxiliary eye (tunneling eye); peripheral eye, bottom edge eye, from inside to outside layer by layer. When the lower steps are blasting, the firing order of each hole is as follows: upper row eye and lower eye; the peripheral eye, from the inside out or from the top to the bottom.

In order to ensure that the blasting network is not initiated at the first explosion, hole millisecond delay blasting is recommended, which delays the explosion from detonator to detonate hole by 5 or 7, ensuring the security of the network. According to the arrangement plan, the holes can be set as per the detonating network diagram, and typical cases are shown in Figs. 2.27 and 2.28.

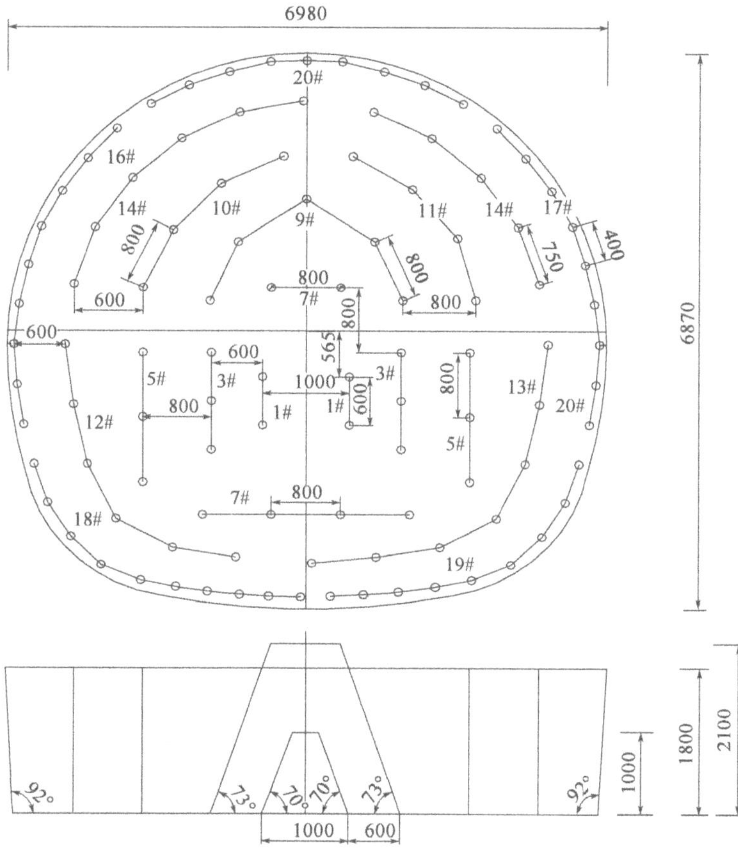

Fig. 2.22. Layout diagram of level II surrounding rock interval tunnel hole (unit: mm).

2.3.3.2.6 The main management system of blasting construction

(1) **The management system of civil explosives**

(a) Construction of temporary storage structures for the fireworks at the site should be approved by the local public security department. Moreover, the temporary storage structures are built strictly according to the relevant requirements. All fireworks shall be stored and managed at the temporary storage point, and any unused fireworks must be recovered and managed by the relevant units each time. It is strictly prohibited to store the fire products privately.

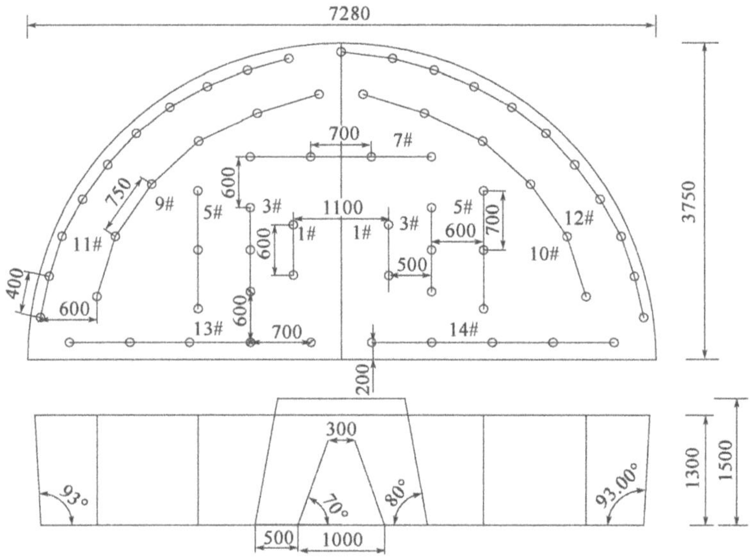

Fig. 2.23. Layout diagram of levels III, IV surrounding rock on the steps of blasthole (unit: mm).

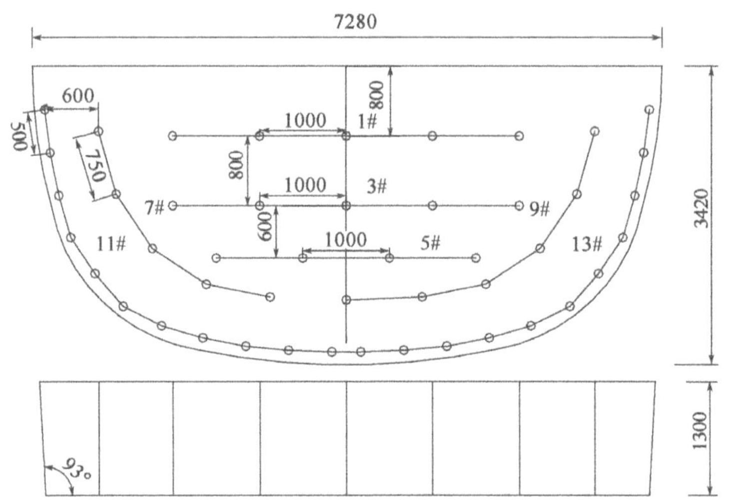

Fig. 2.24. Layout diagram of levels III, IV surrounding rock down the steps of blasthole (unit: mm).

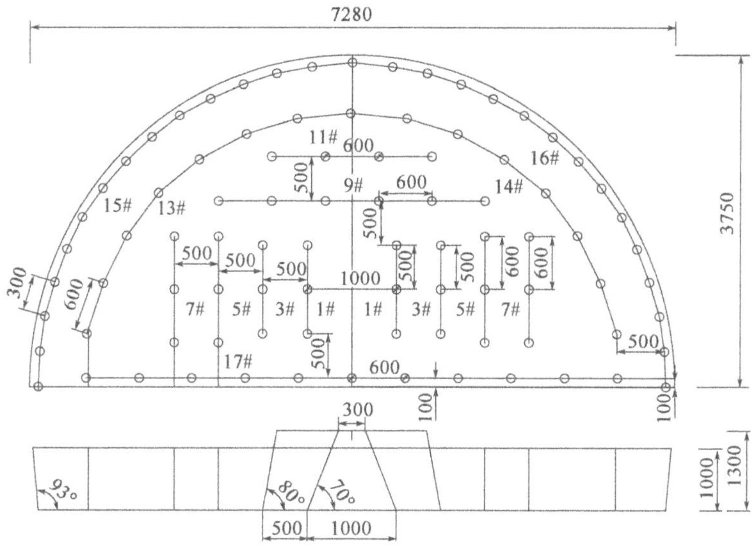

Fig. 2.25. Hole layout diagram of level V surrounding rock up the steps (unit: mm).

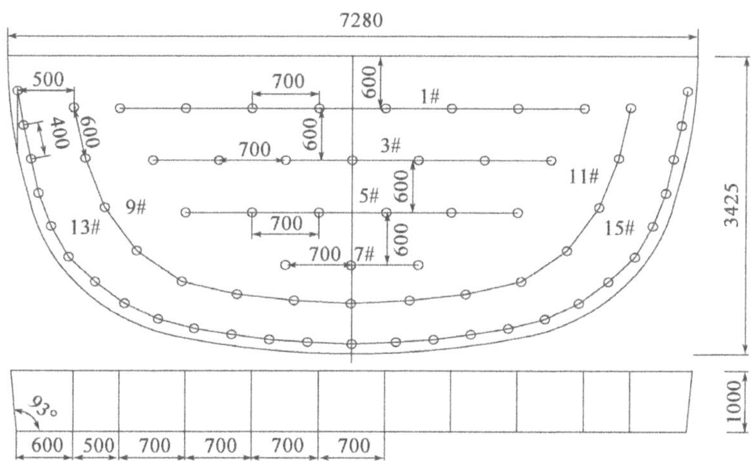

Fig. 2.26. Hole layout diagram of level V surrounding rock down the steps (unit: mm).

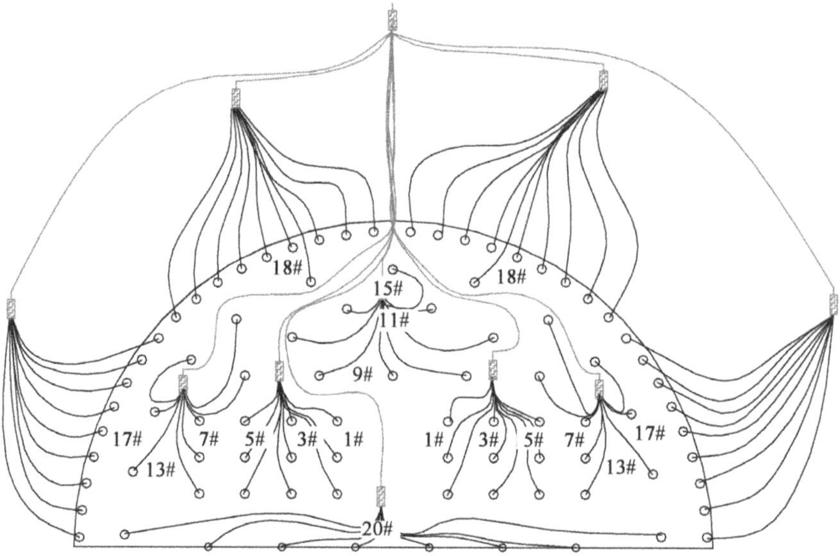

Fig. 2.27. Schematic diagram of step initiation network in a tunnel section.

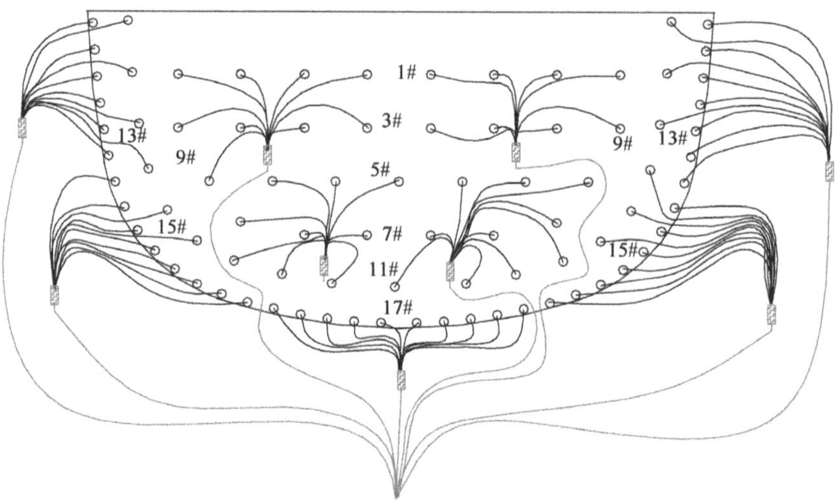

Fig. 2.28. Schematic diagram of starting and blasting network of an underground tunnel.

(b) Temporary storage sites shall be provided with security apparatus or security personnel, and technical preventive measures shall be set up to prevent civil use and avoid the loss or theft.
(c) All fireworks shall be transported and recycled at the temporary storage site. A return registration system should be maintained and everything should be accounted for.
(d) The supervisors should improve the spending, distribution and refund of the explosive items, and make plans based on actual collation, consumption and clearance, in accordance with the day. The procedures are not allowed to be issued publicly. The number of each item shall be counted in person. The handover is clear, and the signature shall be stored separately.
(e) No one is allowed to enter the temporary warehouse and must be admitted to the warehouse only after the approval of the person in charge.
(f) The relevant personnel must master firefighting knowledge, be familiar with all kinds of firefighting equipment and also have practical knowledge of firefighting.
(g) In the case of an accident, one should pay attention to the protection of the site, and report the incident immediately to the security department and the public security organ for investigation and handling. Because the site is not liable, the person who causes the explosive material to be stolen or lost will be held accountable according to the relevant law.
(h) The blasting equipment is managed strictly in the warehouse, and the blasting operation is monitored closely. No explosion equipment is allowed to be taken out. The management and handling of the equipment should be done by one who has been specially trained and has obtained the relevant job certificate.
(i) The handling of fireworks within the site must comply with the following requirements: Detonators and explosives shall be transported separately in a covered container. Within 30 m of the blasting equipment, a minimum of 2 people shall be required to escort the equipment to the work place, and no stopover will be allowed. No more than 20 kg of explosives per person or one carton of original packaging is allowed.

(j) The explosives and detonators used in the working face shall be stored in a special blasting equipment box with separate locks, and should not be randomly scattered. The blasting equipment box shall be placed in an area where the roof is stable, the support is complete and the equipment is not machined. The blasting equipment box should be placed in a safe place and then cordoned off.

(2) **Temporary distribution system of fireworks**

The transportation of explosive articles must comply with the regulations on the administration of civil explosives in the People's Republic of China.

The loading and unloading system with the following conditions must be complied with:

(a) People must understand and obey the commonsense safety rules of operation. It is strictly forbidden to throw, drag or trample on a packing case.
(b) When the loading and unloading personnel are in operation, no chemical fire or nail shoes shall be allowed, and no fire shall be carried.
(c) No loading and unloading can be done during storms or thunderstorms.
(d) During loading and unloading, the vehicle must be turned off, with the brake engaged and the driver must not leave the vehicle.
(e) When transporting explosives, the escorts must be transported with the vehicles, and the vehicles shall not be allowed for use by idle personnel. The vehicles carrying the explosives shall not be allowed to remain in public places or crowded places.
(f) If the vehicle has good visibility of 40 km, low visibility is eliminated. In case of a thunderstorm, vehicles should stay away from the structure of the open field, and during the cold weather, anti-skid measures must be taken.
(g) It is forbidden to transport off duty personnel or an excessive number of personnel in the vehicle. The locomotive and the rear of the vehicle are installed with the special dangerous goods safety mark.

(h) The escorts and the storekeeper must supervise the whole process of handling of the civil explosives.
(i) Vehicles that transport civil explosives must have a transport permit for civil explosive goods issued by the public security department. Explosives must be transported within the effective period and shall not be transported earlier or later.

(3) **System of clearance issued to civil explosion equipment**
(a) The blasting operation unit shall have, according to the design of the blasting, blasting charts and a practical formulation of plans on use of explosives, confirmed by the signature of the technical director. Prior notice should be given to blasting equipment valet parking stations.
(b) The collection of blasting equipment shall be carried out after the technical person in charge of the blasting operation signs the consent.
(c) The blasting equipment provider must hold a blasting demolition operations personnel license, the distributor of blasting equipment must hold a blasting operation personnel license. The personnel keeping, receiving and issuing the explosives must have a certified safety supervisor.
(d) The custodian shall check and confirm that the receiving personnel are the working personnel on duty on the certificate, and confirm that they have been approved and issued the blasting equipment according to the plan. No blasting equipment shall be issued to personnel who have not been identified as working personnel on duty. The blasting equipment shall not be issued beyond the type and quantity as specified in the approved plan.
(e) At the time of receiving the issuance, the number of fireworks should be filled in, and the time of receiving and release of the materials should be recorded by the two parties; the variety, quantity and receival time of the blasting equipment should be acknowledged, signed and confirmed by the parties. This electronic information should be collected and stored.

(f) After the completion of the blasting operation, the remaining blasting equipment should be removed to the temporary storage site in a timely manner.
(g) The custodian shall fill in the accounts of the fireworks in a timely manner, and record the clearance time, the personnel of the two parties, and the variety, quantity and receival time of the blasting equipment, which shall be signed and confirmed by both parties.

(4) **Site transportation system of fireworks**
(a) The demolition should be done according to the blasting capacity of the explosive devices, which should be recorded in person, as the calculations cannot change every single day. The record keeper shall accurately mention the serial number and the issuer's name. The technical leader and safety officer shall supervise the collection, distribution and transportation of the fireworks.
(b) The demolition crew shall go to the demolition site immediately after receiving the fireworks. One cannot carry fireworks in a crowded place.
(c) After receiving the fireworks, the explosive and detonator should be transported separately. No more than 20 kg of explosives per person or one carton of original packaging can be transported.
(d) When the fireworks pass through the shaft, it is forbidden for it to be stopped at the entrance.
(e) It is forbidden to raise or lower the firework products when there is a large concentration of personnel.
(f) Before lifting the fireworks, it is necessary to inform the crane driver and the signal operator first; the signal should be accurate and the operation should be done carefully.
(g) The explosives and detonators shall be placed separately and manned by special personnel after the fireworks are transported to the demolition site.

2.3.3.2.7 Blasting monitoring

Blasting vibration is an ephemeral but very intense event, which not only produces much disturbance in the lives of the residents near the tunnel construction but also causes concerns over the safety of buildings. Therefore, blasting vibration monitoring must be conducted. Any damage caused by the blasting vibration should be strictly controlled. The extent to which monitoring safety is permitted must be assessed, as it provides accurate and reliable data for subsequent construction and guides the subsequent blasting scheme designs, which is an important purpose for the blasting vibration monitoring.

(1) **Monitoring principle**
Due to the rock explosion effect, on the installed monitoring particle sensor particle vibration occurs causing the internal magnetic sensor system to create an air gap thus making the relative movement between coil by means of electromotive force signal, which flows through a wire input variable gain amplifier that enlarges and converts the AD signal, by the clock and trigger circuit. Moreover, through the memory protection signal transmitted by the CPU system input computer, waveform display and data processing software are adopted to improve the waveform analysis and data processing.

(2) **Monitoring instruments**
In the Chinese Academy of Sciences, Chengdu Institute of Measurement and Control, the production of TC-4850 high-precision blasting vibration equipment was an experiment. The instrument has the advantage of being lightweight, waterproof, dustproof, having good pressure, high precision and with wide applications. Using the enhanced 4850 instrument, one can quickly set parameters by pressing buttons and LCD screens at the scene, so as to achieve a fast signal and accurate collection of data. At the same time, the instrument can be easily read at the scene through certain functions, which lend the instrument its characteristic value. It can also be roughly previewed to collect signal waveform. The instrument adapts to different range collections and does not need to do adjusted (Fig. 2.29).

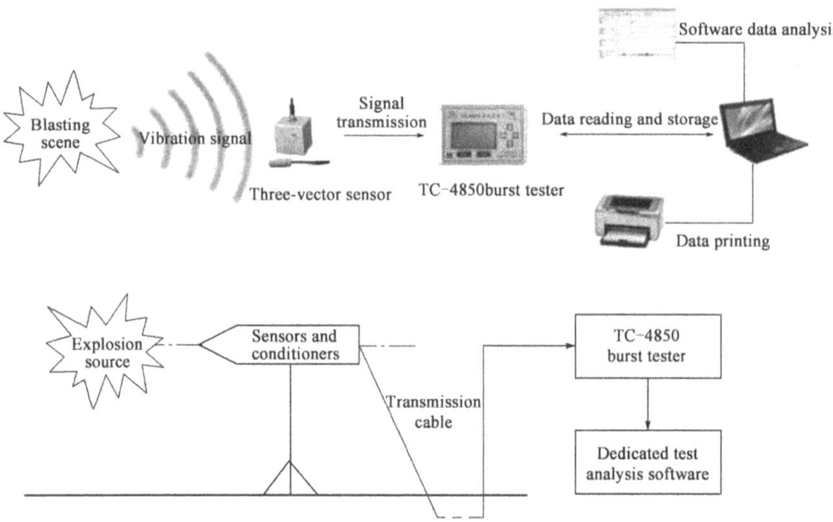

Fig. 2.29. Working schematic diagram of TC-4850 blasting vibration measuring instrument.

It is appropriate to use a separate vibration sensor to measure micro vibration and super vibration. The data recording function is a continuous mode, and the vibration analyzer can display the physical quantity, the main frequency and the recording time all at once. The main technical indicators are shown in Table 2.18, and the composition of the recorder is shown in Fig. 2.30.

(3) **Monitoring and implementation methods**

The real-time monitoring before blasting is according to field survey results and detailed preparation work, and in strict accordance with the working process. According to the excavation blasting scheme and blasting parameters design, for a certain range of vibration around an explosive source, monitoring is conducted of sensitive parts.

In order to ensure accurate and reliable monitoring, examination of the break point near the first object is conducted in a detailed and accurate manner, and then in front of the blasting monitoring system for inspection, testing and calibration. At the same time, the relative position of the object being monitored is determined by having knowledge of the location and

Table 2.18. Main technical indicators of TC-4850 vibration test recorder 8.

TC-4850 vibration test recorder	
The channel number	Parallel three-way
Display mode	Full Chinese LCD display
The power supply mode	Built-in rechargeable lithium battery charging
Sampling rate	1–50 kHz, multiple adjustable
A/D resolution	16 Bit
Frequency response range	0–20 kHz
Acquisition methods	Parallel three-channel acquisition, multi-group cascade
Recorded length	1–160 s is adjustable
Trigger mode	Internal trigger, external trigger
Range	The maximum input value is 10V (35 cm/s)
Trigger record	The continuous trigger record can reach 128–1,000 times
Trigger level	0–10 V (0–35 cm/s) can be adjusted arbitrarily
Storage space	1-m SARM, 128-m flash
Record accuracy	0.05 mV (0.5 mm/s)
Accuracy of reading	1‰
Clock precision	1 month ≤5 s
Data interface	USB 2.0
Battery life	≥60 h
Environment adaptability	–10–75°C (20–100%) RH
Size	168 mm × 99 mm × 64 mm
Weight	1,000 g

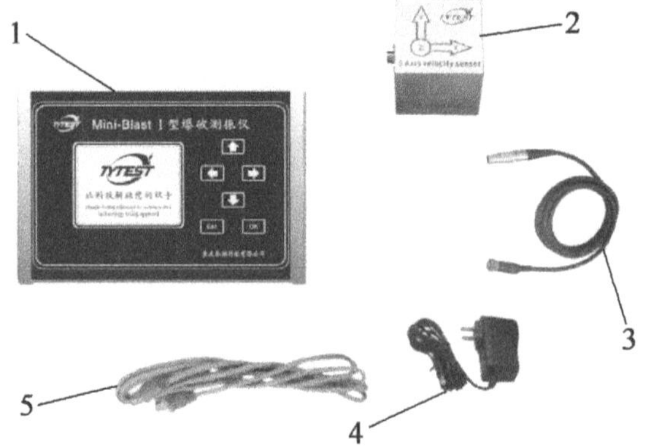

Fig. 2.30. TC-4850 vibration test recorder, (1) blasting vibration monitor, (2) triaxial vibration velocity sensor, (3) speed sensor cable, (4) power charger, (5) network cable.

arrangement of the measuring points, with early access to the site for resettlement, according to the blasting time monitoring.

(1) **Arrangement of measuring points:** According to the design requirements, the blasting vibration measurement points are arranged on the surface for the required monitoring of the structural support column of the building and the side wall of the tunnel. When installing the sensor, it must be installed firmly, otherwise the velocity monitoring data of the particle will be distorted, and the effect of gypsum fixed sensor is generally better. Attention should also be paid to sensor protection, to avoid physical damage caused by blast debris or other objects. The direction of the sensor must also be noted.
(2) **Principle of the arrangement of measurement points:** It depends on the following:
    (a) The location and direction of the section that receives maximum vibration;
    (b) Tracking and monitoring of blasting seismic effect;
    (c) Monitoring of attenuation of the blasting seismic wave.

(3) **The method of measuring points:** In accordance with the principle of blasting vibration and past experience, measuring points in each vertical, horizontal and tangential sensor are arranged and monitored

starting at the bottom of the tunnel and moving to the side, from the nearest point, one at a time. The measuring points of ground structures are arranged at the nearest surface of the blasting center, which is close to the side of the excavation tunnel.

For the underground tunnel, different types of excavations are performed around the hole, with constraints in the upcoming blasting of the adjacent tunnel. The place where the blasting vibration was arranged (near shot point), the impact points in the adjacent tunnel and arrangement of measuring points are shown in Figs. 2.31–2.33.

On the basis of the selection of the structure measuring points, if the foundation is buried under a soil layer, the nearest foundation and solid area are selected as the measuring points.

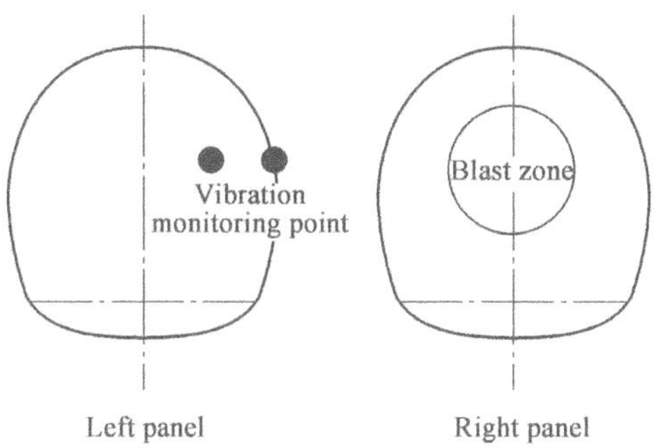

Fig. 2.31. Layout of blasting vibration measurement points in tunnel.

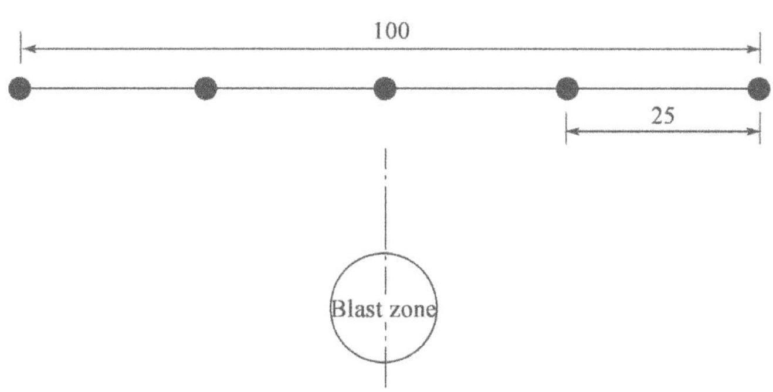

Fig. 2.32. Layout of surface blasting vibration measurement points (unit: mm).

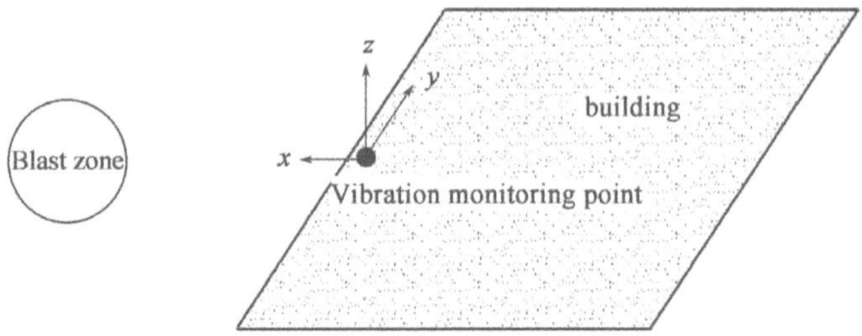

Fig. 2.33. Schematic diagram layout of vibration monitoring point.

2.3.4 *Main excavation method of the tunnel*

2.3.4.1 *Construction method and procedure of full-section method*

In tunnel construction of city metro in hard rock stratum, for levels II, III of surrounding rock, the whole-section method is used, along with the following: measuring contour line unreeling before excavation, drilling and blasting of full-face excavation, early bolting, laying of inverted arch with waterproof layer, inverted arch lining, laying arch and wall with waterproof layer and modification of the second line, and the excavation footage control is 1.5 m per cycle.

The whole-section method is excavated in a step sequence (Fig. 2.34):

(1) Full-section excavation is performed, with initial support.
(2) Building an inverted waterproof layer, with the second lining of the inverted arch.
(3) The arch part and the wall waterproof layer are laid with secondary lining.
   (a) Schematic diagram of construction steps of full-section excavation.
   (b) Full-section excavation method.

2.3.4.2 *Steps method and procedure*

According to the geological conditions, the steps method can be used in levels IV, V, VI surrounding rock for tunnel construction. One can construct a small catheter or bolt in advance according to the rock type

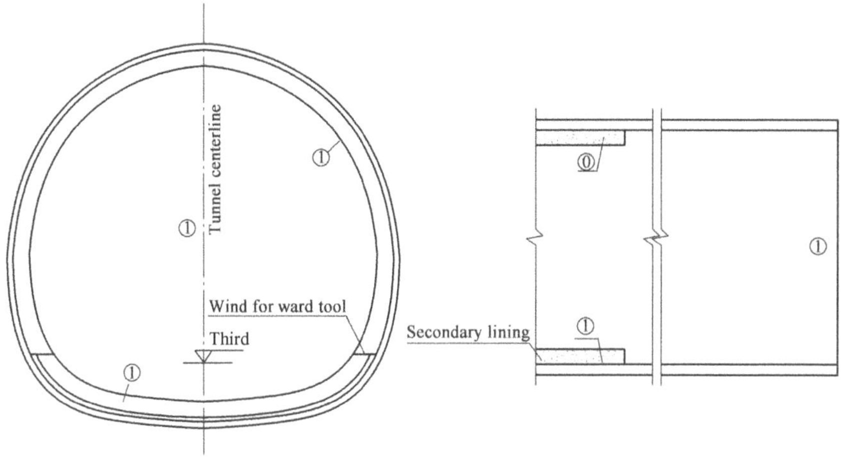

(a) Schematic diagram of full-section excavation steps (b) Full-section excavation section view

Fig. 2.34. Schematic diagram of the whole-section method.

before excavation along the line at the beginning of arch ring, in order to grout for reinforce formation. Then drilling–blasting excavation of the rock mass is performed on the steps, along with spraying of concrete, steel, a hollow bolt and hang net shotcrete. In order to prevent the arch foot from sinking, the arch foot is equipped with the anchor bolt, or a method of increasing the arch foot is used to set up the steel frame of the lower step. The bolt and the shotcrete are sealed and treated. When the upper steps are excavated at 3–5 m (when the steps are not larger than 3 m), the lower steps are also excavated. When the surrounding rock is in poor condition, it is necessary to reserve some core soil for the safety of the excavation.

The construction procedure of the step method is shown in Figs. 2.35 and 2.36.

(1) The stages of the steps method are as follows:
  (a) To excavate the steps in the tunnel and to support the steps in the early stage.
  (b) To excavate the steps under the tunnel, and to support the steps in the early stage.

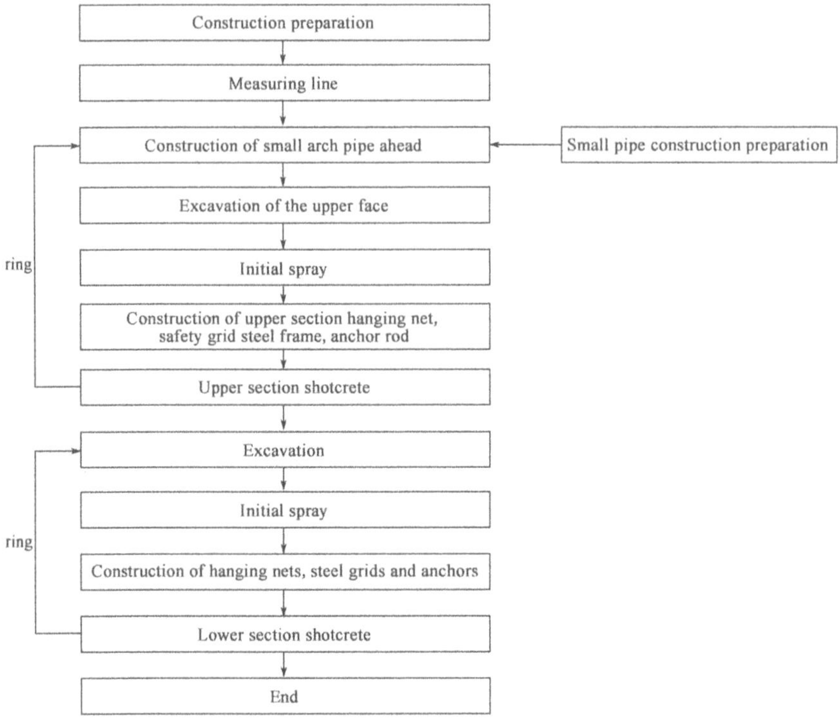

Fig. 2.35. Construction process flow chart of step method.

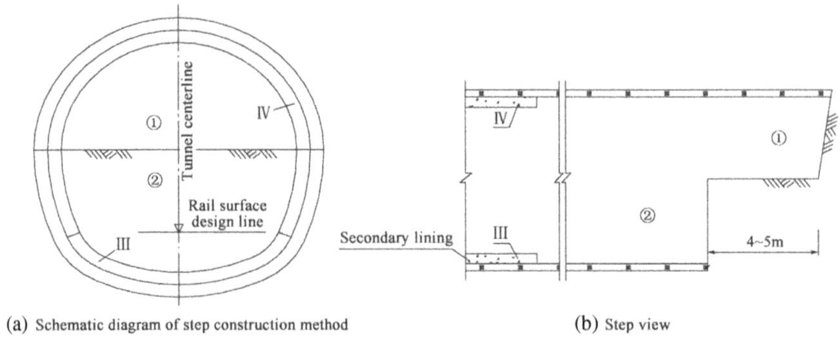

Fig. 2.36. Schematic diagram of construction process of step method. (a) Construction parameters of the step method, (b) section of the step method.

(c) Laying an inverted waterproof layer and applying the secondary inverted lining.
(d) The arch and the side wall waterproof layer and the secondary lining are fixed.

(2) The control of section block and step length is done by the step method.

Under normal circumstances, the step length is controlled within 6–8 m, with worse geological control at the 5-m mark. For the water-rich formation or for constructing steps in the sand, the weak formation step length should be shortened appropriately for the steps according to the blasting excavation conditions. For digging a replacement gun or by artificial pick processing, which is done for effective operation of equipment, steps clearance should not be less than 3.5 m.

The excavation cycle is carried out according to the design and specification requirements.

The lower steps are excavated by blasting and artificial cooperation. Excavators are used, and the excavated material is loaded into a dump truck to be shipped out.

2.3.4.3 *Median septum excavation (CD method)*

The CD method is suitable for the shallow buried and adverse geological section of the tunnel excavation. According to the category of the surrounding rock before excavation, along the line at the beginning, an arch ring for a small catheter is fixed in advance. Then grouting reinforcement is done. Then excavation is done for the stairs on the left side of the rock by drilling–blasting. Before beginning to spray the concrete, steel structures are installed temporarily on the vertical baffle plate using hollow bolts and net-suspended spray concrete. To prevent the foot arch, arch foot and temporary vertical clapboard from locking in the foot bolt, measures must be taken. Then the excavation is done for the steps on the left side of the rock mass with shotcrete use after erection of the steel frame and temporary partition, with bolts, down the steps. Finally continuing right up the stairs on the right side and down the steps on the right side of the primary support, unilateral steps of 3–5 m are excavated (short upper steps are

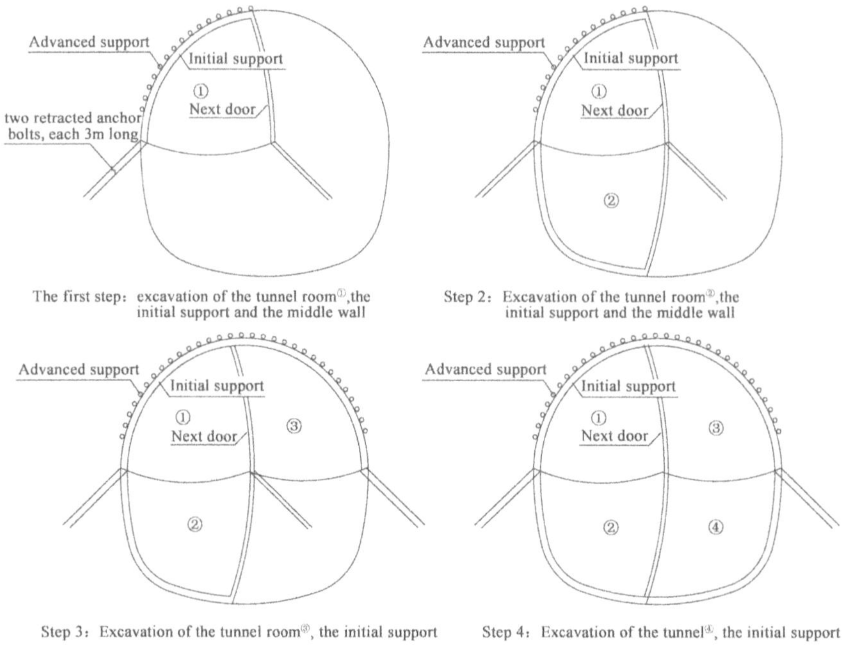

Fig. 2.37. Schematic diagram of the construction process of the CD method.

constructed when the excavation is not more than 3 m). Each cycle of excavation is 0.5 m.

(1) Step sequence of CD method (Fig. 2.37):
   (a) Excavate the first step of the tunnel, and construct the primary lining and temporary partition board of the upper step.
   (b) Excavate the second step of the tunnel, and construct the primary lining and temporary partition board of the lower step.
   (c) Excavate the third step of the lower steps, and construct the primary lining of the lower step.
   (d) Excavate the fourth step of the lower steps, and construct the primary lining of the lower step.
   (e) The temporary partition plate is removed and the waterproof layer is laid.
   (f) The arch part and the wall's waterproof layer are laid for the secondary lining.

(2) The control of section block and step length of the CD method

In general, step length is controlled within 3–5 m. In the case of poor water formation in the wall rock stratum, the step length is shortened appropriately. The conditions of blasting excavation require a replacement gun or artificial pick processing. For excavation in level V surrounding rock, the common steel frame spacing is 0.5 m.

2.3.4.4 *Intersecting method* (*CRD*)

The CRD method is usually applied in large sections and long-span tunnel construction. According to the rock type before excavation, along the line at the beginning, an arch ring for a small catheter is fixed in advance. Then grouting reinforcement is done, and excavation is done for the stairs on the left side of the rock by means of drilling–blasting, with initial shotcrete and steel, and temporary installation of vertical and horizontal clapboard, a hollow bolt and hang net shotcrete. To prevent the foot arch, arch foot and temporary vertical clapboard from locking in the foot bolt, measures must be taken. Then excavation is done on the left side of the rock mass under the steps, with concrete spraying after erection of the steel frame and temporary partition, with bolts, down the steps shotcrete closed processing. Finally continuing right up the stairs on the right side and down the steps on the right side of the primary support, unilateral steps of 3–5 m are excavated (short upper steps are constructed when the excavation is not more than 3 m). Each cycle of excavation is 0.5 m.

(1) The process of CRD excavation (Fig. 2.38):
   (a) The first step of the tunnel is to be excavated, and the first step has to be supported by the temporary horizontal partition plate and vertical partition plate.
   (b) Excavate the second step of the lower steps, and construct the primary lining and temporary partition board.
   (c) Excavate the third step of the lower steps, and construct the primary lining and temporary horizontal partition board.
   (d) Excavate the fourth step of the lower steps, and construct the primary lining.

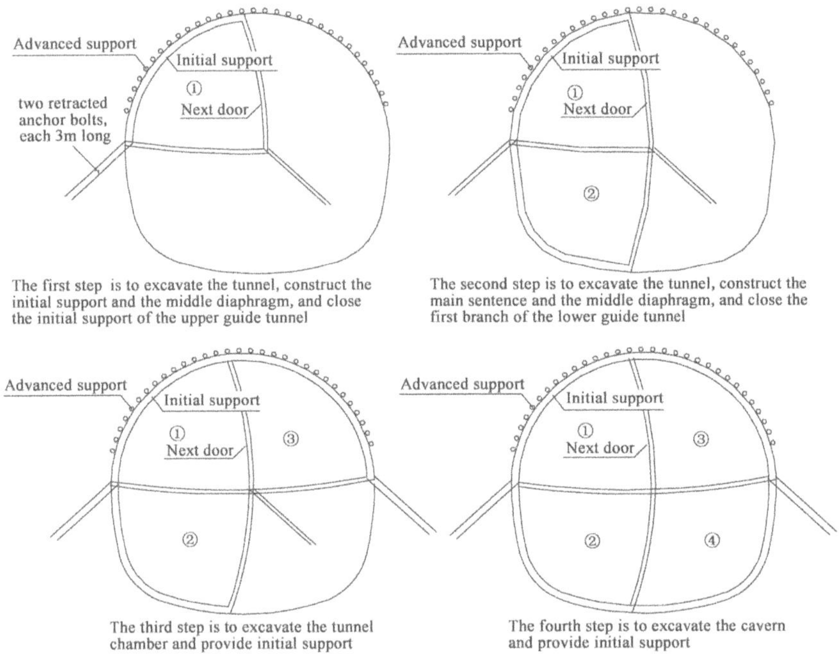

Fig. 2.38. Schematic diagram of construction process of the CRD method.

(e) The temporary partition plate is removed and the waterproof layer is laid.
(f) The arch part and the wall's waterproof layer are laid for the secondary lining.

(2) Section block of CRD method and step length control

Under normal circumstances, the step length is controlled within 3–5 m. In the case of poor water formation in wall rock stratum, the step length is shortened appropriately. The conditions for blasting excavation require a replacement gun or artificial pick processing. For excavation in level V surrounding rock, the common steel frame spacing is 0.5 m.

2.3.4.5 *Central drift method*

In the construction of the urban metro, the excavation method is usually applied in special areas such as cross-tie lines and small-distance tunnels. As shown in Fig. 2.39, excavation is done according to the category of the

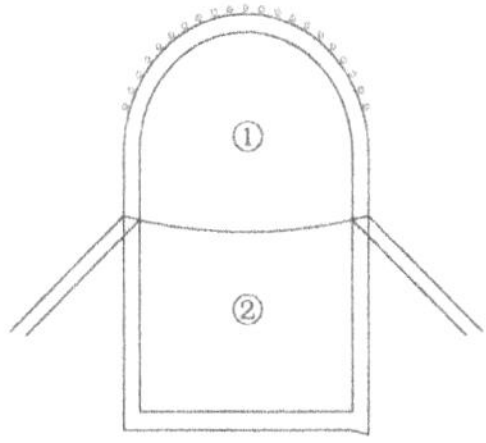

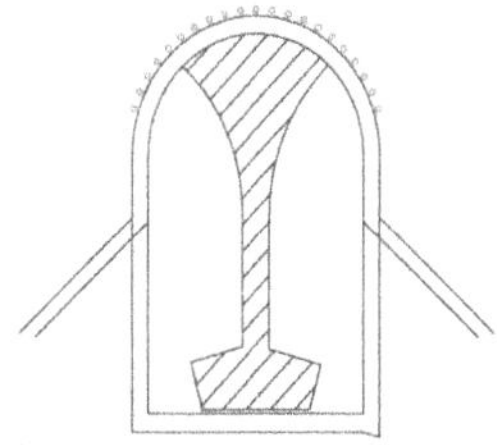

First, the guide tunnel is supported in advance, the guide tunnel is 1/2 in the step method, and the upper and lower steps are staggered by 3 ~ 5m. After the excavation is completed, initial support is applied and the middle guide tunnel is closed

2. After the middle guide tunnel is completely penetrated,the lower and top waterproof layer of the middle partition wall is laid, and the second partition structure of the middle partition wall, the connection between the middle partition wall and the middle hole, and the side tunnel second liner are preset with a reinforced connector and a later waterproof layer Laying joints

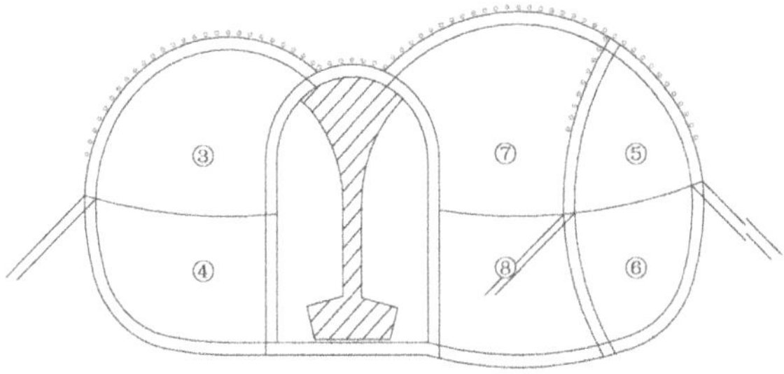

3. After the partition wall has been placed, the main method will excavate the main holes 3 and 4 on the left side, and the CD method will excavate the main holes 5, 6, 7, and 8 on the right side.

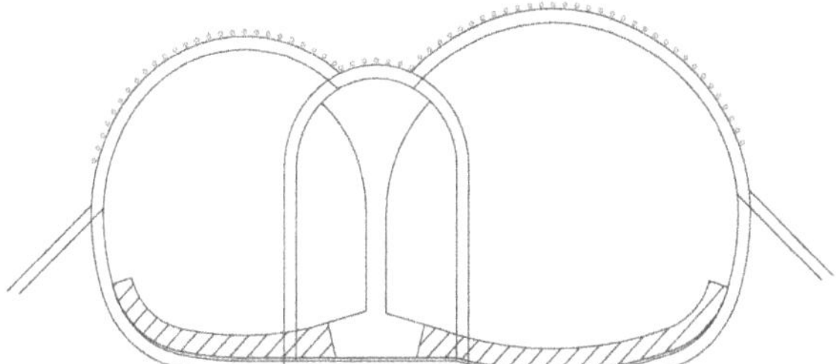

Partially remove the middle partition wall (step length 6 ~ 8m), lay Yanggong waterproof layer and apply Yang-gong secondary lining

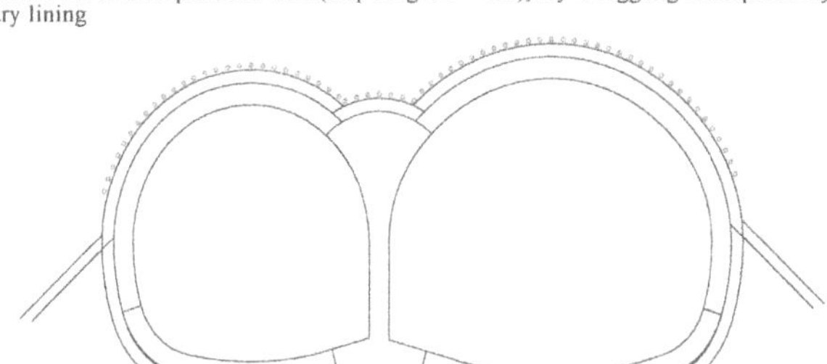

V. Continue to lay the waterproof board of the arch and apply the secondary lining of the arch to form a closed second lining. The excavation of the next cavern can be carried out after the construction of the second lining of the continuous arch section

Fig. 2.39. Schematic diagram of the construction process of the guide hole method.

surrounding rock in the middle of the multi-arch tunnel before setting a middle drift along the early pilot tunnel lining; an arch ring for a small catheter is fixed in advance. Then grouting reinforcement is done, and by the steps method the early middle drift excavation and support are completed. In the middle of the tunnel, the lower part of the partition wall and the top waterproof layer are laid. After the middle wall casting, steps on the left side of the main section of excavation (small section) and on the right side of the main section of CD excavation (section is larger) are constructed. After the even arch section's linking, a partial break in the door (step length is 6–8 m) is created, with an inverted arch waterproof layer and arch of the secondary lining. The steps of a single hole chamber are 3–5 m, and the length of each cycle is 0.5 m.

2.3.5 *Key points of steel arch support technology*

(1) When the steel frame is made, the connection of the components is the key process, which shall be carried out in strict accordance with the relevant regulations to ensure the quality of all the different kinds of welds. The following requirements shall be noted in the process:
  (a) During processing, it is necessary to standardize the production strictly according to the design requirements, make the sample platform, with lines, check and test the same, and mark the number to ensure the accuracy of production.
  (b) The steel frame is fabricated by rebar processing. All the steel frames are welded to the joint plate during processing, so that the steel frame can be connected securely.
  (c) The connection of the component is the key process of steel frame processing, and it should be carried out strictly in accordance with relevant regulations to ensure the quality of all the different kinds of welds.
  (d) After the steel frame is processed, it should be tested, and if its allowable error meets the requirements of the specification and it gains the supervisor's approval, it is put into use.

After the steel frame is processed, it should be put on the cement surface to test it, and the allowable error is shown in Table 2.19.

Table 2.19. Steel frame processing and assembly inspection standards.

No.	Project		Allowable deviation (mm)	Frequency of checking		Checking method
				Scope	Number of checks	
1	Steel frame processing	Sagittal height and length of Arch frame	+20, 0	Nature	1	With a ruler
		Wall frame length	±20		1	
		Cross-section dimensions of wall frames (high, wide)	+10, 0		2	
2	Steel frame pre-assembly	Altitude	±30		3	
		Width	±20			
		Torsion resistance	20			

(2) During the steel frame installation, one needs to accurately measure the first point, line and the horizontal pile to ensure the accuracy of its installation to meet the requirements of the outlined design (Table 2.20). According to the design of the spacing between steel connections and steel bars, the steel frame near the surrounding rock has to meet the design requirements of a 3-cm clearance protection layer. It is strictly prohibited to fill the steel frame with gravel or keep stones behind it. Steel and welding are firmly attached to the end of radial anchor design.

(3) The steel frame should be wedged tightly into the surrounding rock. When the steel frame has a large gap between it and the initial spray layer during the installation, the block should be set as shown in Figs. 2.40–2.42.

2.3.6 *Key points of shotcrete support technology*

Shotcrete support is used in the concrete jet as it will advance based on the proportion of the mixture. After adding accelerator, the concrete is

Table 2.20. Steel frame processing and assembly inspection standards.

No.	Project	Allowable deviation (mm)	Frequency of checking		Checking method
			Scope	Number of checks	
1	Steel longitudinal	±50	Every nature	3	Inspection with certificate and a steel ruler
2	Steel frame transverse	±30			
3	Elevation deviation	±30		2	
4	Vertical degree	5‰		3	
5	Steel frame cover thickness	–5 mm			

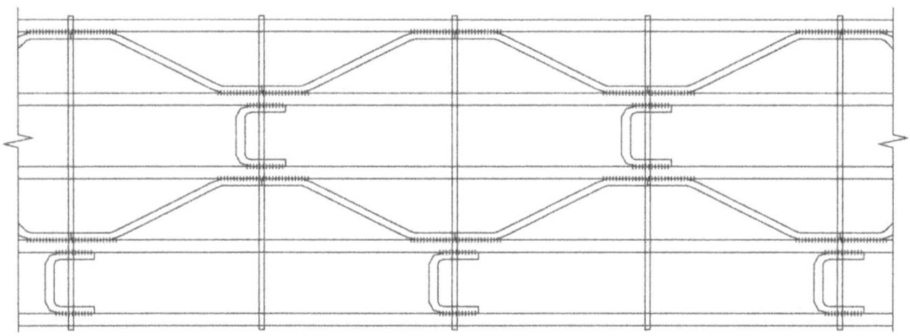

Fig. 2.40. Grid steel frame expansion diagram.

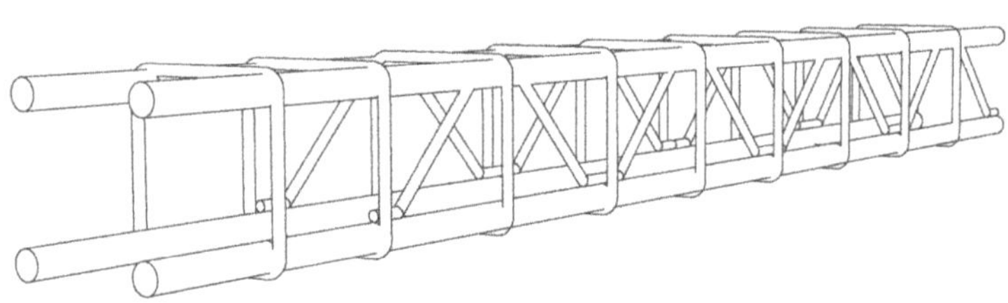

Fig. 2.41. 3D stereogram of grid steel frame.

applied through a concrete jet nozzle to the wall surface, leading to rapid solidification and forming a layer of the supporting structure, between the support and the surrounding rock. The solidification time changes according to different water addition times. The shotcrete is usually divided into

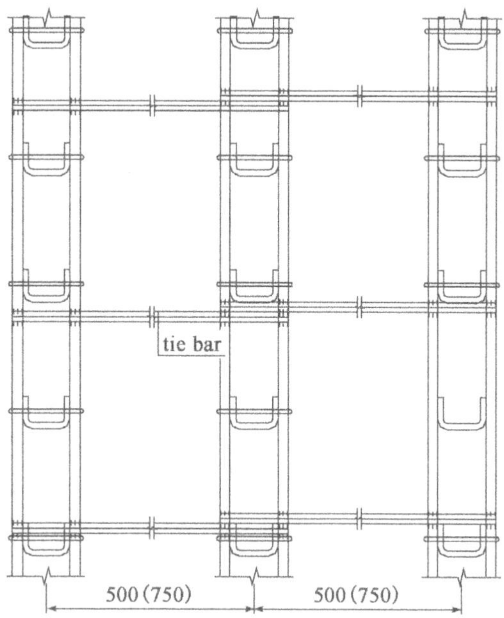

Fig. 2.42. Vertical connection diagram of grill steel frame (unit: mm).

dry spray, tidal spray and wet spray, which are the three construction methods. The construction methods and related characteristics of shotcrete are shown in Table 2.21.

(1) The raw materials of shotcrete should meet the following requirements:
   (a) Cement: Select P. O42.5 cement, the performance conforms to the existing cement standards.
   (b) Coarse aggregate: 5–10-mm gravel is used to meet the specification requirements. The cumulative weight percentage of aggregate gradation should meet the standard requirements.
   (c) Water: Drinking water.
   (d) Accelerator: Good quality; accelerating agent comes into play, at the beginning of net cement slurry; final setting time should be according to the test batches of the actual use of the raw material, the mixing ratio and parallel comparison experiments, with an allowable deviation plus or minus 1 min.

Table 2.21. Comparison of shotcrete construction and related characteristics.

Project	Dry spray	Tidal spray	Wet spray
Water addition time	Mix the water with the dry mix at the nozzle.	Add some water when making the mixture, and add water at the sprinkler head according to the mixture ratio.	Mix all the water in the mixture.
Concrete quality	Quality depends on the operator level and is not easy to control.	Mixes well, good quality.	High requirement for mixed concrete and easy to manage injection quality.
Work difficulty degree	The field mix system, the feeding limit is lower.	The field mix system, the feeding limit is lower.	The equipment occupies a large space, the limit is higher.
Technology and operation	Spray on time, manual operation.	Spray on time, manual operation.	Spray not timed, manipulator operation.
Springback quantity	More	Less	There are less hard rocks, more soft rocks, less large sections and more small sections.
Fault treatment and maintenance	Easy	Easy	Troublesome
Stive	More	Less	Less
Micro sweat on human body	Excessive dust, damage to respiratory tract.	Damage to respiratory tract.	Damage to the skin.

(2) The concrete spraying machine should have good sealing ability, conveying continuous and even feeding capacity, and should meet the concrete construction needs.
(3) The accumulative error of the raw materials is cement, accelerator plus or minus 2%, sand and gravel plus or minus 3%.
(4) The rock face should be prepared and cleaned before spraying.
    (a) Before spraying, clean up the surface to be sprayed, remove the surface pumice, stone slag or accumulation, dig out the undercutting part and flush the spray surface with a high-pressure air nozzle. For wet mud rock, high-pressure air is used to clean the rock face.
    (b) After the preparation of the sprayed surface, the spray thickness mark should be set on the anchor bar.
    (c) Geological data should be collected and organized before shotcrete spraying, and there should be cooperation with supervisors and design geological engineers to make the geological drawings.
    (d) The spraying shall be carried out immediately after the excavated surface is exposed. The operation shall meet the following requirements:
    (e) The shotcrete operation should be segmented and divided, starting from the top to the bottom, forming a firm movement; then the gap between the steel frame and the rock wall is sprayed first, after which the parts between the two steel frames are sprayed.
    (f) The spray concrete is stratified and the thickness of the jet is determined according to the injection site and the design requirement; the first layer should be sprayed just before the next layer is sprayed.
    (g) The resilience value of shotcrete should not be more than 15%.
    (h) Wet shotcrete is possible due to the work of wind pressure, so the appropriate distance between the nozzle and the spray (the wet spraying machine specification parameters shall prevail) is maintained, in order to avoid compressed air from being driven away from the concrete mix and to increase the resilience value of the coarse aggregate.
    (i) The spray nozzle should be kept vertical to the surface. If the spray surface is covered by grating or a steel mesh, the nozzle can be tilted slightly, but should not be less than 70°.

(j) The storage time of the shotcrete mixture should not exceed 20 min.
(k) When the design thickness of shotcrete is greater than one injection's thickness, it should be sprayed in layers.
(l) A minimum time interval between two injections should be maintained and the previous layer should be wet before spraying the next layer. In 2h after shotcrete spraying, water must be sprayed. The curing time and the number of water sprayings are modified according to the cement variety and air humidity, but in any case, the curing time should not be less than 14d.

(5) In the shotcrete lining section, in order to meet the requirements of the spray surface level, it should be noted that the spraying operation area should be fully illuminated. The nozzle should always be kept at a right angle, with suitable distance and injection pressure. The concrete should be sprayed at a thickness wherein it does not flow, and can be repeatedly sprayed to the specified thickness. After the injection, the thickness should be confirmed and the surface treatment should be done.
(6) The surface of the shotcrete should be compacted and flat, without cracks, shedding, leakage and other phenomena, and the deviation is allowed to be plus or minus 3 mm and the chord ratio should not be greater than 1/6.
(7) Supplementary methods for seepage and surface construction are as follows: When the sprayed surface rock has leakage or water seepage, it should be dealt with, in line with priorities: first the row, then the block. For wet shotcrete on the rockface, it is appropriate to reduce the water–cement ratio and wind pressure. When the jet sprays concrete at a high pressure, when the wind blows on the rockface immediately after water injection, leakage or water seepage could occur, especially when water is sprayed from a distance, as opposed to when you spray gradually from a shorter distance.
(8) During blasting operation, there should not be less than 3h between the shotcrete cycles.
(9) The temperatures of the sprayed concrete and mixture in the jet shall not be less than +5°C.

(10) The compressive strength of the sprayed concrete specimen should be appropriate for the construction conditions, and the sprayed concrete board and the specimen cannot be made of normal concrete. Then the following rules are cut out for standard or non-standard compressive strength specimens: the same mixture ratio should be used for each injection for a length of 20 m, concrete volume should be less than 100 m^3, strength of the arch and wall specimens should not be less than 100 m^3 and these should be made for one group of compressive strength of specimens. When the material or mix ratio changes, the test pieces should be made separately.

(11) The hammering method is applied to check the spray layer on the surrounding rock and the bonding between spray layers. For spray layer thickness, at every 10-m (20 m) range, check one section and each section from the vault center line. Each 2 m of the borehole check one point, cross-section. The checkpoint jet thickness should be more than 60% of the design thickness. The minimum design thickness is not less than one-third that of the design thickness, the total of which is averaged in order to be qualified.

2.3.7 *Grouting after the initial support*

In order to ensure that the primary support and surrounding rock are close grained and to prevent deformation of the surrounding rock, while at the same time improving the quality of waterproofing, every 3 m or so in the vertical direction along the tunnel arch ring, a spacing of 1.5 m is advisable and a side wall ring at a spacing of 3 m or so is suggested, with plum flower arrangement. Grouting depth at the beginning is 0.5 m. At the beginning of each pilot tunnel with a closed loop for 3–5 m, there should be timely grouting, as shown in Fig. 2.43.

The embedded grouting tube usually takes a ϕ 42 steel pipe, buried 3–5 m under the surface. Afterward, grouting for the cavity between the surrounding rock and the initial support is done in a timely manner. It is recommended to use a grouting pump water–cement ratio of 1:1 slurry, with grouting pressure P = 0.3–0.5 MPa, as long as there is a drip- and wet-stained surface, with repeated grout injection until no longer wet stained, to seal water.

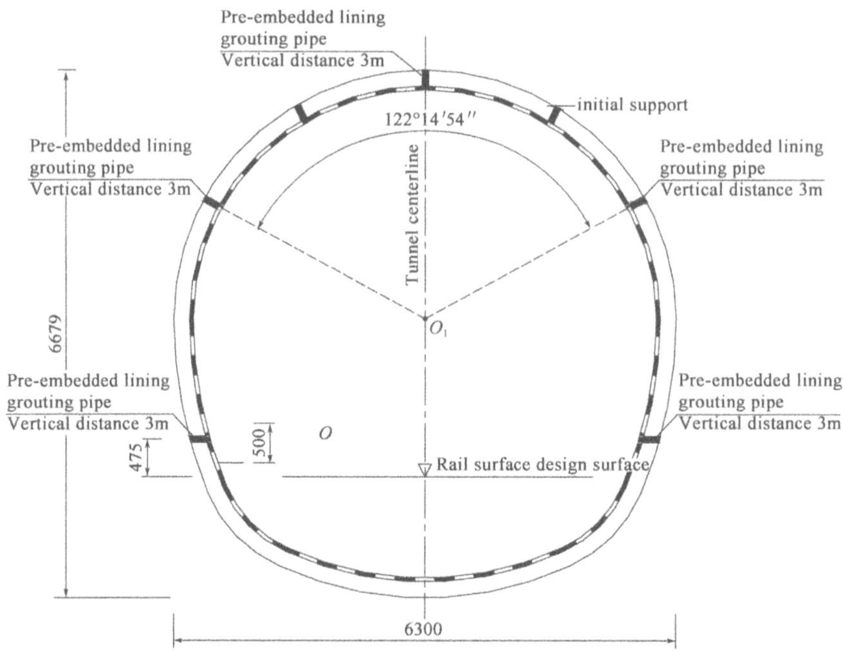

Fig. 2.43. Pre-embedded grouting pipe layout (unit: mm).

2.3.8 *Summary*

The drilling and blasting method of tunnel excavation in the hard rock stratum, along with the supporting technology, is relatively mature, but in the urban construction surrounded by a busy environment, there is a large probability of technical difficulties taking place. One should attach great importance to the protection of the surrounding buildings and control the blasting vibration velocity. The focus should be on the following:

(1) In city construction, when influenced by the environment, at a narrow construction site, the layout should adhere to the principles of compaction. A reasonable shaft area should be 1,000–1,500 m^2. A conditional area should first be chosen to plan the inclined shaft in the hole. The slag from the slope is more than twice that of the shaft. If there is poor geological development, the slope underground geological section (such as arch sand) should not be more than 100 m, and one should not choose a plan involving the inclined shaft in the hole.

(2) The method of drilling and blasting in the city should be carried out according to a proper design of blasting, which should be "multi-perforated, with less charge and blasting", and it is recommended to use the automatic vibration speed monitoring method to implement the management system of the fireworks.
(3) The selection of the construction equipment is very important, especially the slag equipment of the shaft which should meet the requirements, and the slag equipment in the inclined well is also very important. The ZL50 side flip loader can be used.
(4) The supply of fireworks is one of the key factors affecting the construction efficiency of urban metros. One should strengthen the organization and strive for the blasting of each working face in the morning and evening.
(5) The choice of the method has a great influence on working efficiency. The method should be chosen according to the geological situation and adjusted in a timely manner. If the condition of surrounding rock is at levels II, III smooth blasting can be used for the whole cross-section. The penetration control in the 1.5–2-m range. If the geological condition is bad, one should choose the method of the up and down steps, with a method of "small step to run", in which there will be controlled penetration and controlled explosion. One should read the "ten eight guidelines" quickly.
(6) In tunnel construction, the support should be timely. The grill steel frame is recommended to be concentrated in the processing plant. Not only can this save the field work space but it can also guarantee processing quality and improve the operation efficiency.
(7) At the beginning of the grouting behind the segment after the tunnel excavation cycle, the holes are filled and one has to focus on grouting seepage in important areas. For the subsequent two linings, the construction is done against a good working face.

2.4 Waterproofing construction technology of drilling–blasting tunnel

2.4.1 *Waterproof system*

At present there are many all-inclusive tunnel waterproof types, which prevent rows from leaking and provide drainage control combined with three types. There are many pros and cons to each kind of waterproofing and drainage way, and there is no particular method that is adapted to the regional scenario. There are successes or failures in each case, so to strengthen the process control is the key to waterproofing in construction. Three waterproof lines are set up in the tunnel:

(1) **First line of defense:** Initial support + back grouting.
(2) **Second line of defense:** The flexible waterproof layer. The waterproof layer is made of a non-black non-asphaltic polymer self-adhesive film waterproof plate (coil) (Y P S 1.5 mm-GB/T 23457-2009).
(3) **Third line of defense:** The structure self-waterproofing. The secondary lining adopts the C45 and P10 waterproof reinforced concrete, and a secondary grouting is made between the second lining and the waterproof layer.

The waterproof system of the excavated structure is shown in Fig. 2.44.

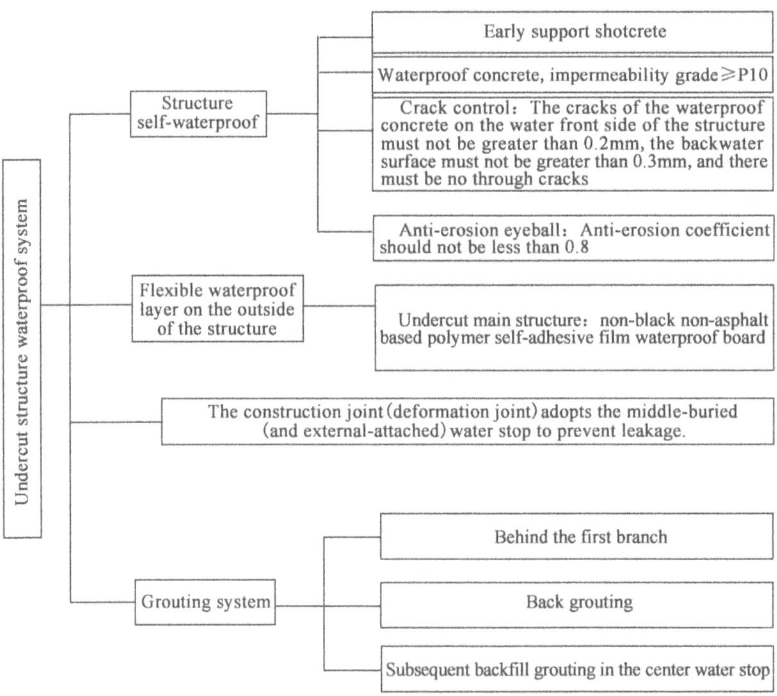

Fig. 2.44. Waterproof system diagram of the excavated structure.

2.4.2 *Waterproof base treatment*

Base processing is an important working procedure before the construction of the waterproof board (coil), mainly including seepage treatment and surface roughness. If the process cannot be done well, it will directly affect the construction quality of the waterproof board (coil). At the beginning, seepage is considered a systems engineering issue, and specific treatment plans should be formulated according to the size of seepage.

【Engineering project:】The case of water leakage treatment in a certain interval of Qingdao Metro Line 13.

2.4.2.1 *Steel frame disposal measures*

The steel section is in a medium-water-rich area, with net-like connected cracks, and early-bolting shotcrete thickness of more than 200 mm. The waterproof partition settings are as follows: At the end of the construction area, in a water-rich area (serious leakage location) at both ends, every 50–100 m a waterproof partition wall is set according to conditions, along with a waterproof curtain along the tunnel ring, namely, along the normal direction two ring drillings are set, with hole spacing of 0.4 × 2.0 m and hole depth of 4.0 m.

2.4.2.1.1 Drilling and grouting

(1) **Borehole and embedded grouting pipe**
   (a) *Arch part*: The hole is set within 120° of the arch, 4,000 cm deep and the longitudinal hole is 2.0 m.
   (b) *Side wall*: The upper and lower sections are drilled with a hole position of 1.0 m, the longitudinal hole is 3.0 m and the hole depth is 80 cm, with staggered setting on both sides.
   (c) Embedded note phi 32 steel tube is used as a slurry pipe, with a steel pipe tail 40-mm range setting thread, in the hole after using anchoring; the required intensity is reached after installing the control valve for grouting, concrete as shown in Fig. 2.45.

(2) **Grouting:** The test operation of the grouting pipe system: After the equipment is in place, the system will be tested for water absorption

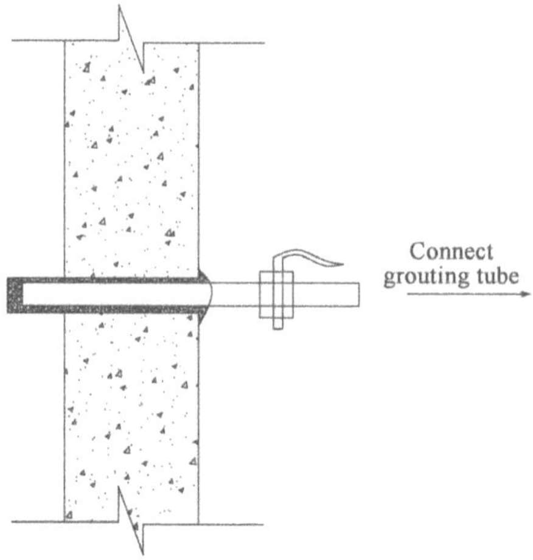

Fig. 2.45. Drilling and embedded grouting pipe diagram.

by the final pressure 1.5–2 times, and the wind, water and electricity will be connected. The ability of the piping system to withstand pressure will be checked, along with whether there is leakage and whether the pipeline connection is correct. It will also be checked whether the equipment is performing normally and a test run will be performed. The time taken is usually 20 min.

(3) **Grout slurry:** For continuous grouting, 1–2 slurry tanks are provided.
   (a) *Grouting material and proportioning cement*: With high activity and a production date not more than 3 months away, ordinary Portland cement of P. O level above 45 is used. The water used is drinking water, the water temperature is not less than 6°C and the pH should be in 5–8 range.
   (b) *Grouting ratio*: The cement ratio is 1:0.8–1:1.5.

(4) **Grouting order:** After the first injection hole, according to the principle of precipitation funnel, from the vaults, in case of a string of pulp, based on the interval of the holes or the pore pressure, the adjacent hole's slurry flow might jam after pressure grout injection until the correct pressure is ensured.
(5) **Grouting pressure control:** The grouting pressure is controlled by the oil pressure of the grouting pump, and it is controlled at 0.3–0.5

MPa. In order to prevent a high crushing speed and pressure that is too fast to return to the pulp and slurry, intermittent grouting is repeated.

(6) **Modifying the order:** In order to prevent slurry formation, the drilling and grouting press up, down, left and right in the holes in a staggered order. The stop valve must be closed in time after the grouting finishes. The grouting stops for 2 days, according to the leakage of the tunnel and the regional specifications.

2.4.2.1.2 No steel frame disposal measures

No steel section of general design is used for the supporting form of levels II, III surrounding rock, which is given priority in dealing with underground fissures, joint micro development and water cracks, with the following plugging order: arch department, the sidewall, then the end of the tunnel. The early-bolting plugging method is as follows:

(1) **The leak is stopped at the point of concentration:** In an area with a small amount of water, from where the concentration of water leaks out, the leak is plugged, and in an area with a large amount of water, which leaks from a scattered area, the leak is stopped by deep-hole grouting.
(2) **The plugging agent is used to plug the leak:** After removing the loose concrete layer in the water outlet, the water pressure should be reduced by using cloth or cotton yarn.
(3) **Deep-hole grouting prevents leakage:**
    (a) The layout of the grouting hole at the concentrated outlet point is around the 1.0-m range and the hole depth is 2.0 m; the hole layout is shown in Fig. 2.46.
    (b) Grouting
        (I) *Grouting materials and proportioning cement*: Ordinary Portland cement with high activity and a production date of less than 3 months prior to usage is applied. A water–cement ratio of 1:1 is used, with the water being drinking water, at a temperature of not less than 6°C and a pH of 5–8.

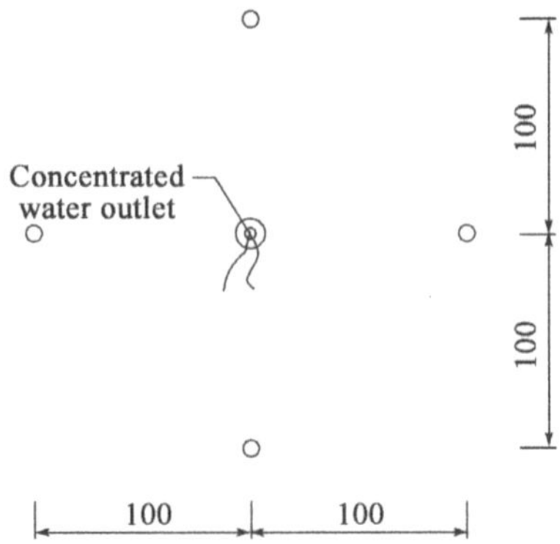

Fig. 2.46. Hole layout deep-hole grouting (unit: mm).

(II) *Grouting control parameters*:

1. *Grouting pressure*: The final pressure is 0.3–0.5 MPa and the grouting lasts for 5–6 min.
2. *Diffusion radius of single-hole grouting*: $R = 0.8$–1.8 m. The effective diffusion radius is 60–80 cm.
3. *Single-hole grouting*: The final hole is pressure controlled.

(4) **Tunnel bottom inverted arch cleaning:** After the removal of the waste residue from the tunnel bottom backfilling, the inverted arch is rinsed clean and sprayed with water pointed at the loose weak concrete, identifying the cracks in the surrounding rock. There are ~32 buried drilling pipes (tail with thread), which are washed with water first and then filled or sprayed with concrete making it waterproof and closing the inverted arch again. After being concreted at a strength of 60% or more, the steel pipe at the end of the hose is demolished, using an epoxy cement mortar blocking pipe. The water point messy surface infiltration or surface flow is arrested by drilling, grouting or plugging as required following the deep-hole grouting plugging technology.

(5) **Base-level acceptance:** A pre-construction conditional acceptance needs to be done of the base surface before the waterproofing construction, which will mainly include the following: the early clearance size, transfixion measurement and the surface roughness meeting the design requirements, no obvious leakage phenomenon at

the beginning, seepage water processing standards generally without a clear design and specification requirements only at the beginning, two lining waterproofing layers, and extensive seepage only in the early stage of processing.

2.4.3 *Construction of waterproof board (coil)*

[Engineering example:] The construction case of a waterproof plate (coil) on Qingdao Metro Line 13.

2.4.3.1 *EVA waterproof panel construction*

Material specifications for the waterproof layer: geotextile cushion layer (general design) of 400 g/m^2 and flashing of 1.5 mm. To reduce the flashing between the longitudinal weld and waterproof board according to the length of the inverted arch and arch wall, one can use a coil scale, at 1:0.8 for the control.

The fixed geotextile is equipped with a nail gun, a hot melt washer and a waterproof plate, which are welded together; the waterproof plate is welded with a double-seam welding machine and an electric soldering iron is used to repair it.

2.4.3.1.1 Buffer laying

The inverted arch is paved with an empty paving method, while the side wall and the arch roof waterproof panel adopt the EVA gasket fixation method.

Buffer layer material is not less than 400 g/m^2 unit weight geotextile.

With masonry nails and the waterproof board well matched with the hot melt gasket material (EVA), the buffer layer is fixed on the base. The masonry nails are fixed with a nail gun, the arch fixed point spacing is 400 mm, the arch wall protection distance is 800 mm and the plum flower layout is formed. Low base-level areas should have added protection points, so as to avoid breaks in the waterproof layer of the lining concrete. The nail depth shall not exceed the gasket plane, otherwise it will easily pierce the waterproof layer.

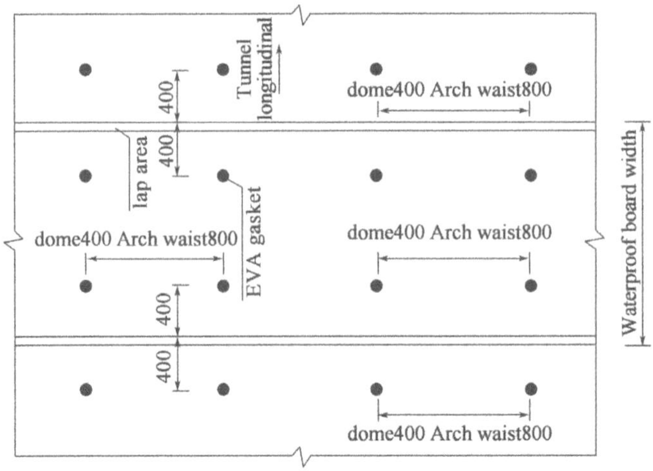

Fig. 2.47. Fixed schematic diagram of geotextiles (unit: mm).

Laid non-woven fabric is fixed to the desired location with a hot melt gasket and nails (material specification: hot melt thermal 70-mm diameter and 5–6-mm thickness, metal gasket 20-mm diameter and 2-mm thickness, nail length 25–30 mm). The nail gun is used for fixing the non-woven fabric bottom-up in a fixed cycle, with a width not less than 50 mm. The special hot melt gasket and nails are used to decorate the plum flower form, at an arch span of 0.4 m with a sidewall of 0.8 m. The non-woven and spray concrete surface form a dense paste, and should be laid smoothly, with no uplift or fold. After installation, one should check whether the hot melt gasket is in good condition and determine whether it can meet the installation requirements of the waterproof panel. If it is not suitable, the hot melt gasket should be repaired (Fig. 2.47).

During construction, one should carefully check whether the nails are out of the gasket or not, and whether there is excessive use of the hammer.

2.4.3.1.2 Waterproof board

(1) The method of fixing the waterproof board will be used to reduce the number of T-shaped welds and cross welds, and reduce manual welding to ensure the waterproof effect.

(2) The waterproof board is manually welded on the plastic circular gasket by the hot melt method. The welding should be firm and reliable, and the waterproof plate should be removed when pouring and vibrating concrete.
(3) During the waterproof board fixing, one should pay attention not to pull too tight or else the board may bulge. A good waterproof board shall conform to the base level of the concavo-convex fluctuation, and maintain a natural, smooth level, without breaking, so as to not affect the two linings of concrete or flashing outside the round gasket.
(4) While laying the waterproof board, the waterproof plate should be fixed at a predetermined position, and then a manual electric welding machine should be used for heating, at a hot melt pressure voltage of 220 V with the power at 500 W and a heating temperature of 0–450°C. It is fixed in a bottom-up manner with the hot melt. At the beginning of the electric high-pressure welding, the heat is lower; half an hour after the peak, normal-pressure welding of each hot melt gasket should be conducted for 3–5 min. The head of the pressure welding machine should be at the optimum temperature of 300–400°C. An insulation cloth cushion layer should be placed between the manual hot melt pressure welding machine and the waterproof board to prevent heat damage to the waterproof board. The waterproof board welding should be made in the special fixed non-woven hot melt gasket.
(5) As for the circumferential directions and longitudinal seam construction, as a result of the waterproofing of the whole pack, and inverted arch lining, the construction of the circular and longitudinal seams is done separately. The welding machine for the waterproof plate adopts an automatic double-seam method; the temperature control part adopts automatic constant temperature PID control. The speed control part adopts pulse width modulation (PWM) with an automatic voltage-stabilizing circuit, DC servo motor drive, the output torque and stable walking. The following specifications are maintained: a predetermined welding speed, voltage (V): 220 V, frequency (HZ): 50 HZ, power (P): 800 W, welding speed (V): 0.5–5 m/min and the heating temperature (T): 0–450°C. The welding joint ring to the single

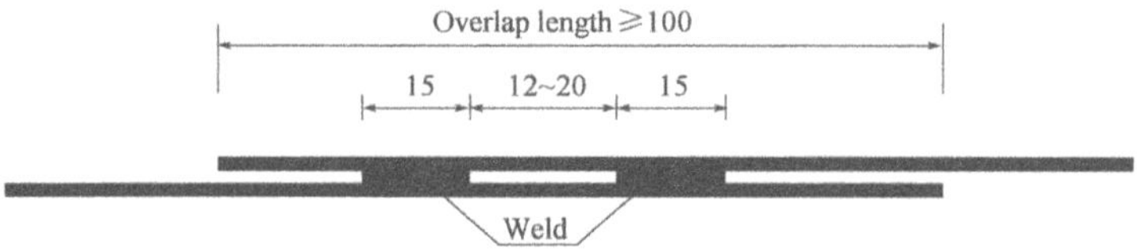

Fig. 2.48. Welding diagram of waterproof plate (unit: mm).

channel needs to be run for 25–30 min (cycle length 28 m or so), during which the operators need to control the temperature. The actual applied support is adjusted to meet the demand and the optimum temperature is between 350°C and 400°C. The lap width is 100 mm, the width of the two welds is not less than 15 mm and the cavity is between 15 and 20 mm. There is an air pocket between two welding seams, which is used for tightness testing of the instrument and detection of welding quality. In addition, the presence of welding dust on the surface of the waterproof board should be avoided. When welding, one must also pay attention to the accuracy of the lap at a width of 10 cm and ensure that it is not skewed. To smoothen the welding, top-down circular welding must be done, which will also ensure joint compactness (Fig. 2.48).

(6) Inspection of the waterproof plate's weld quality must be done. The waterproof board has hot melt welding seams with double seams, at a lap width of 10 cm. After gas detection, to block the airway, at one end and then the other, a pressurized air tightness test instrument is used, with an inflation pressure of 0.25 MPa, maintained for not less than 15 min, with an allowed pressure drop of 10%, which will continue to decline. If any leak is found, comprehensive manual repair welding will be performed.

2.4.3.2 *Construction of waterproof rolling material*

A flexible waterproof layer is adopted for the underground tunnel project before paving with a black asphalt base since the glued polymer film (YP S 1.5-mm GB/T 23457-23457) is a high-density polyethylene sheet. With the adhesive polymer film, there is a special performance requirement for the granular layer, which is sticking prevention so as to realize

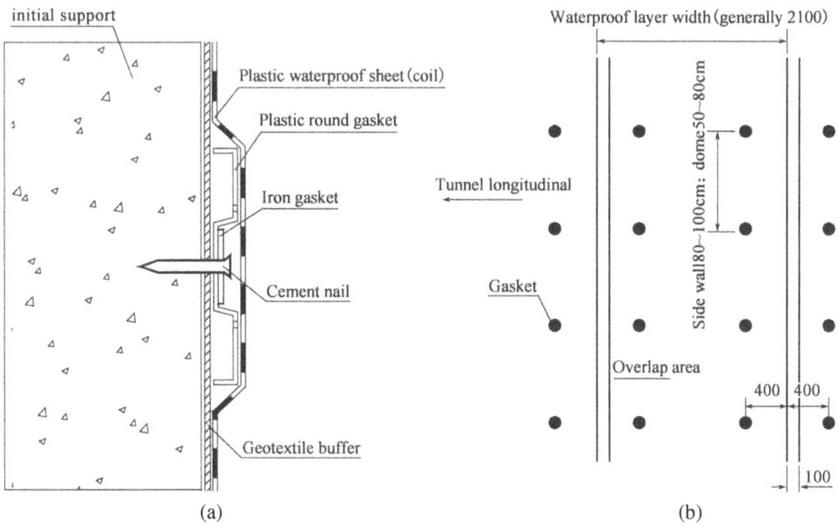

Fig. 2.49. Fixed method of waterproof rolling material and layout plan of waterproof rolling material (unit: mm).

integration with the concrete structure of asphalt base polymer before paving the waterproof coiled material. The initial setting of the cement slurry in the concrete is under the effect of pressure. Through the creep, the relative permeability layer is prevented from sticking, thereby forming an effective interpenetrating bond and great intermolecular force. After concrete curing, the gap between the waterproof rolling material and the main body of the structure is sealed to the maximum extent, eliminating channeling completely. The "external anti-interior sticking method" is adopted, the plane part adopts the hollow paving method and the facade is laid by the mechanical fixed method. The laying sequence is the inverted arch and the side wall first, then the arch top and the two side walls (Fig. 2.49).

2.4.3.2.1 Construction process of waterproof rolling material

The basic processing for laying of the reinforced waterproof rolling material is followed, with the pressure edge acceptance of the waterproof protection layer during concrete pouring.

The requirements for waterproof layer are as follows:

(1) The base surface of the waterproof layer should be cleaned, and the required level of smoothness should be satisfied: *D/L* is less than 1/50 and *D* is between the adjacent convex surfaces. Maximum depth, *L*, and minimum distance are required between the adjacent two convex surfaces, and the concave and convex parts should be smooth and flat. If all the requirements are not met, the bulging parts should be chiseled, along with a 1:2.5 leveling of cement mortar. The pit part shall be filled with cement mortar. The base surface shall be clean, flat and firm, and shall not have loose material, sand or any peeling phenomenon.
(2) Any uneven parts adopt a 1:2.5 cement mortar processing ratio; if the base level is in a poor condition, it can be laid first with a 400 g/m^2 of geotextile cushion layer.
(3) The base surface may be slightly damp, but there should be no clear water flow; otherwise, water shutoff treatment or temporary drainage should be carried out.
(4) All the negative angles are made of 1:2.5 cement slurry for 5 cm × 5 cm of obtuse angle where *R* is greater than 5 cm round corner, and the positive Angle is 2 cm × 2 cm, or the round angle of *R* is greater than 2 cm.

2.4.3.2.2 Construction of waterproof layer

(1) The waterproof layer adopts a self-adhesive film waterproof rolling material, where the separation film close to the first side is not torn off; the isolation film with the surface layer of the cast-in-place concrete structure should be torn off before pouring concrete.
(2) The waterproof head adopts the mechanical fixation method early on, with a fixed point 2 cm from the coil edge, at a distance not more than 50 cm, with a nail not less than 27 mm long; the gasket waterproof layer is firmly fixed on the surface at the grassroots level, wherein the gasket is not less than 2 cm in diameter, to avoid loss when pouring concrete.
(3) The effective lapping width of the two adjacent coils is 10 cm (not including the nail hole). The upper pressure is required to be overlapping. When connecting, the insulating film on both sides of the waterproof rolling material is required to be torn off. Short side lap joints should be staggered by more than 1 m.

(4) After the waterproof layer is laid, the reinforcement layer shall be laid on all construction joints and deformation joints, and the width of the construction joints shall be 50 cm. Deformation joints' strengthened layer width is 1 m; the construction seam reinforcement layer and waterproof layer are self-adhesive, and each side of the deformation seam is 10 cm. The insulating film on the surface of the inner waterproof layer should not be torn off (that is, the waterproof layer and the reinforcing layer should not be pasted).
(5) The coils should be sealed with polysulfide building sealant, and the width should not be less than 10 mm.
(6) The applied waterproofing materials may produce a spark, so one should set up a temporary mobile protection board or asbestos board or take other measures in order to avoid sparking or burning of the reinforced waterproofing materials.
(7) Excluding the air under the coil, the roller should be pressed firmly, and the surface of the coil should not be distorted, crumpled or bubbling. After the side wall rolling material is finished, the end of the coil should be fixed. In low-temperature construction, the waterproof rolling material and base surface should be properly heated and then laid.

The inverted arch is laid with a waterproof layer of coiled material isolation membrane, and immediately afterward a 70-mm-thick C20 fine stone concrete protective layer is casted. After the completion of the waterproofing of the lateral wall, in the subsequent steel bar construction, one should pay attention to the protection of waterproofing materials and avoid bad waterproof layering. If the layers are damaged, one should arrange for personnel to use the same waterproof materials to conduct repair work. If the damaged area is full of sticky patches, a patch adhesive layer should be used in the damaged edge at a minimum distance of not less than 10 cm, which can reinforce cast-in-place concrete.

2.4.4 *Construction joints and deformation joints*

2.4.4.1 *Construction joints*

The construction joints are divided into circular construction joints and longitudinal construction joints, and the waterproof structure of construction joints is shown in Fig. 2.50.

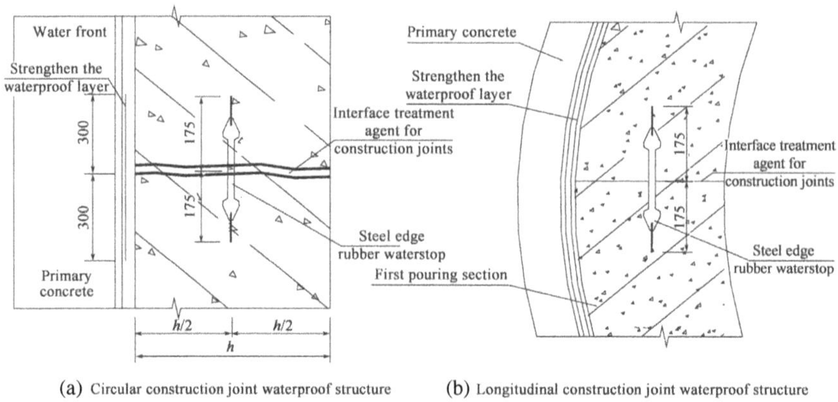

Fig. 2.50. Waterproofing construction diagram of construction joints (unit : mm). (a) Waterproofing construction of circular joints. (b) Waterproofing construction of longitudinal joints.

(1) An interval of 10 m between the construction joints and the corresponding setting of the construction joints in the second lining of the platform is guaranteed.
(2) There should not be any shear force or bending moment between the horizontal construction joints or the plate and side wall of the junction. The side wall and inverted arch above the intersection point should stay 300 mm away from the wall; the wall has reserved holes. The distance to the construction joints from the edge of the hole should not be less than 300 mm.
(3) Before the vertical construction joints are concreted, the surface should be cleaned up, the concrete interface primer or the cement base permeable crystalline waterproof coating should be applied and concrete should be poured in time.
(4) Before the concreting construction of the level joints, the surface laitance and debris should be removed with a brush or concrete interface treatment agent, and then net pulp is laid, with cement base material such as crystalline waterproof coating, with a 30–50-mm-thick layer of 1:1 cement mortar, and concrete in a timely manner.
(5) In the middle of the construction joints, a steel side rubber hose stopper and a grouting pipe are set. All the rings are equipped with a

600-mm-wide additional waterproof layer on the outside of the construction joint, and the waterproof material is selected for the waterproof plate (coil) in such a way that it can be bonded with the main structure.

2.4.4.2 *Deformation cracks*

The setting of the deformation joints is shown in Fig. 2.51.

(1) The interval of the deformation joints is set according to the drawing, and the general interval is set along the junction of the station and the cross channel and the normal line of the tunnel. The width of the deformation joints is generally 20 mm.
(2) In case of leakage, it is difficult for the deformation joints to be repaired, and the deformation joints should be equipped with an auxiliary waterproof layer to enhance the treatment of the waterline.

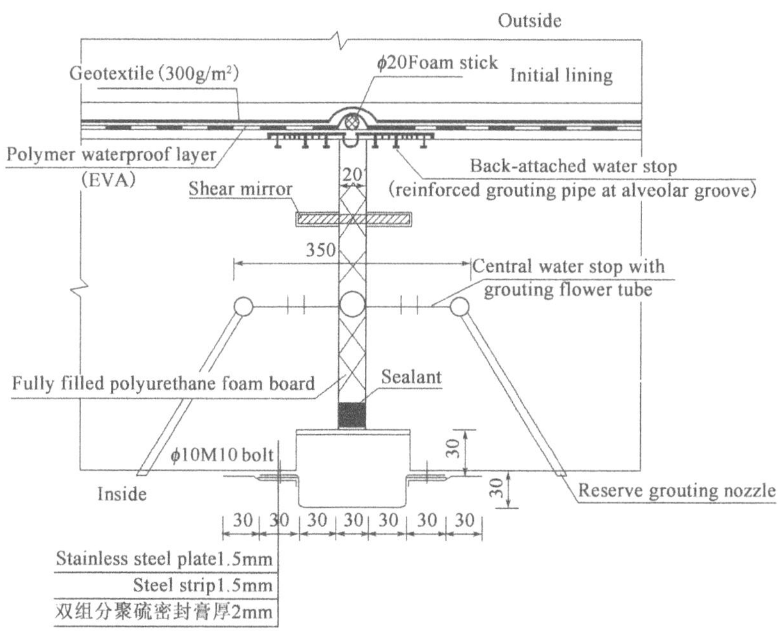

Fig. 2.51. Waterproof structure of deformation joints (unit: mm).

(3) One should arrange for a water-stop belt on the outside of the molded concrete at the deformation joints, and use the back hose surface between the rack and the mold to build a waterproof concrete layer to seal off the water. At the same time in the back of the outer rack, inside the hose, on the wing root fixed grouting pipe, one can use pulp out of the grouting tube to evenly fill in the gap of the hose and concrete parts, thereby achieving the goal of sealing out water.
(4) In the middle of the deformation joints, the rubber stop belt with the grouting pipe (center air hole in the center form) forms a closed waterproof line.
(5) The arch of the deformation joints and the inner wall of the side wall are set up with a stainless steel flume, and a small amount of water seepage is introduced into the drainage ditch and discharged into the station or interval sewage pumping station.

The inside of the deformation joints is sealed with sealant. The sealing paste is required to be closed along the deformation joint ring. No break point is allowed in any part, so as to avoid the phenomenon of water channeling.

2.4.5 *Summary*

Waterproofing is the key link in the construction of underground excavation.

(1) One has to pay special attention to any early signs of leaking and perform a close inspection of the construction area. One should also ensure that at the beginning of the dry base, there is a precondition of waterproof board (coil) construction. Therefore, the beginning of a job is one of the most important links in tunnel construction; the main aim of the design is to ensure that the project is completed safely. At the beginning of the construction, a unified quantization standard is established, along with a system of checking the base conditions. Each section of the work area shall be no less than 100 m in length, a professional team of experts will be chosen for water plugging and the most stringent measures will be adopted to ensure early-stage water plugging.

(2) Waterproof board (coil) quality is dependent on the operation personnel and the quality of the work. Therefore, in waterproofing construction, one must strictly implement a traceable quality system, when each waterproof joint is managed in a real-name system, with proper management of the operating personnel, and maintenance of records of each seam test result, thereby improving the workers' sense of responsibility and fundamentally improving the quality of waterproofing.
(3) Protection of waterproof board (coil) should be done with great attention to detail. The destruction of the waterproof board (coil) tends to occur in the process stage, such as damage of steel binding, damage of the waterproof board (coil), concrete damage of the plate (coil) and so on; therefore, one should strengthen process control to achieve a good-quality waterproof board (coil).
(4) The choice of waterproof board (coil) is also very important; it is suggested to choose self-adhesive polymer material.

2.5 Secondary lining construction technology of tunnel

2.5.1 *Preparation for lining construction*

The drilling and blasting method is conducted using a template car, with tunnel secondary lining according to the first inverted arch after the order of the implementation of the second line, with the arch forming a wall lining, thereby reducing construction joints and improving the waterproofing ability of the second lining. The inverted arch and arch wall longitudinal construction joints should be set on the rail surface elevation at 100 mm, near the secondary lining of the longitudinal section, at a length of 20–25 mm, based on interval curve radius, with the length of each section being 9 m, 10.5 m and 12 m.

2.5.1.1 *Construction preparation*

(1) After the initial support of tunnel excavation is completed, technical analysis of the measured data in the early stage will be carried out in time. The feedback information of the test results will have the following conditions:
   (a) There is a significant decrease in the displacement rate.
   (b) The net space convergence is less than 0.15 mm/d.

(c) The displacement that has occurred accounts for more than 80% of the total displacement, which indicates that the initial support deformation of the tunnel has been basically stable, and the secondary lining construction can be carried out.

(2) Inspection of the materials and the new equipment will be carried out, along with selection of the commodity concrete manufacturer selection and other relevant preliminary work.
(3) According to the design drawing, the lining reinforcement is processed in time.

2.5.1.2 *Clear bottom and base surface treatment*

According to monitoring value, the primary support closed loop tends to be stable, with no vault subsidence upon checking the tunnel. The waterproof board (coil) is measured for base treatment. Before base treatment, it should achieve a solid, smooth, coordinated layer with no obvious concave and convex objects. By dividing the backfill grouting due to the water plugging, one can manage the bottom digging process and repair the base, highlighting the mutilation caused by sharp objects or the chisel used for flattening the concrete bulge.

2.5.1.3 *Waterproof board (coil) and construction of geotextiles*

2.5.2 *Secondary lining construction technology*

[Engineering example:] The second lining construction case of Qingdao Metro Line 13

2.5.2.1 *Construction of waterproof protective layer of fine stone concrete*

The inverted arch on the waterproof board (coil) should be laid with a fine stone concrete cover, in order to prevent steel bar damage (Fig. 2.52). As it is applied to the commercial concrete on the earth's surface using a concrete pump truck in the operation, the waterproof board (coil) on the wall at the edge of the inverted arch is pressed 1 m along the vertical single direction in the tunnel. Furthermore, construction personnel cannot

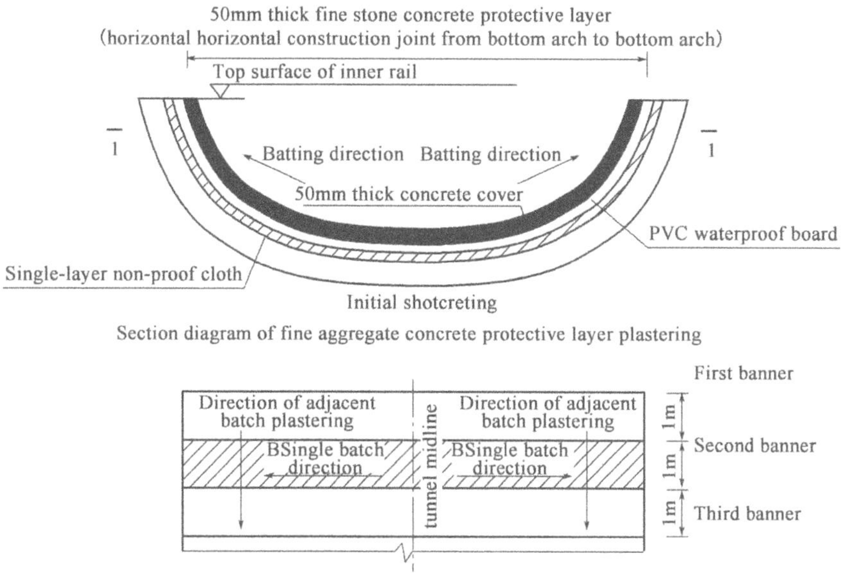

Fig. 2.52. Schematic diagram of the protective layer of fine concrete.

wear shoes directly on the waterproof board (coil) above or on the layer of geotextile laid for temporary protection.

2.5.2.2 *Lining reinforcement processing and construction*

(1) A second lining reinforcing bar material is used.
(2) In reinforcement processing, the first set of inverted reinforcements is adopted on the surface precast.
(3) The remaining two reinforcing bars are used to lay a steel plate on the lined inverted arch concrete, and the steel bars are measured on the steel plate and welded to it in sequence. In steel processing, a swing steel bar bending machine is used in the tunnel; at the same time, it will be reinforced in the shaft-hoisting system at the bottom of the hole. In the process, the steel bar bending machine is used to prebend the steel bar, and then the steel bar mold is used to shape the steel bar manually. The steel processing plant in the tunnel is randomly changed along with the lining of each group.
(4) After the completion of the inverted waterproof layer and the protective layer, the reinforcement is carried out and the steel reinforcement

is applied first, and then the arch wall reinforcement are applied after the inverted arch concrete is removed.

(5) The reinforcement of the inverted arch is done using artificial direct ties, and the reinforcement of the side wall arch is done with a multi-function easel.
(6) The link of the ring to the main bar and the longitudinal distribution of the reinforcement are mechanically connected.
(7) The joint of the reinforcement should be set at a lower force. The reinforcing steel joint should be staggered, and the area of the joint should not be more than 50% of the total 35D area.
(8) The control of the protective layer of the inverted arch and the spacing between the two layers is fixed. The lower layer is made up of the same grade strength concrete, and the distance between the two layers is controlled by the horse bench.
(9) The construction of the side wall and the reinforcement of the arch are made by using a simple scaffold, and then the concrete is poured.
(10) The embedded grouting pipe shall be prepared according to the design requirements when the reinforcement is fastened, and the grouting should be filled after the lining is finished. For the reinforced protection layer construction, a protective layer (wide) of 50-mm length and ×40–50-mm thickness with two linings of the same grade strength of concrete block are utilized at a set distance of 1000 mm (ring) and ×1,000 mm (vertical) with a staggered quincunx. For concrete block prefabrication in the earth's surface, after checking if the concrete block's strength meets the requirements the prefabricated wire is embedded into it, and steel binding is fixed tightly.

2.5.2.3 *Construction of lining reinforcement connection*

2.5.2.3.1 The process flow

(1) Steel wire head processing and requirements: cutting material, steel bar end face, strip rolling thread, silk head quality inspection, cap protection (if necessary, sleeve protection), silk head quality spot check, storage to be used.

(2) Rebar connection: The steel bar is in position, and the protective cap or sleeve protection cover is screwed down. The screw head is screwed into the sleeve joint to tighten the self-check mark, and the construction inspection is done.

2.5.2.3.2 The construction method

(1) **Cutting material and joint construction:** Steel bar straightening is done in advance in accordance with the requirements, and transverse incision should be along the vertical axis, with a reinforced flat end. Unauthorized use of grinding wheel cutting or other specialized cutting equipment is forbidden.
(2) **Ribbed and rolled threads:** Stripping the rib thread straight, the operator self-checks the quality.
(3) **Cap protection (with sleeve protection if necessary):** Cap protection (with sleeve protection if necessary) prevents the thread from being knocked or contaminated.
(4) **Quality inspection of silk head:** The quality inspection department conducts spot checks on the quality of the silk head.
(5) **Store for use:** According to the specifications and type of classification, storage is planned and executed for different materials.
(6) **Rebar connections:** The docking is reinforced but does not freely rotate or is not very convenient. To advance some or all of the screws on the sleeve into a connected steel screw thread, the reinforced or reverse screw sleeve connection is turned to the desired location. Finally, the connection is made by means of a steel bar that is mutual to the top locking sleeves. For reinforced steel that is completely unable to rotate, such as a bending steel bar, one must adjust the steel internal force in areas, such as the construction joints, the post-pouring belt, lock nuts and connecting sleeve, so that it can be screwed into the extended thread in advance, then screwed into another bar end thread, with a locking nut sleeve. Finally, the application has either a positive or negative effect on the thread teeth and sleeve, and can loosen or tighten the two reinforcements in order to achieve a lock connection effect.

Tightened joint separated from the unscrewed joint: The length of the single-side exposed wire should not exceed 2P (P is pitch), and quality inspection shall be conducted on the finished joint.

2.5.2.3.3 Thread head inspection

(1) Thread profile and diameter: Visual inspection is done with a thread ring pass. When the tooth is finished, the width of the tooth shape is 0.3P and the cumulative length of the incomplete thread is not more than 2 threads. The ring gauge can be threaded smoothly, and the ring stop gauge should not exceed 3P (P is the pitch).
(2) Length of silk head: With a caliper or special gauge, the length is measured and it should be 1/2 the length of the standard sleeve.
(3) The operation personnel shall use the ring and ring rules to inspect each of the 10 threads, and visual inspection shall be carried out one by one. The surface of the silk head shall not have any damage or corrosion as that affects the performance of the joint.
(4) Inspection of the joint should be carried out by means of random sampling by an inspector, and any working class scale wire head can be chosen for a test batch, with a random check of 10%, and no less than 10. If the qualification rate is less than 95%, the other end should draw the same amount of wire inspections. If the total qualified rate of both inspections is not less than 95%, the batch is considered to have qualified products. Otherwise, one should stop the processing equipment including the batch processing wire heads, find out the reason and solve the issue. At the same time, one should individually test this batch of steel wire heads, check if any are eligible for use and analyze the unqualified wire heads.

2.5.2.3.4 Joint site inspection and acceptance

(1) **Inspection of appearance quality:** The length of the single-side exposed wire should not exceed 2P when the joint is tightened.
(2) **Tightening torque test:** The tightening torque should not be less than the relevant provisions as given in Table 2.22.

Table 2.22. Tightening moment of steel joint with straight thread rolling.

Bar diameter	≥16	18–20	22–25	28–32	36–40
Tightening torque	100	200	260	320	360

(1) Tensile strength test of unidirectional tensile strength

Joint inspection is done of 500 parts under the same construction conditions, using the same materials and grade, with the same form and specifications, because less than 500 are tested as a test batch.

For each inspection lot, joints shall be given to the structure of randomly in the three joint specimens (randomly chosen under special circumstances in the processing of finished steel wire heads) to check their uniaxial tensile strength. When the tensile strength of the three joint specimens is in accordance with the general technical specification of the steel mechanical connection, the acceptance criterion is satisfied.

If the tensile strength of one specimen does not meet the requirements, all six specimens in the same batch should be inspected again. If the strength of even one specimen in the reinspection does not meet the requirements, then the acceptance criterion is dissatisfied.

There are three types of unilateral tensile test failures in steel bars and sleeves. As long as the strength requirement is met, any failure form can be judged to be qualified.

After the sampling test is taken, the steel bar in the original joint is used for the same type of reinforcement as bonding, connection or welding and other mechanical connections.

After the inspection of the unqualified joint batch, a solution should be put forward for further study of the design and supervision.

(2) Product protection

The rebar connection is put into the box and stored in the warehouse. It must not be stored in open air to prevent it from getting wet due to rain. The processed silk head is provided with a protective cap or sleeve to prevent dirt damage. The processing equipment is placed in a rainproof tent. The protection of the motor and electrical parts must be ensured, especially on rainy days as they are easily damaged by damp and moisture.

There is great need for the construction of a post-cast strip, which is required in pouring concrete before the use of the reinforced connection device. To prevent the slurry from forming in the quick-release connector during construction, which would stain the cotton bar, one must use the protective cover.

2.5.2.4 *Inverted lining construction*

2.5.2.4.1 Inverted construction conditions

According to monitoring, the tunnel's bottom supporting closed loop tends to be stable with no vault subsidence. After checking that the tunnel clearance meets the requirements, the waterproof board (coil) is measured for the tunnel bottom base treatment and laid to achieve a solid, smooth coordinated layer, with no obvious concave and convex shapes. These are divided into teams at the beginning of backfilling grouting process before dense water plugging is done using the water plugging dug handle to pick up the bottom that is filled with spray. Base level highlights the mutilation caused by sharp objects or concrete bulge chisel flat base with equal steps for processing.

2.5.2.4.2 Laying of inverted waterproof board (coil)

After the base surface has reached the required state, it is necessary to install the inverted waterproof plate (rolling material), and then the next procedure can be carried out after satisfactory inspection has been conducted. The waterproof panel (coil) shall be protected from damage during the laying process.

2.5.2.4.3 Waterproof board (coil) protects the bedding

In order to prevent the lining of the bottom plate from slipping, the waterproof sheet (coil) is laid on the waterproof plate (coil), and a concrete layer of 7 cm is laid along with a layer of fine stone. The cushion concrete is used by artificial hand to push the rubber wheel cart and artificial stall. During the construction process, one should pay attention to the protection of the waterproof panel (coil). During the operation, the construction personnel should pay attention to the light release and avoid the waterproof plate (coil).

2.5.2.4.4 Inverted rebar connections

The ring is bound to the longitudinal reinforcement of the column, and the stirrup is used to bind and position it. According to the middle line and elevation, the site is positioned. The bottom of the steel bar shall be provided with sufficient concrete blocks to ensure the thickness of the concrete cover at the bottom. In the process of the placing the inverted lining, one should avoid welding the waterproof plate (coil), so as to avoid burning the waterproofing plate (coil).

2.5.2.4.5 Inverted formwork support and water-stop installation

After checking the clearance of the lining reinforcement, the base plate formwork is mounted. The base plate and the corner of the side wall shall be welded with the template bracket to fix the wall template. If necessary, the transverse brace shall be provided for reinforcement.

Before the closing of the plug template, the ring is installed in the construction joint to the water-stop belt. The requirements of the water-stop belt are satisfied and it is fixed firmly with the plug template. The construction joints before the installation of the waterstop will first have a lining and a clean reserve tank, applied with adhesive to paste it in the groove of the joint, where the joint parts will be reinforced with high-strength screw fixation.

2.5.2.4.6 Inverted concrete pouring

In order to ensure the compactness and liquidity of the inverted arch concrete, a concrete slump of 16 cm + 2 cm is provided with symmetry from the two sides during concrete pouring, using an artificial plugged-in vibrator. An internal vibrator which lights up when the vibration is too intense is used, so that it does not destroy the waterproof layer and waterstop. During floor concrete pouring, to facilitate control at the bottom of the radian, one needs an inverted arch radian consistent steel model with plaster as the standard to control the radian, along with ample construction preparation. The inverted arch is a bearing structure of concrete, with a strength of 70%, which can be dismantled. Immediately after dismantling it, one should cover the sprinkler with geotextile. The waterproof concrete should be cured for not less than 14d.

2.5.2.5 *Construction of arch wall lining*

2.5.2.5.1 Laying of arch wall waterproof plate (coil)

The waterproof layer is made of waterproof board (coil) and a whole package of geotextile. The construction process is detailed in the earlier section on waterproofing construction.

2.5.2.5.2 The lining of the arch wall is banded

According to the line and elevation of the tunnel, the lining reinforcement of the arch wall can be used to avoid the unstable impact of the waterproof plate (coil). The steel bar should be bound as far as possible to reduce spot welding; if it must be welded, it should be on the waterproof board (coil) or asbestos board insulation layer, so that the waterproof board (coil) does not burnt out. Between the steel bar and the template, there should be sufficient quantity of the concrete cover block of the required strength to ensure the thickness of the protective layer.

2.5.2.5.3 Concrete construction of arch wall formwork

(1) **Formwork platform support system**

The arch wall concrete formwork uses custom line steel, based on profile line balancing and the time limit for the project, with each line being equipped with a template for cars to prevent enlarged second-line faults, and ensure the quality of secondary lining construction. The selected length is 9 m, with a bottom die adopted using a hydraulic cylinder telescopic template to apply the hole curve section. In order to ensure that the car is within the panel and check the strength and stiffness of the supporting system, the car panel has to have an 8-mm-thick steel plate, along the longitudinal beam and girder steel support. In order to prevent car floatation and inward displacement, the car walk utilizes a 24 kg/m standard rail, with a walking speed of 6–8 m/min and motor power supply of 380 v/50 Hz, with the car brake setup checked suitably.

There are 5 rows and 15 perfusion windows in each row. The arch mouth infusion set perfusion tube and concrete are used to facilitate delivery pipe connection, with the rest of the perfusion being used to cover the plate, which is flexibly opened and closed for both the perfusion and the vibrating mouth. All infusion ports and car joints shall be strengthened, and the joints of the platform shall be tight to ensure the effect of secondary lining. The plate vibrator is attached 1 m above the template on the foundation setup. Six units are used for micro vibration to increase concrete compactness. The car tunnel template support elevation is as shown in Fig. 2.53. The elevation drawing of the section tunnel template platform is shown in Fig. 2.54.

(2) **Car positioning**

As the end of the inverted arch or tunnel backfill concrete meets the design strength, with the assembly line rail shuttering trolley in the hole, the measurement of tunnel clearance is done, with the midline and control elevation point being released. After assembly of the car into the tunnel section, which has been laid out, it is centered first and then the

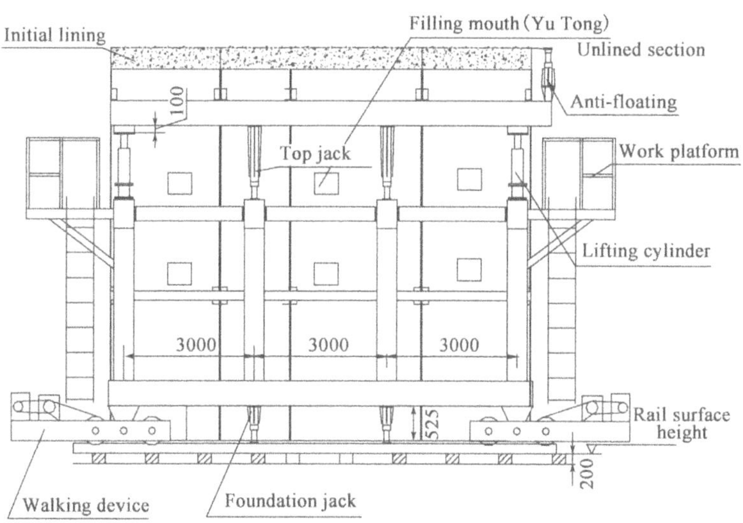

Fig. 2.53. Side view of the tunnel formwork platform (unit: mm).

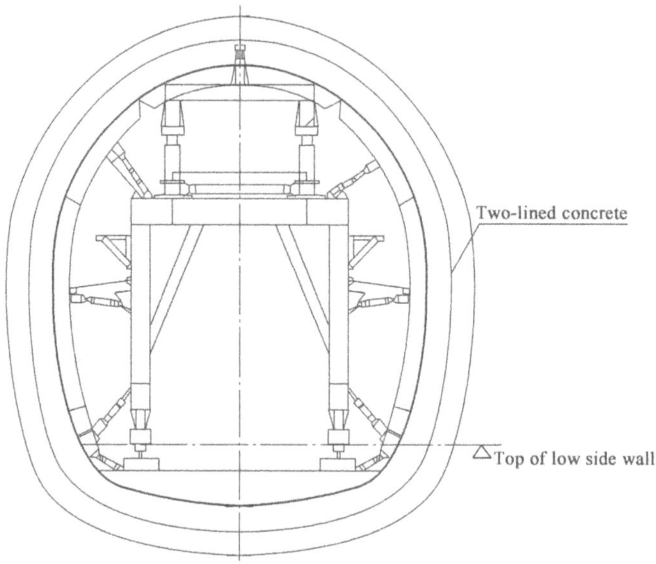

Fig. 2.54. Tunnel formwork platform support elevation drawing.

height and width are adjusted according to the design parameters (whole tunnel 30 mm, outside obligate sedimentation value), satisfying the clearance requirement. The car is adjusted after supervision until it has passed the review of concrete placement; the car positioning process is shown in Fig. 2.55.

(3) **Allowable error for each size**

Height: +10 mm, –0 mm. The deviation value error of the middle and line of the trolley: 10 mm. Clearance error: +20 mm, –0 mm.

(4) **Concrete construction of arch wall**

Taking the C45 waterproof reinforced concrete as the arch wall concrete, the concrete permeability grade for P10 utilizes a slump of 120–160 mm and a 5–20-mm grade coarse gravel aggregate in order to ensure appropriate concrete liquidity. Concrete pouring should be carried out in accordance with the car set on both sides of the window when layered, in symmetric perfusion.

While pouring concrete, the greatest concrete drop height cannot exceed 2 m. The height difference of concrete before and after the car

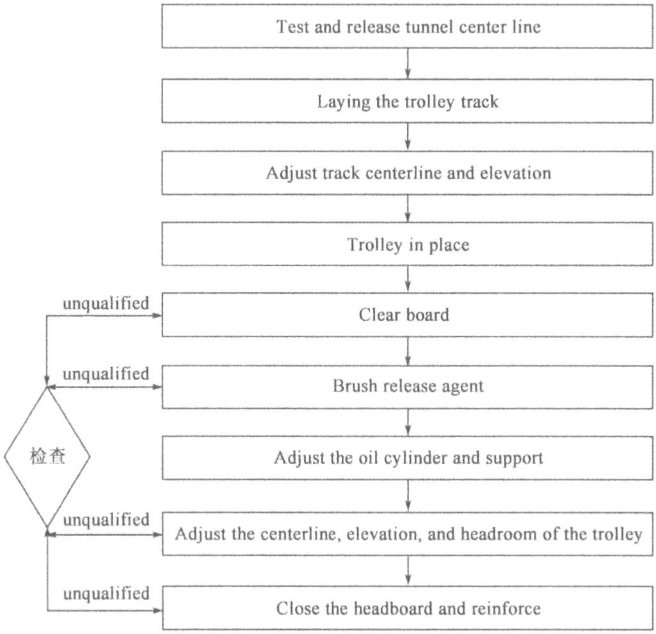

Fig. 2.55. The positioning flowchart of the template vehicle.

cannot be more than 0.6 m, and the concrete height cannot be more than 0.5 m. It is strictly prohibited to unilaterally cast from a height greater than 1 m. The concrete may not directly scour the car template surface flow to the pouring position, preventing segregation of concrete. In the concrete pouring process, it is strictly prohibited to add water. Concrete pouring, tamping and vibrating are conducted with an attached vibrator, along with a plugged-in vibrator. The vibrating shall not touch the waterproof layer, steel bar, the embedded parts or the plug plate.

The concreting shall be continuous at the position of the side wall, and the concrete shall be poured into the arch after the proper interval. In the arch parts, the pumping concreting extrusion method is adopted. The construction method is as follows: First, the casting near the lining has to be completed at one end of the second line, then, in turn, by means of backward extrusion the casting car parts and car on the other side arch concrete are completed. For pouring concrete on the arch, the arch department's reserve observation vehicle is used to plug the holes, using a powerful searchlight to look at the arch. When the concrete pouring arch has turned, the conveying should be stopped immediately to avoid the car from being crushed. In order to ensure the lining of the concrete arch part, on the installation 5 root injection pipe ends are reserved (ϕ 42 steel pipe, 90 cm long, with cotton closed) to meet the required strength of the lining concrete, with two rounds of backfill grouting behind the lining.

Concrete pouring is done by pump feeding into the duct connected to two lining trollies on the vault operating platform. A ϕ200-mm soft tube is used to transport the concrete from the vault conveying operation platform to the blanking window of each car. The lining shutter window has 5 layers of material on each side for the three windows, overall in the plum flower form layout. The casting principle is from the first layer of the blanking window on the left side of the short wall casting, with concrete materials from three windows. When concrete height reaches 50 cm, a mobile hose is used on the right side of the short wall after the first layer under the window. On the left side of the concrete and on the right side of the concrete, one uses a plugged-in vibrating rod. The relative height should not exceed 50 cm on both sides of the casting concrete.

According to the strength growth law and construction experience of concrete hardening in the hole, concrete stripping is usually done after 24–36 h; concrete curing can be done immediately after ripping, using a personnel sprinkler, with a curing time of not less than 14d. After the car has been demolded, the surface of the car should be cleared in time to deal with a surface coating release agent before the next step; the release agent uses artificial besmear to brush. Release agent pollution should be avoided when concrete reinforcement and stripping adhesion are conducted in the car.

See Fig. 2.56 for the construction process of the second line.

(5) **Precautions for concrete construction**

Before the concrete placing, one must check the template (or car), steel, embedded parts, reserved holes, construction joints, deformation and waterstop, and remove impurities in the template after the acceptance of concealment but before concreting. In the process of concrete pouring, it is necessary to observe the template (or car), support, waterproof plate (coil) steel reinforcement, embedded parts, reserved hole, etc., and find any problems that should be dealt with in time.

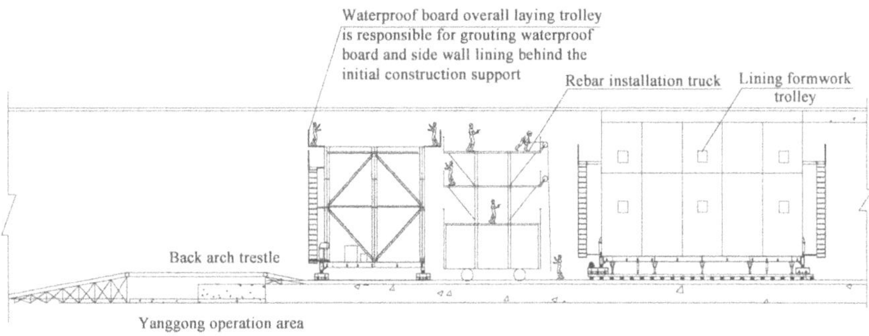

Fig. 2.56. Construction flowchart of the second line.

One should also check whether the concrete pouring in the vault is compacted by using the arch filling hole.

2.5.2.6 *Backfill grouting construction*

Tunnel model building construction is greatly influenced by the limitation of concrete lining, especially the vaults. There is usually a hole behind close-grained concrete, which greatly affects the whole strength of the lining concrete. At the same time, the grouting anti-seepage function is important to a certain extent. Therefore, in our tunnel construction, one must use compacted backfill grouting behind the lining concrete in the hole to ensure that the whole strength of the lining concrete is maintained along with the waterproof function. Three grouting short tubes are preset for each 3 m at the beginning of the tunnel, and the side wall is 2 and the inverted arch is 1. The embedded grouting pipe is made of ϕ32-m steel tubes, and the second lining surface is exposed for 3–5 cm, as shown in Fig. 2.57.

2.5.2.6.1 Construction preparation

(1) One should push the self-propelled easel to the grouting position and stabilize it. Then, one should install the air pipe and water pipe to the grouting position, install the grouting equipment and carry out the grouting test to ensure smooth grouting.

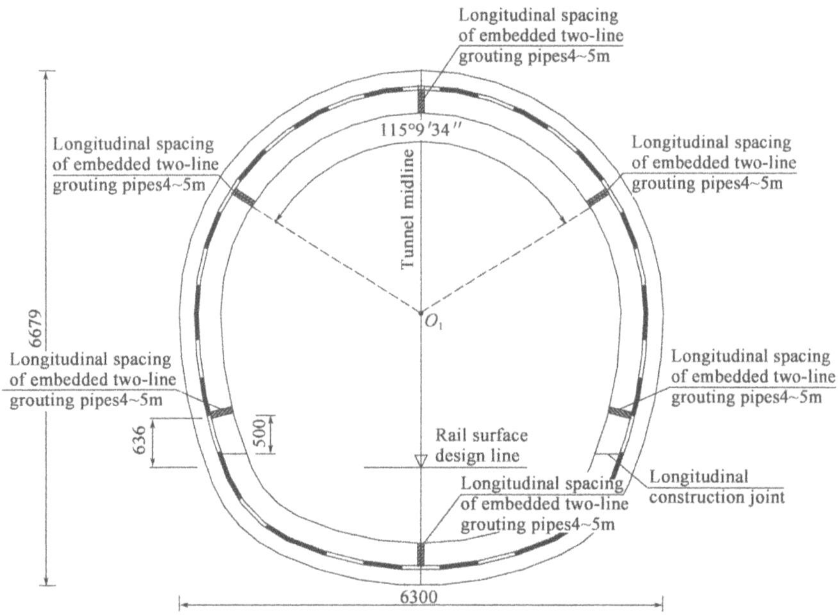

Fig. 2.57. Pre-burial grouting pipe layout (unit: mm).

(2) One should clean and repair the reserved grouting hole to ensure its smoothness.

(3) The slurry (in a water-rich location, use double-liquid appropriate) is mixed when there is good irrigation. The water–cement ratio should be controlled to 0.5 to 1.0, with an anchor solid strength of not less than 25 MPa. Consolidation is done after compressive strength, tensile strength and good stability. The slurry is mixed in strict accordance with the guidelines issued by the laboratory; there should be no water in the process of filling slurry.

When the concrete strength reaches 70% of the design requirement, grouting operation can be carried out. The grouting is divided into two steps: First, pressure is applied at 0.2–0.3 MPa to perform grouting. Then, after grouting is finished, the standard end pressure is reached at the constant pressure time of 5 min.

(4) Each sequence of hole grouting is infused according to the order from low to high. When the first hole is lowered into pulp, the neighboring holes are drained, forming a string of slurry, which will

continue to the grouting until the end of the hole can be blocked. The same grouting order is followed from the arch to be symmetric on both sides so as to ensure the quality of the work.

(5) Grouting should be carried out continuously. If the grouting is interrupted, it should be restored as soon as possible. After the grouting is restored, the injection rate will be much lower than before the interruption, and the slurry will be stopped quickly, and the hole should be cleaned and refilled.
(6) The grouting process should be monitored and recorded to ensure the quality of grouting.

2.5.3 *Summary*

The second lining construction is the last part of the excavation interval construction, and it is necessary to pay special attention to the maintenance of the base surface during construction. In addition, due to the particularity of geology in the Qingdao area, special attention should be paid to the durability index of concrete by strictly controlling the protective layer of reinforced concrete. The relevant points are as follows:

(1) Since the waterproof concrete is an important line of defense, the underground waterproofing system's quality should be improved, in addition to the quality of concrete itself, as it also strengthens during vibration, based on a reasonable setting of the plate vibrator's number and position, needing continuous improvements in the process of construction and parameter optimization. Vibration or vibrating deficiency can cause leakage, concrete imperfection or honeycomb surface pitting Vibration of newly poured concrete may lead to damage to the concrete sand line, slurry leakage, or serious damage to the template car.
(2) Since waterproof concrete is an important line of defense, the underground waterproofing system needs to improve the waterproof ability. First, one should strengthen the quality control of the concrete itself. If there are suitable conditions the construction unit should set up a self-built concrete mixing station, so that the quality can be controlled. The chosen concrete mix should have strong comprehensive

strength and be from a good manufacturer. A set of quality management measures should be established, for example, a supervision inspection station, selective examination of silt content and gradation, choosing a good water-reducing agent, etc.

(3) In order to strengthen the vibration, the number and position of the plate vibrator are set up reasonably in the template platform, and the configuration parameters should be optimized through continuous improvement in the construction process. Vibration or vibrating deficiency can cause leakage, concrete imperfection or honeycomb surface pitting. Vibration of newly poured concrete may lead to damage to the concrete sand line, slurry leakage or cause serious damage to the template car. Double-lining constructions generally face more difficulties, such as a long concrete pouring time, a strong focus on the supply of concrete in the city and setting it aside in advance when necessary, an on-site mixing station, avoiding cold joints caused by concrete supply problems and many other quality problems.

(4) Details such as slow work, incomplete construction joints, short construction joints, sidewall sinking and roughness all attract attention, and must be taken seriously. The workers must do a good job. According to the requirement of horizontal construction joints and circular construction joints, the water-stop steel edge must be embedded in the middle.

(5) Double-lining construction of most processes is likely to aid in the waterproofing. Damage should be avoided in the steel binding template, end seal welding, template car, construction joints, etc.; therefore, construction process control is the key.

(6) Quality control of the slurry filling is the key; a fresh pouring of concrete is influenced by gravity. It is not easy to fill concrete even though there is space; therefore, strict grouting is a must, which will guarantee that the vault is close grained and ensure that the important links of the construction joints are not leaking.

2.6 Typical engineering project cases

2.6.1 *Underpass (side through) building*

With the increase in construction of the urban rail transit system, the presence of tunnels under existing buildings is becoming increasingly common, which leads to a higher requirement for their safety. The tunnel excavation construction technology directly affects this safety, thereby becoming greatly in demand, especially due to the technology's high significance and low impact on the surrounding environment.

For the change mechanism of the upper structure near the pile foundation in tunnel excavation, domestic and foreign scholars have carried out extensive research from various aspects, along with calculations, analyses and engineering countermeasures. The experimental studies include centrifugal model tests and field tests. Mortoit (1979) studied tunnel construction through the indoor model test and the influence of pile foundation on the bearing capacity and settlement, and found that the impact on the tunnel construction of pile foundation is great. He draw the following conclusion: In the soft underground soil, the tunnel's influence on the existing adjacent pile or upper part of the pile is the main problem. Lee (1994) reported on an engineering example of a tunnel with seven new layers between the two layers of the basement of the building's pile foundation. Despite the estimated total long-term stratigraphic subsidence, the possible effects of the pile foundation are not clear. Hergardent *et al.* (1996) studied tunnel construction by means of the centrifugal model test and the influence on the end bearing pile. In experiments, the test tunnel, located in three different positions, measured the settlement of pile foundation and the surface subsidence. They also studied the tunnel construction and the influence of the bearing capacity on the pile foundation, and found that when the horizontal distance of the tunnel is less than 0.25 times the tunnel diameter, the pile will be affected severely; the distance from the tunnel when the pile reaches is than 2 times the diameter is almost unaffected. For building damage caused by tunnel excavation, foreign scholars have conducted studies in a systematic and detailed manner, and older research articles are available from the 1950s; to date, there have been some preliminary results and evaluation criteria, of which the following is a simple introduction. In the past the study of subsidence damage was generally divided into three basic types: The first kind was the empirical method, based on the case history of compilation and research of the definition of damage and the easily measurable or experienceable relationship between parameters. The second method considered the principle of structural engineering, in order to obtain the permissible limit by means of the selected size and material parameters, and deduced the limit value of actual cases compared with history. The third method can be summarized as the combination of numerical calculation and case-based statistical damage criteria.

For building subsidence control technology, from the engineering example of metro construction, there are a lot of experience for reference. B. P. Kassap (1992–2000) and others reported on the Boston I-interstate 93 under the north tunnel segment and the metro Red Line station engineering. The tunnel is located in Boston, Atlanta street and Sayin Mo street cross-section, which, in turn, is superimposed on a station hall floor, with the MBTA tram underground station and Red Line station, a total of three layer in the structure. The Red Line station was founded in 1914 and the project uses the subsurface excavation method, with a structure sole-plate of about 19 m. The surface of the station is located in the gray mud soil and clay, and the station is a reinforced concrete structure, which is still intact. Only local, visible water seepage occurred, with cracks and degradation due to age, before the renovation, and the state of the structure after comprehensive assessment was said to be too weak to bear complete construction due to an increase of stress and strain.

A new tunnel was constructed for the four-lane road tunnel, located in the Red Line south station directly, with the upper supporting structure 1.5 m from the station floor. In this the method of underpinning construction was adopted. Initially, on one side of the excavation construction of two shafts, the depth is about 35 m. Under the existing metro station, two pilot tunnels were planned along the direction parallel to excavation, with 18-m-deep soil grouting reinforcement, forming two reinforced soil walls. Then three layers of excavation within the walls was conducted, and a reinforced concrete wall was poured, as it had a direct effect on the fresh rock mass. In the pilot tunnel adjacent to the existing metro station in a vertical direction, at a certain interval, excavation is conducted using the post-tensioned method with pouring of a reinforced concrete beam, which provides support to the concrete wall. In this way, underpinning system is developed. In an area of $(25 \times 36)\ m^2$, the tunnel excavation is conducted. In order to guarantee the normal operation of both the station and affiliated facilities thereof, 24-h monitoring of construction process is done. The structure-bending deformation value is 6 mm and the outer limit is 10 mm.

According to the basic design specifications of the building foundation, the allowable values of the building deformation are shown in Table 2.23.

Table 2.23. Control standards table for main existing construction settlement.

Floor type	Junction type	Basic type	Early-warning value (mm)		Maximum settlement specified in Code (mm)
			Subside (mm)	Bias	
Multi-storied building	Brick concrete structure	Isolated strip footing, foundation	25	0.0015	30

Urban metro tunnels are mostly built in busy areas, and in limited urban spaces, where the tunnels often pass under some existing buildings. Tunnel construction will inevitably produce disturbance to the rock, destroying the balance of the original rock soil mass, with in situ stress redistribution, causing a settlement and deformation of the stratum, which will all affect the safety of the aboveground adjacent buildings. Therefore, it is necessary to strictly control the settlement and deformation of strata and ground buildings.

[Engineering example:] Qingdao Metro Line 3 is under the Wujiang range of high-rise buildings

2.6.1.1 *Project background*

The left line of Qingdao Metro Line 3 is under six high-rise buildings, and the right side is under four high-rise buildings, and the crossing of buildings is shown in Table 2.24.

This section introduces the case of the holiday inn, which is built on relatively poor surrounding rock.

2.6.1.2 *Location and geological conditions of the tunnel under construction*

(1) The holiday inn has four underground floors, with 28 floors above ground, a frame structure, raft foundation + anti-floating bolt, and was built in 2010. The left and right lines of the area are all underground, as shown in Figs. 2.58 and 2.59.
(2) Geological data: The design for level IV surrounding rock, according to the top-down principle with other related information is as shown in Figs. 2.60 and 2.61.

Table 2.24. Building below the Wujiang interval.

Item	Name of buildings	Underpass building mileage	The surrounding rock grade	The length of the crossing	Buried depth	Design excavation method	Basic mode
Left line	Public resource trading hall	K7+648-K7+700	III	52	23.5	Full section	Raft foundation + anti-float bolt
	Holiday Inn	K7+736-K7+816	IV	80	26	The steps method	Raft foundation + anti-float bolt
	Xinguidu district building	K7+844-K7+860	III	16	27.6	Full section	Reinforced concrete column base
	Building 2, Weidong Shangcheng community building	K7+919-K7+939	III	20	27.6	Full section	Raft foundation + anti-float bolt
	Building 3, Weidong Shangcheng community building	K7+968-K7+988	II	20	27.2	Full section	Raft foundation + anti-float bolt
	Building 4, Weidong Shangcheng community building	K7+031-K7+051	II	20	27.2	Full section	Raft foundation + anti-float bolt
Right Line	Public resource trading hall	K7+670-K7+705	III	35	25.1	Full section	Raft foundation + anti-float bolt
	Holiday Inn	K7+748-K7+816	III	68	26	Full section	Raft foundation + anti-float bolt
	Building 1, Weidong Shangcheng community building	K7+877-K7+897	III	20	26.5	Full section	Raft foundation + anti-float bolt
	Building 2, Weidong Shangcheng community building	K7+948-K7+968	II	20	27	Full section	Raft foundation + anti-float bolt

Fig. 2.58. Holiday Inn location.

Fig. 2.59. Basement.

2.6.1.3 *Main measures taken in blasting excavation*

According to the geological characteristics of the Qingdao area and the relevant construction experience, the blasting vibration velocity should be controlled within 0.5 cm/s, which will also control the blasting vibration speed. The measures taken at the site are as follows:

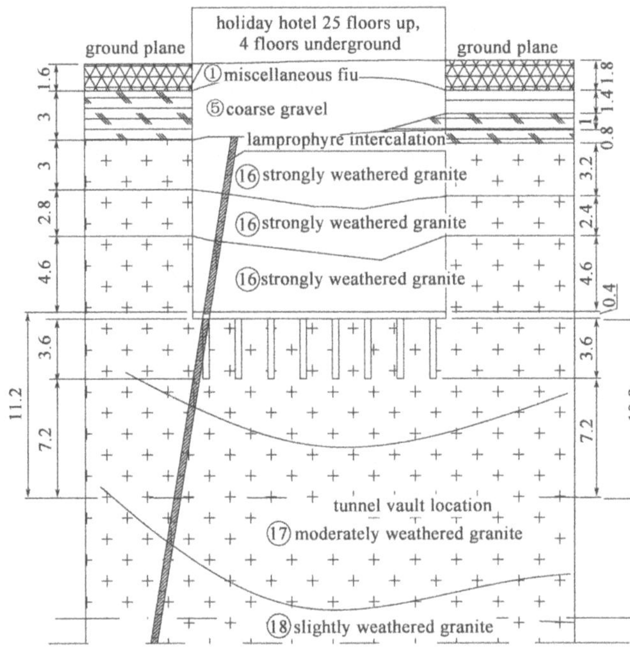

Fig. 2.60. Geological profile of holiday inn (size unit : mm).

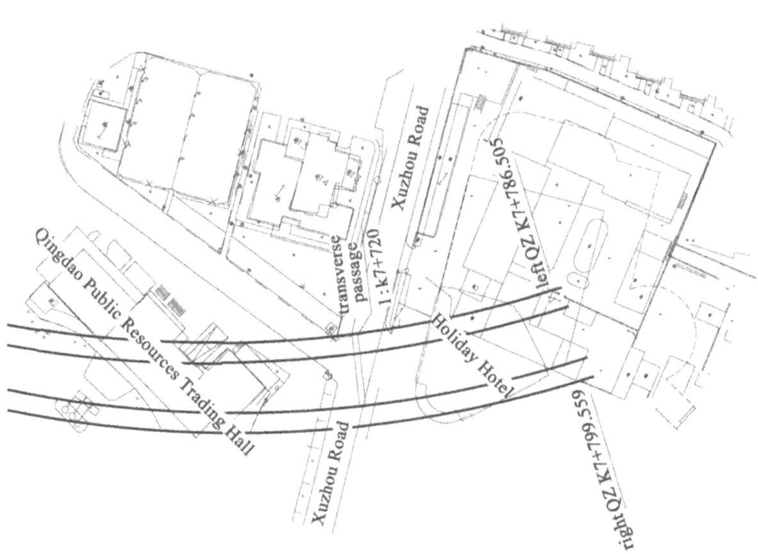

Fig. 2.61. The schematic diagram of the relationship between the holiday inn and the tunnel.

(1) **Bench excavation:** The holiday inn has a tunnel roof plate distance of only 11.2 m. In order to reduce the influence of blasting vibration on buildings, and fully considering the construction progress and efficiency, under the circulation speed of 0.5–0.9 m, the YT-28 blower gun drilling and the excavation steps method are used, with a bench height of 3.2 m, steps 6–9 m in length; the drilling and blasting method of excavation is utilized, with small excavators, grilled slag, dump trucks, two 10-t improved derrick slag (skip 3 m^3). Construction is done according to the surrounding rock conditions using different drilling depths, with reasonable control over the explosive load, improving the blasting efficiency of the surrounding rock, minimizing the overbreak and waste quantity. Reasonable adjustments in the construction process are made, involving formation drilling–blasting, muck transportation and supporting the assembly line. After excavation, spray anchor bracing and monitor measuring can be done in a timely manner.

(2) **Hollow large-diameter straight-hole cut:** Utilizing a geological drilling rig, one should dig at a drilling diameter of 150 mm and drilling depth of 30 m in hollow straight lines around the cylinder using the 8 auxiliary eye drilling bit with 4 layers with the center hole at the center points where the deep eyes are set at 1.0 m.

(3) **Peripheral eye relief hole:** The peripheral eye perforation spacing is 35 cm, which increases the vibration reduction effect; the peripheral eye is also used as the vibration reduction hole.

(4) **Peripheral eye compartment charge structure:** The peripheral eye is equipped with the structure of the detonating cable spacer to improve the blasting ability.

(5) **Strengthen the hole blockage:** When a plane flies from the International airport, it causes the same amount of perforation as the blasting vibration. Therefore, depending on the length of this period and the amount of perforation charge, the practical application proves that in the tunnel stemming jams and tunnels, compared to conventional blasting without backfilling, the current method of blasting can save 20% of explosives, as it adopts a stemming congestion charging structure and fundamentally weakens the blasting vibration.

(6) **Initiation sequence and interval time:** A reasonable, scientific deployment of the perforation detonating sequence and time interval

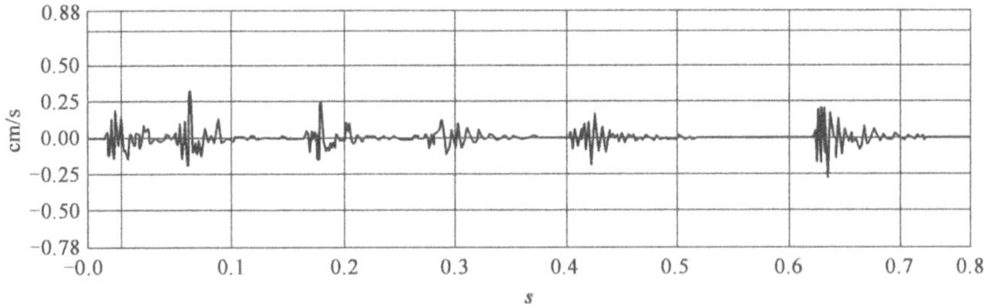

Fig. 2.62. Monitoring waveform of blasting vibration velocity.

can not only improve the quality of blasting but also reduce the blasting vibration effect and is conducive to protect the structure of the building. The millisecond detonator tube uses a non-electric initiation system first with a time interval of less than 50 ms and with a later time interval of more than 50 ms. For tunnel blasting excavation, in practical blasting, the interval after firing is greater than 50 ms.

(7) **Strict control of excavation footage:** The step method is used to construct the surrounding section on the left side of holiday inn. The steel frame spacing is 50 cm and the circular input is 0.5 m, which corresponds to the steel frame spacing.

(8) **Monitoring of blasting vibration velocity:** Blasting time shall be measured properly through monitoring and adjusting the hole charge; the same detonator initiation dosage and firing interval time can be used. The aboveground buildings and the basements of the building should be monitored and blasting parameters should be adjusted by monitoring the data (Fig. 2.62).

2.6.1.4 *Major measures to protect surface buildings*

2.6.1.4.1 Tunnel excavation and support

Under the tunnel, the construction strictly follows the principle of "pipe lead, strict grouting, short footage, weak blasting, strong support, fast sealing and measuring". The construction strictly abides by the "first line, then grouting, then digging, grouting section, digging a section, supporting a section, closing a section". Construction is in strict accordance with the design drawing of the building, with proper protection measures undertaken for content and pipeline, excavation and support. Excavation is done by the steps method, which supports and strengthens the existing level 1–2 surrounding rock. When necessary, one can change the grid steel frame for the steel section support, so that the force is held steady. At the same time, a feasible construction method is prepared for the construction of concrete structures and pipelines, so as to ensure their safety and reliability.

2.6.1.4.2 Reinforcement of building grouting

The sleeve valve pipe ring (vertical) is set to the spacing of 1.5 m, with the plum flower form layout. Grouting adopts a 1:1 single-grout ratio, with arch grouting hole distance at a minimum of 3 m, thereby preventing stress damage to tunnels. Grouting should be conducted before testing, controlling the grouting pressure at 1.5 MPa for the final test. The grouting pressure should not be excessive; in the case of high grouting pressure, it causes the ground to lift up and endangers the safety of the building. The building reinforcement elevation map is provided in Fig. 2.63, the building reinforcement elevation map of the tunnel in Fig. 2.64 and the building reinforcement plan sketch in Fig. 2.65.

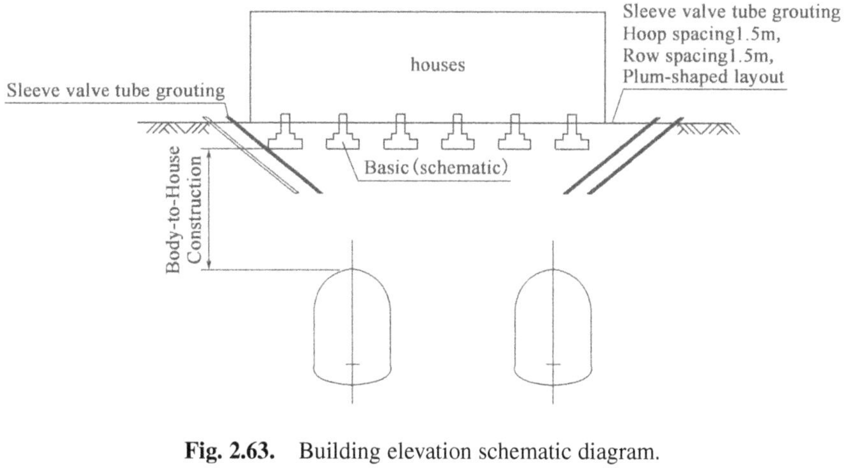

Fig. 2.63. Building elevation schematic diagram.

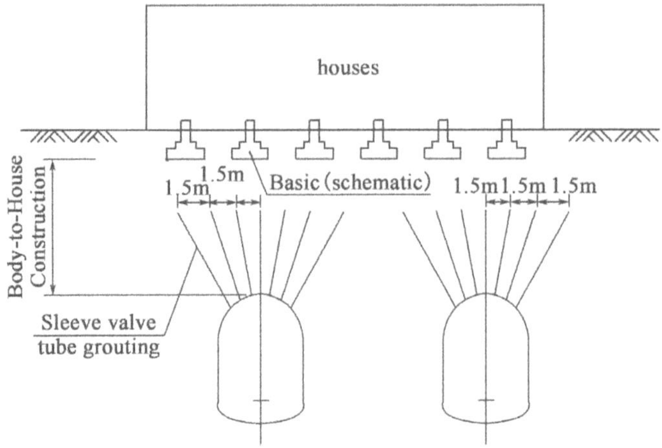

Fig. 2.64. Building reinforcement facade of the tunnel.

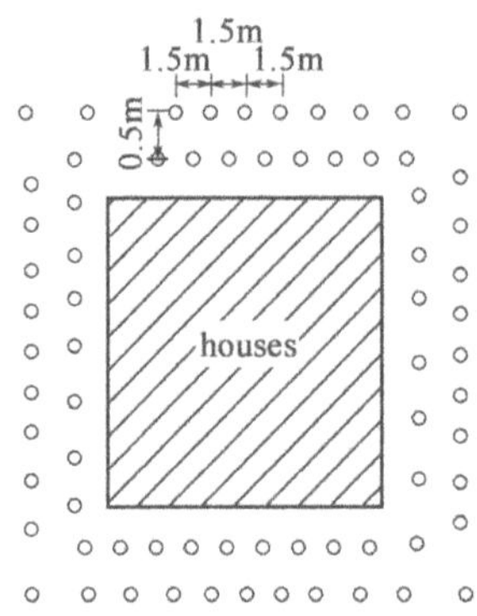

Fig. 2.65. Schematic diagram of building reinforcement.

2.6.1.4.3 Grouting water plugging

(1) **Blocked palm surface grouting**

Underground interval is set in the water area of groundwater status to II, III level (grade II water percolating capacity is set as 10–25 L/min at 10 m and grade III water percolating capacity is set as 25–125 L/min at 10 m) or more, using radial grouting to perform waterstop measures. Tunnel water inflows through a hole, and if it is measured to be large in other areas, it is controlled by the radial grouting water plugging measures.

After grouting for water plugging, the state of the groundwater should be less than level I or no water at all for tunnel lining construction. If the construction has to be conducted in the presence of water, it can be near borehole drainage so as to decrease water pressure; at the same time, water plugging and grouting should be done. Groundwater status should be at level I (grade I water percolating capacity as 10 L/min at 10 m) for using shotcrete closed water plugging measures in construction.

(2) **Full-section deep-hole grouting to plug water**

Deep-hole grouting has many constraints. After the grouting, the hole is filled with pressurized slurry, which leads to extrusion of the section fissure water in and around the area. One must ensure that the surrounding rock is of a certain strength with close-grained paste filling, which helps in the consolidation with the rock mass.

(3) **Radial filling grouting stops water**

The surrounding rock has good structural integrity and based on the location of a small section and deep-hole grouting, if the local effect is not satisfied, the tunnel excavation contour will be set outside the ring to stop water from entering into the deep hole. After the tunnel excavation, along the contour, the spacing is set to 60 cm, with a 350-cm small catheter radial circular layout. The whole section is washed using pure water after slurry grouting, with the water pressure in the area aiding consolidation.

(4) **Design parameters of grouting**

(a) The effective diffusion radius of the single hole for the slurry is 2.0 m.

(b) The effective radius of grouting reinforcement is 3.5 m outside the lining.

(c) The normal grouting pressure is 1.5 MPa, and the final grouting pressure is not greater than 3.0 MPa, and the construction is adjusted according to the on-site grouting test.

2.6.2 *Down (side) pipeline*

In underground tunnel construction, the methods used are varied based on the different regional hydrological conditions and the characteristics of the soil. The pipeline system in the city is multifarious, with different pipe diameters, pipe types, embedment depths, and so on being used. Many bodies of research in this field are confined to a certain type of geological condition or a certain kind of pipeline, and the analysis of factors affecting the deformation and stress of the pipeline is not specific.

Ariman's study shows that the limit of the weight is very small relative to the surrounding soils, and soil deformation will increase the additional stress and deformation of pipeline. Stress, strain and deformation of the pipelines are caused by the deformation of soil around them.

Singhal analyzes the ductile iron pipes of different diameters and creates formulas for the bending moment, torque and force of the pipeline. He also obtains the pipeline displacement formula as follows:

$$E_P I_P\left(\frac{d^4 W_P}{dy^4}\right) + KW_P = Kf(y), \quad K = 2K_\infty,$$

where E_P is the pipeline elastic modulus, I_P is the moment of inertia of pipeline section,

$$K_\infty = 0.65\sqrt[12]{\frac{E_s d^4}{E_S I_P}} \cdot \frac{E_S}{1 - v_s},$$

where K_∞ is the foundation reaction modulus, d is the diameter of the pipeline, E_S is the elastic modulus of soil layer, v_s is the Poisson's ratio of soil layer.

In addition, many domestic scholars have done a lot of research on the existing pipelines, the details of which are summarized in Table 2.25.

Urban metro tunnel construction causes major problems for various municipal utilities pipelines, especially the gas, water, oil and other pressure-based pipelines. Pipeline settlement due to tunnel construction is extremely easy, but vibration control indexes are overrun, causing harm to the surrounding residents and leading to adverse social impact. But in limited urban space, the cost of removing all the pipes above the tunnels is even more excessive, so tunnels under construction now often pass under some municipal public pipes.

Table 2.25. Domestic numerical simulation research summary.

Author	Year	Description	Analysis software
Wu Bo, Gao Bo.	2002	A three-dimensional model of tunnel–soil–pipeline is established to simulate the actual construction.	ANSYS
Wu Bo	2004	The centrifuge implementation is combined with numerical simulation.	ANSYS
Wei Gang	2004	The influence of roof pipe construction on pipeline deformation is studied, and the influence of the parameters of the pipeline and soil layer along with soil strengthening measures is analyzed.	—
Jiminpeng, Meng Xizhang	2005	The influence of tunnel construction on pipelines is analyzed, considering the mutual influence of the tunnel, soil layer and pipeline; the distance of tunnel and pipeline, shield construction, grouting and other factors are also analyzed.	ANSYS
Jiang Xinliang	2005	Two-dimensional finite element dynamic simulation analysis and comparison with measured data.	ABAQUS
Bi Jihong	2006	The stress and horizontal, vertical displacement of the pipeline are analyzed under the influence of pipeline materials, burial depth and elastic modulus of soil layer.	ABAQUS
Luo Jianjun	2007	The interaction model of soil–pipeline–tunnel is used to analyze multiple pipelines.	FLAC3D
Wu Weiyi	2008	This paper establishes the coupling model of "tube–soil–structure" and analyzes the law of force and deformation of pipeline in a metro project in Shenzhen.	FLAC3D
Veckay	2008	The influence of driving force, excavation sequence and stress release rate is considered.	ADLNA
Ma Keshuang	2008	The influence of excavation steps, internal support and burial depth of pipeline on pipeline displacement is studied.	ABAQUS
Xi Taiyue	2010	Considering the fluid–solid coupling effect, a comparative analysis is made of the three situations of non-precipitation, dynamic precipitation and normal precipitation.	FLAC3D

The channel tunnels constructed under the worn pipeline and those under the buildings adopt similar technical measures, but one should pay attention to the tunnel and any underground operation should make calculations strictly ahead of time when reinforcing the relative position of the catheter and the pipe; a safe distance from the device is not less than 2 m. In addition, in the construction of gas, oil and other pipelines, blasting construction shall be controlled by micro seismic control.

Micro seismic control blasting usually adopts the following technical measures to control or reduce the blasting vibration velocity and the vibration effect: (1)The blasting speed shall not be greater than 1.5 cm/s. (2)Blasting is monitored by calculation, determining a reasonable blasting scale and the correct blasting design and construction, and adjusting the dose of explosive charge. One can make full use of the explosives, according to the objective requirements of blasting, determining the maximum allowable dose for minimum seismic effect. (3)Differential blasting, according to the principle of superposition of stress wave, uses equal charges, wherein the technology can make the blasting seismic waves' energy disperse in time and space, making the main phase stagger, thereby effectively reducing the blasting vibration intensity, in general by 1/3–1/2. (4)One must use a pre-splitting blast or ditch damping. Between the blasting area and the protected objects, drilling must be conducted in advance for 1 or 2 rows as intensive shock holes, or one can use pre-splitting blasting to form a certain width of fissure or excavation damping groove, all of which can perform obvious shock absorption effects, thereby reducing the general vibration by 30–50%. (5)To strengthen the monitoring of the blasting vibration effect, a vibration instrument is adopted to at the network room, which measures the stress state and provides more accurate data for security check computation.

2.6.3 *Small-spacing tunnel*

Two single-line spacings are generally used when the tunnel excavation hole diameter is small, commonly known as the small-spacing tunnels. The Qingdao northern metro line 13 consisting of two small-spacing tunnels can be considered as an example, having a tunnel clear distance between 1.2 and 6.52 m. The two-line tunnel construction influences each other. The small-spacing tunnel vault has a depth of 4.6–6 m. The surrounding rock rating for the ultra-shallow depth excavation tunnel is class VI.

To ensure the safety of the tunnel during construction, the following measures are needed to reduce strata disturbance, improve the stability of the perturbed formation, especially for the grouting reinforcement and protection of the surrounding rock cylinder, and strengthen the constraints and the stability of the surrounding rock.

(1) In the process arrangement, first the right line of the tunnel is constructed, and then the left line of the tunnel, which is more than 30 m away, is constructed.
(2) Before the whole body tunnel excavation is carried out in vault 120 within two rows of ~42 small ducts, grouting is performed with L = 4 m, ring and longitudinal spacings of 0.4 and 2 m with the corresponding driving angles of 12° and 30°, small intraductal pure water slurry grouting, grout water to cement ratio of 0.5:1 and a grouting pressure control of 0.5–1 MPa.
(3) Tunnel excavation is done by the circular steps method, in strict accordance with the design method of the excavation process. During blasting excavation, one should take seismic control measure before the splitting blasting and the static blasting measures when necessary, especially for the buildings near the left line tunnel. The excavation sequence and the determined lap time will have a great influence on the stability of the rock mass. If the tunnel excavation is conducted in sections, it would shorten the exposure time of the rocks. The closed loop should be used as soon as possible, using shotcrete when there are necessary constraints and adding a temporary inverted arch, which is essential to the construction and structural safety. In the excavation, a circular excavation length control within a common steel frame spacing is utilized.
(4) Tunnel construction is done in the first side wall rock using a grouting pipe embedded at ϕ42, with grouting tube spacing of 1.0 × 1.5 m, L = 3.5 m, in the plum flower form layout. Drilling depth is determined based on the thickness of the area between the rock mass. The support is clipped in the rock mass grouting reinforcement, using cement grouting slurry. Installation of grouting tube in done using the

cement grouting pipe orifice and linen, silk, and it is connected to the drill hole wall extrusion in full; the grouting tube is checked for pulp and fixation. Grouting cement solidification should have enough strength. After the grouting, timely inspection of the grouting effect should be done, ensuring that it meets design requirements, or else grouting holes must be added.

(5) When necessary to clip something in the rock mass, one can use tensile anchor reinforcement. The pulling anchor is of ϕ25-mm rebar, with each root reinforcement adopting single spacing, and any form of extension is prohibited. The bolt from the vertical direction along the tunnel line is at a spacing of 75 cm, each row has 4 roots, at a spacing of 1.5 m. The anchor end pad with a concrete leveling layer is a must, and the concrete strength should measure up to 100% for prestressed anchor tension.

(6) After the construction of the left line tunnel, spandrel, on rock pillar side to the rail surface at a line height within the scope of the arch wall parts, is advanced to add a small tube or hollow grouting anchor auxiliary construction. Attention must be paid to the construction of the tunnel as the anchor shall not be stretched into the early construction period to the beginning of the outer contour line. At the beginning of the construction of the tunnel, the loop must be closed as soon as possible.

(7) There is vast scope for primary support in the arch department during the construction using a molten steel tube sprayed ~42 times and embedded at 120°, which is used as a grouting pipe, in a section with a wall thickness of 3.5 mm, length of 4,000 mm, at a distance 400 mm × 2,000 mm, where the plum flower form is a decoration. The early-bolting closed loop, after a certain length, is placed behind the line pressure slurry. If the groundwater leaks out of more locations after excavation, the lining and backfill are placed at those locations until grouting can be performed on the surrounding rock.

(8) Due to the narrow spacing of the tunnel, hole excavation has a great disturbance on the surrounding rock, so one must strengthen the monitoring of the surrounding rock. In the process of the tunnel construction, mainly due to surface subsidence and vault subsidence, peripheral convergence in surrounding rock, blasting vibration and

anchor internal force, observation and monitoring in a timely manner is necessary for proper informational feedback measurement, so as to ensure the safety of the tunnel. Before the construction, a specific construction monitoring plan should be worked out. The changes in the inside and outside of the hole should be mastered at any given point of time, and the measures should be taken in a timely manner if the ground settlement extent and the sinking of the arch top in the hole exceed the warning threshold.

2.6.4 *Broken horse-head door*

The main vertical and horizontal channels into the inclined well are where the job is first carried out, including the horse-head door and the stress of the ingate break which changed the original structure. If there is improper construction, then it is easy for deformation to form in the original structure, for surface subsidence to increase and even for serious structural failure to form. Therefore, it is necessary to choose reasonable construction methods and construction sequences according to the characteristics of the structure, and give necessary reinforcement and support to the structure.

Overall construction technology: In front of the main construction, an advanced geological forecast is completed, of the channel, interval, wall, first-line mileage period of ingate, right-line mileage ingate and left-line ingate.

In order to complete the left line and the right line of the small distance to the horse-head gate, certain specific opening sequences are conducted as detained in Fig. 2.66. See Fig. 2.67 for sections.

2.6.4.1 *Specific measures*

(1) Geological radar is mainly used in prediction, and the complex sections of the surrounding rocks are drilled in advance.
(2) The segmental block is divided into the initial blocks of concrete in the transverse channel, and the grid steel frame is erected synchronously. When chipping, one must proceed according to the horizontal channel construction step.

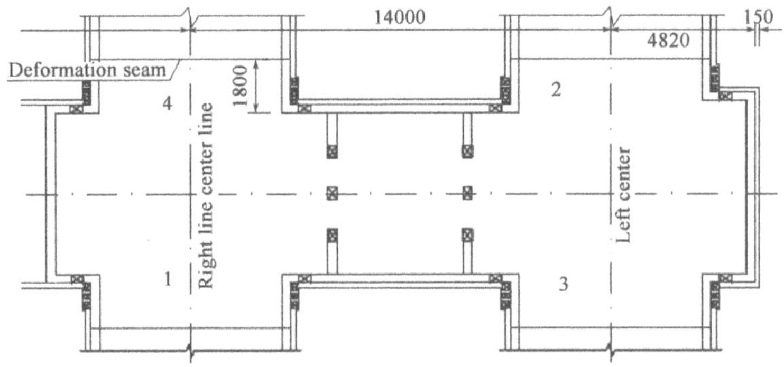

Fig. 2.66. Front door opening sequence diagram (size unit : mm).

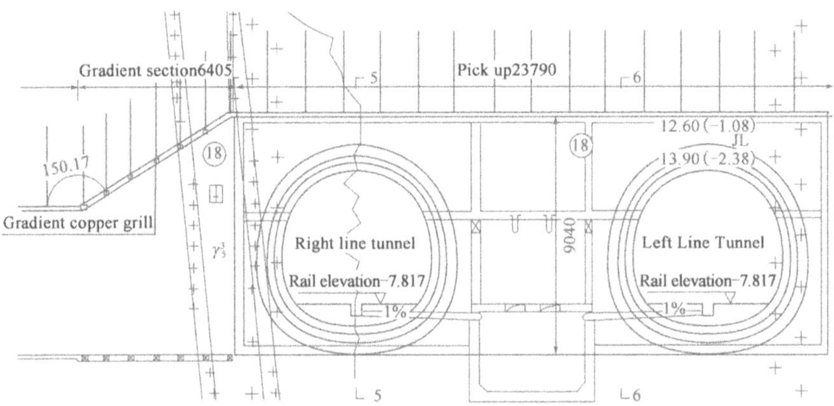

Fig. 2.67. Section profile of the front gate and transverse channel (dimension: mm).

(3) The horse-head door into the front tunnel is set with 3 trusses at a spacing of 300 mm.
(4) In the front of the tunnel, steps of 5–6 m can be used to excavate the shotcrete of the upper steps before proceeding to the lower steps.

2.6.4.2 *Main construction process*

2.6.4.2.1 Measurement

According to the elevation of the horse-head gate, the measurement gives out the position of the horse-head door, then gives out the center-line position, then returns the coordinate point to the two sides, and then draws the excavation contour line of the entire channel.

2.6.4.2.2 Removal of horse-head door of the steps

In order to ensure ingate safety of construction, one should use drilling–blasting to perform the excavation. Ingate breaking is done on the steps by first causing the ring to explode. Concrete is applied along the outer contour line break, before being cut out at the beginning of the horizontal channel to retain the contour line of the reinforcing steel bar and by erecting a trunk line tunnel grid steel frame. It is reinforced with the passage of the cross around the break and the reserved steel welding is retained firmly. The flow of shotcrete is stopped before it gets set. Three trusses are set in the opening section of the horse-head door and the anchor pipe is set to strengthen the support of the cross mouth.

2.6.4.2.3 Break down of the entrance to the lower steps

When the distance reaches 5 m, excavation of steps and the supporting closed constraints, ingate break of the next steps and cutting down of the steps near the ingate area at the beginning of the horizontal channel are required, with a dense grid steel frame row and three common channels. In the next step of the excavation process, an appropriate guarantee is placed on the stability of the excavation slope with gratings installed throughout.

2.6.5 *Summary*

The wear under buildings and pipes in the construction of small-spacing tunnels is considerable. The horse-head door is a breaking process in urban metro tunnel constructions, which faces many problems. In the process of construction according to the different geological conditions and the surrounding environment, one has to achieve safe construction. Through multiple engineering practices, the construction in such a situation, in different stages of the design and progress, involves the following:

(1) For both pipelines or buildings above the tunnel, detailed investigation and survey should be conducted, and units designed on the basis of the information gained, with special construction schemes and technical solution to ensure that drilling–blasting excavation under adjacent buildings or pipelines is safe and reliable.
(2) Controlling the parameters during drilling–blasting excavation is the key in construction. The area should be analyzed before the initiation of construction, with situations like surface subsidence being controlled with specially formulated measured and advanced blasting parameters adopted along with automatic monitoring means, which is the process of dynamic adjustment.

Bibliography

[1] Wang Mengshu. Construction technology of tunnel and underground engineering in China. Beijing: China Communications Press, 2010.
[2] Jin Feng. Development and exploration of the construction technology of the drill and explosion method for tunnel tunneling. *Building Materials and Decoration*, 2016 (45): 270–271.
[3] Zheng Zanyi. Simple analysis of the construction technology of secondary lining of tunnels and solutions to common problems. *Jiangxi Building Materials*, 2016 (20): 135–136.
[4] Liu Changyu, Zhang Qing, Wang Zhuo. Analysis and grouting treatment of tunnel secondary lining vaults. *Municipal Technology*, 2016 (05): 170–173.
[5] Liu Xiaowei, Liang Shasha, Guo Yimin. Analysis of the impact of underground drilling and blasting excavation on existing road. *Building Materials and Decoration*, 2016 (35): 248–249.
[6] Wu Liangyong, Li Xiahong. Development and application of tunnel boring and blasting construction technology. *Building Materials and Decoration*, 2016 (27): 267–268.
[7] Gao Guibin. Study on construction technology of rock tunnel drilling and blasting. *Heilongjiang Science and Technology Information*, 2016 (18): 229.
[8] Wang Liwei, Xi Jian, Zhang Luming. Analysis of construction safety of underground tunnel under existing buildings. *Traffic World*, 2016 (11): 79–81.

[9] Yu Yongyan, Gao Yongtao. Impact of drilling and blasting method of subway tunnel excavation on underground pipeline. *Engineering Blasting*, 2015 (04): 6–10.
[10] Zou Qingqi. Research on the construction technology of existing buildings under the tunnel of shallow and dark excavation. Xi 'An University of Science and Technology, 2015.
[11] Li Yingjie. Application of tunnel excavation scheme and lining construction method in tunnel engineering. *Journal of Changzhou Institute of Technology*, 2010 (05): 11–15.
[12] Li Fan. Research on construction risk management of drilling and blasting subway cross-sea tunnel project. Qingdao University of Technology, 2015.
[13] Zhang Tian. Impact analysis of tunneling and blasting vibration on nearby natural gas pipelines. Chongqing Jiaotong University, 2014.
[14] Zhao Hongpeng. brief analysis on construction technology of small spacing tunnel. *Henan Science and Technology*, 2014 (09): 64.
[15] Sun Jianlin. Research on the impact of underground tunnel penetration under the drilling and blasting method. Central South University, 2014.
[16] Fu Xiaohai. Study on the impact of metro tunnel construction on existing buildings. Beijing Jiaotong University, 2014.
[17] Dong Xiaoyu. Preliminary study on construction technology of tunnel waterproof project. *Chinese Residence* (Late September), 2013 (09): 120–121.
[18] He Kangli. On the construction technology of secondary lining of tunnel. *Urban Architecture*, 2000 (14): 89–122.

Chapter 3

Open-Cut Station Construction Technology

3.1 Overview of open-cut station construction technology

The excavation method is used for the foundation pit under the protection of a certain supporting system, and then the construction method of underground engineering is applied to the main structure of the foundation pit. It is a priority that the metro station is built using a stable method, compared with other techniques; the Ming excavation method achieved progress in constructing the structure of the foundation pit, which can be applied to different soil layers. The aspect of the envelope structure is conducive to the methods of manual excavation pile, underground continuous wall, i-steel pile, steel sheet pile circumference II and so on. Among them, the artificial dig-hole pile wall is used to achieve the goal of water and soil retention, which makes it suitable for the deeper underground construction environment; it can be utilized in a square or circular cross-sectional form. The construction intensity is higher in this than other methods. The underground continuous wall method is suitable for a soil environment of saturated sand, which can effectively block the seepage of groundwater and control the soil pressure.

The advantages of the Ming excavation method are as follows: (1) Design convenience: The force of the supporting structure and the anchoring system of the excavation method are clear, making it easy to choose a reasonable design scheme and parameters. (2) Facilitating rapid construction: Under normal circumstances, the open-cut construction site would require a more open excavation method, with a larger scope of work and more people, equipment, materials, machines and tools, thereby aiding rapid construction. (3) Easy to control construction safety, quality and schedule: Much of the excavation method of construction and operation process can be directly observed and examined, facilitating inspection, whereby safety hazards are easy to find, security measures are easy to formulate and implement, and emergency rescue conditions are better. (4) Low cost and wastage: In this case the wastage quantity and project cost are low, as the excavation method, mining method and shield method require relatively small staffing and simple equipment, with relatively high construction efficiency. It is the first method to be considered in metro station construction because it is better than other construction methods.

At the same time, the Ming construction excavation method has the following disadvantages: (1) Demolition work affects traffic: In the open-cut method that is used for urban construction, under normal circumstances, there is a need to demolish buildings, change lines and trees, and traffic control when necessary. (2) Weather conditions: It is difficult to excavate in cold regions or during windy, foggy, rainy, snowy, freezing weather conditions, which can also be dangerous. (3) Greater environmental impact: The Ming excavation method results in noise, dust, sewage, vibration and so on, having more influence on the environment; in addition, precipitation may lead to groundwater levels drastically changing, ground subsidence, inclined buildings and underground pipeline damage. (4) Tendency for foundation pit instability and failure of the whole: In poor geological and complicated environments, improper design or construction could cause slope sliding or overall buckling, leading to disaster and significant casualties.

The core steps of the construction include lowering the underground water level, supporting the slope, excavation, and construction and usage of waterproof technology. The hard rock region has a particularity of construction process, so the engineers need to fully understand the features of the metro foundation pit construction from three aspects, management, technology and personnel, all of which need to respond in a timely manner to emergent problems that may occur, so as to promote the development of the whole surface of the metro construction.

3.2 Excavation construction of open-cut station

3.2.1 *Setting of construction transportation ramp*

The construction transportation exit ramp is generally set in the location where transportation is convenient. The following factors should be considered when setting up the layout:

(1) The slope of the area and the ability to transport vehicles on the ground after excavation should be considered.
(2) The slope should be convenient for the vehicles to turn around, considering the road traffic conditions around the foundation pit.
(3) It is necessary to provide temporary support on both sides of the ramp.
(4) It should be convenient for earthwork excavation and loading.
(5) The slope of the road has an influence on the subsequent excavation and the construction convenience of the ramp construction.
(6) When the site construction is allowed, the ramp can be extended to the outside of the foundation pit to obtain a longer slope distance for ease of usage.
(7) A circular driveway is built at the bottom of the slope. When there is no circular lane, the vehicle will return on the ramp and go down to the road.

The ramp can be arranged in the direction of the long side of the foundation pit, and can be arranged along the corner of the foundation pit. For the base pit which can meet the slope of the vehicle, the slope can be arranged on the long side of the foundation pit. This arrangement will affect the excavation and provide support to the soil and stone on the side of the ramp, but only the stability of the slope side is considered, which is conducive to the safety of the transportation vehicles. For the small foundation pit that does not meet the vehicle climbing grade in the direction of the long side, the slope is generally arranged in the diagonal direction of the foundation pit to obtain the longest slope distance. This arrangement has little effect on the support construction, but it is necessary to consider the slope stability after excavation.

In order to reduce the influence of the ramp's setting on the construction period, a spare ramp can be set up at the end of the foundation pit to solve the temporary transportation problem during the support construction of the slope.

3.2.2 *Excavation sequence of foundation pit*

Quaternary soil–rock combinations are abundantly found during groundwater excavation of the foundation pit surrounding the cutoff curtain, requiring the setting up of the inside of the foundation pit with a dewatering well inland in the water, or a water pit during the excavation under convenient conditions. Temporary drainage measures shall meet the discharge requirements of groundwater, rain or stormwater and construction water.

The excavation of the foundation pit is generally carried out by the top-down stratification and subsection support flow operation. For the three-level foundation pit with less strict displacement, the excavation method of simple support can be adopted after the excavation of the slope to the basement. For excavation conditions of foundation pit stability or instability, work should be conducted according to the geological characteristics of the pit and the likelihood of damage, in a top-down manner, with segmented job-hopping and supporting time, especially for unordered extensive excavation; further, blasting operations are strictly prohibited.

The thickness of the surrounding soil of the foundation pit during excavation should be suitable for the setting of the anchor pole position, and the control system should be controlled to 30–50 mm under the anchor rod, meeting the requirements of the construction machinery. According to the requirements of the project schedule, excavation in the middle of the surrounding area can be performed to create a working face for the supporting work. As there is no weak soil layer during rock excavation, when conditions of excavation progress faster, the debris can be set aside to meet around the foundation pit slope stability and anchor construction machinery operations after the excavation of the first pit. According to the progress of the support operations, the surrounding soil to be excavated is divided into layers and subsection supports.

3.2.3 *The construction is open to all laws*

In the foundation pit of soil and rock, an excavating machine is used to excavate the surface soil and then the rocky part. The mechanical

Fig. 3.1. High power bulldozer with cracked soil.

excavation method is generally adopted for hard rock and weathered soft rock. With a large power bulldozer, the rock is broken into pieces, as shown in Fig. 3.1, and then loaded onto the bulldozer. It is possible to use the "chisel crack method" to excavate upon considering the weathering degree of the rock, the inclination of rock formation and joint development, and the manpower required for the splitter, and then confirming after a field test. The technology required and the experience of the equipment operators are closely related.

For the weathering rock, blasting excavation must be carried out, generally in the middle of the foundation pit for the initial blasting, forming a certain flattened surface, with terraced blasting excavation utilized for the surrounding area. Control blasting or static blasting must be used when near supporting structures and slopes.

After blasting the rock slope surface or basement, usually the hydraulic crushing hammer is installed in the excavating machine.

For the surrounding environment the use of explosives to perform blasting is not allowed; vibration, shock, or static blasting are also not allowed which could cause broken rock to be flung about in a slingshot manner requiring high security and vigilance to monitor, leading to relatively high cost.

3.2.4 *The case of construction opening*

The formation of the Huanghai east road station of Qingdao Metro Line 13 is a typical "earth rock assemblage". The main structure of the station is 160 m with an effective platform width of 11 m and an effective platform length of 80 m. The station is equipped with 2 platforms, 2 wind pavilions and 1 fire service outlet. The bottom plate is buried at a depth of 14.3–15.9 m, the roof covering is relatively uniform and the covering thickness is about 3 m. The main structure of the Huanghai east road station is constructed using the open-cut excavation method. The width of the foundation pit is 24.48 m, the width of the standard section is 21. 7 m, the depth of the foundation pit is 16.95 m and the lowest portion of the foundation pit is at a depth of 16.36 m, as shown in Fig. 3.2.

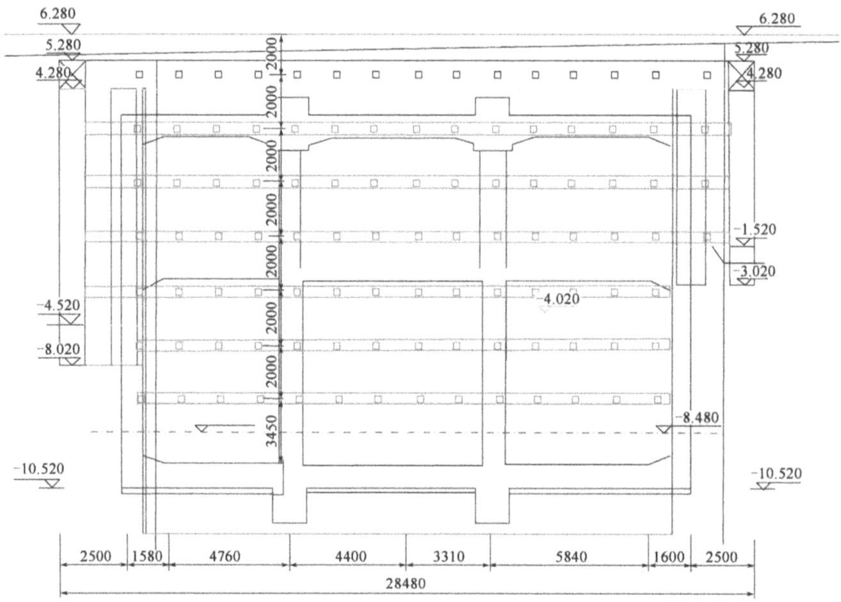

Fig. 3.2. Transverse profile of foundation pit expansion section (unit: mm).

3.2.4.1 *Excavation preparation*

(1) As stipulated in the design of technical standards, geological data and the detailed information on the surrounding buildings and underground

pipelines, it is important to perform a strict and careful analysis on the deep foundation pit construction organization design (including environment monitoring) and construction procedures, as one may encounter during the excavation foundation pit instability, mud flow or water seepage or phenomena such as quicksand, requiring forward-thinking emergency measures and reservation of related supplies in advance. A ground drainage system should also be prepared in the foundation pit.

(2) In accordance with the design requirements, steel wire is purchased and prepared, and a steel waist girder is prepared for excavating, transporting and discarding, so as to ensure a continuous flow of work.

(3) The buildings surrounding the foundation pit should be investigated, and a detailed report of the foundation pit, surrounding buildings, ground and underground pipelines should be prepared. A plan should be drawn up to control the process and provide adequate protection, which includes the setting up of monitoring points, tests of the initial data and debugging of the testing instruments.

Equipped with sufficient excavation and transportation machinery, mechanical testing, maintenance and other work should be performed to ensure the normal operation of machinery.

3.2.4.2 *Excavation and enclosure at the top of the foundation pit*

(1) In crown beam excavation, chiseling is done on the pile head. In the model casting crown beam, the beam in the reserved phi 110 steel pipe is anchored to the cable construction and the crown beam intensity at the rear end is set at 90%, thus meeting the design requirements of the tensile anchor cable. This is achieved by positioning the crown beam after retaining the wall.

(2) The construction of the crown beam is carried out in sections of 30 m or so at a time. After blasting, excavators and artificial combination are used to excavate.

(3) After the structure is completed, the excavation of the soil in the foundation pit is started.

3.2.4.3 *Construction of crown beam*

3.2.4.3.1 Top beam construction arrangement

The drill hole pile is equipped with a steel-reinforced concrete crown girder and is connected to the whole body, and the cutter head is 1000 mm × 1000 mm (width × height).

The construction of the crown girder is arranged after the completion of the bored pile, and the construction of each section shall be considered according to 30-m sections. A wood mold is used. The steel bar is processed in the field and transported to the site by the concrete conveying truck. The vibrator compacts the ground and the water level is maintained.

3.2.4.3.2 Construction process of the crown beam

See Fig. 3.3 for explanation of the process.

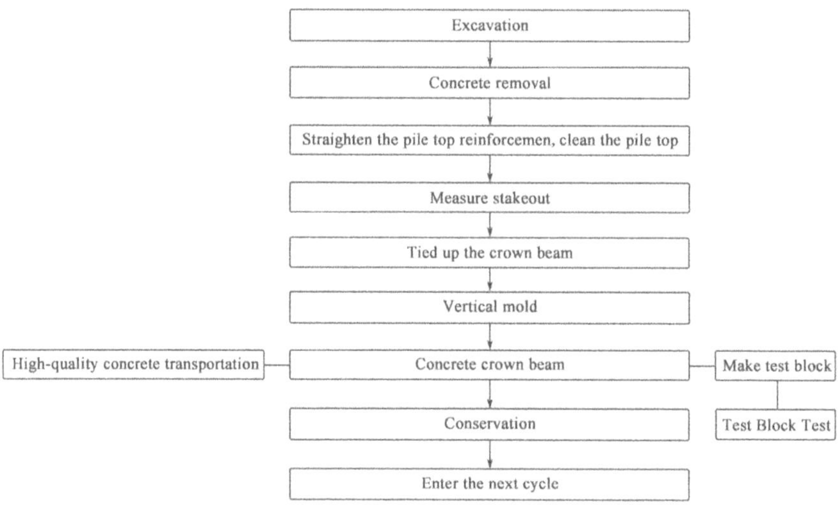

Fig. 3.3. Process flowchart of crown beam construction.

3.2.4.3.3 Construction methods and technical measures of canopy beams

(1) After construction of each section of the bored pile is completed, the concrete of the pile head is chipped.

(2) To remove the remaining soil and floating pulp, the pile top is reinforced and aligned, and then cleaned with clean water.
(3) In accordance with the design requirements, girder steel reinforcement is required, and the length of the main bar and the main bar of the crown beam are reserved.
(4) The model is reinforced.
(5) Each section of the crown beam is concreted one at a time. After the concrete is poured, the water should be poured in a timely manner and the curing time is not less than 7d.

3.2.4.4 *Excavation of main foundation pit*

3.2.4.4.1 Excavation sequence of foundation pit

The station excavation is divided into nine sections, and the construction section of the station excavation is shown in Fig. 3.4. After the envelope structure has been closed, the station is sliced and dug from one end to the other. The time and space effect should be fully taken into account in the excavation, the excavation should be divided into small sections and the bolt should be set in time to be completed in 6 h.

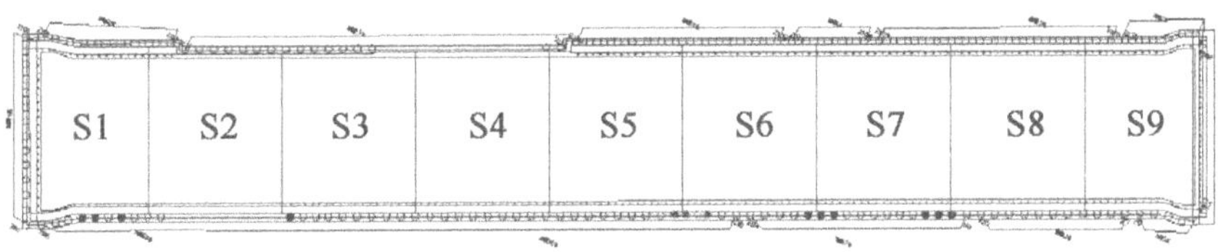

Fig. 3.4. Schematic diagram of block construction of Huanghai east road station.

3.2.4.4.2 Excavation schedule

Excavation of a single working face, with continuous operation for three shifts is conducted in 950 m^3/d.

3.2.4.4.3 Excavation of mechanical equipment and labor organization

Excavation is carried out using 300 excavator segmentation steps, below 8 m with two digger firms to dig to the basement above 30 cm using artificial stripping, and the last remaining distance using a 20-m long-arm excavator and auto cranes for construction. Considering the influence of the weather, mechanical efficiency and other factors, excavating, discharging the mechanical equipment and labor force are equipped with 150% of the average excavated amount. The ability to export soil is dependent on the transportation time, the amount of excavated debris, etc.; it is equipped with a vehicle for the purpose.

3.2.4.4.4 Excavation procedure of foundation pit

The excavation of the foundation pit follows the principle of "vertical stratification, horizontal section, top-down and anchor excavation", which is divided into nine sections from west to east.

The specific steps are as follows:

- **Step 1:** Excavate the first layer of earth to the bottom of the crown beam, remove the pile head, form the crown beam and retaining wall, reserve the steel pipes and set the first bolt. See Figs. 3.5 and 3.6.
- **Step 2:** The height of the excavated step is about 1.5 m. After the network of the layers is finished, the excavation will continue to the second cable. At 0.5 m, the second anchor cable is set. See Figs. 3.7 and 3.8.

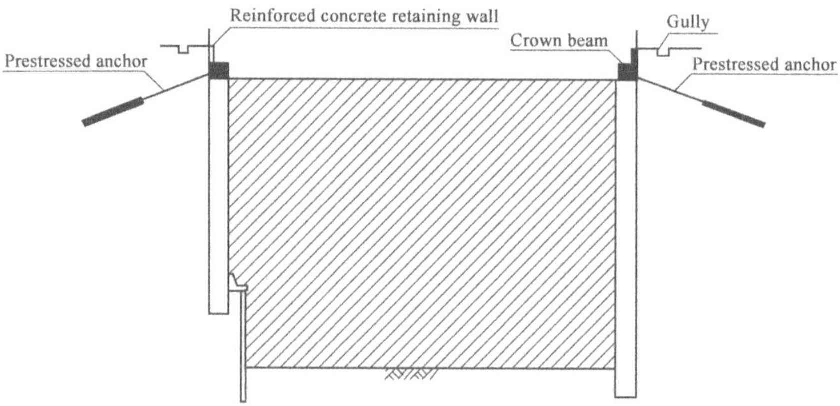

Fig. 3.5. Cross-section diagram of step excavation of the foundation pit.

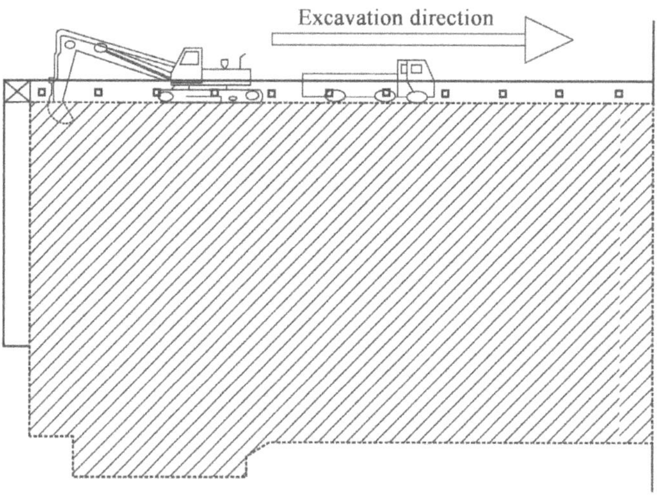

Fig. 3.6. Schematic diagram of earthwork excavation.

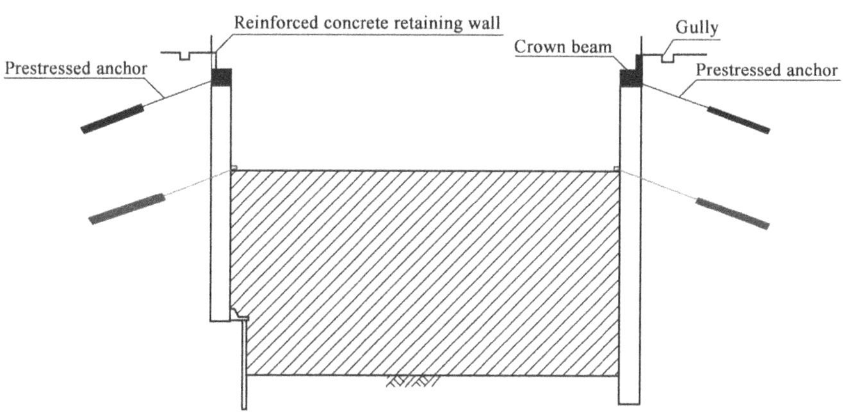

Fig. 3.7. Cross-section diagram of excavation of the foundation pit.

- **Step 3:** The excavation height is about 1.5 m. After the network spraying is finished, the excavation will continue down to the third anchor line. At 0.5 m, the third anchor cable is set. See Figs. 3.9 and 3.10.
- **Step 4:** The above method is continued to excavate the platform of the locking foot, using the steel pipe pile and the lock foot beam, and setting the fourth anchor cable. See Figs. 3.11 and 3.12.

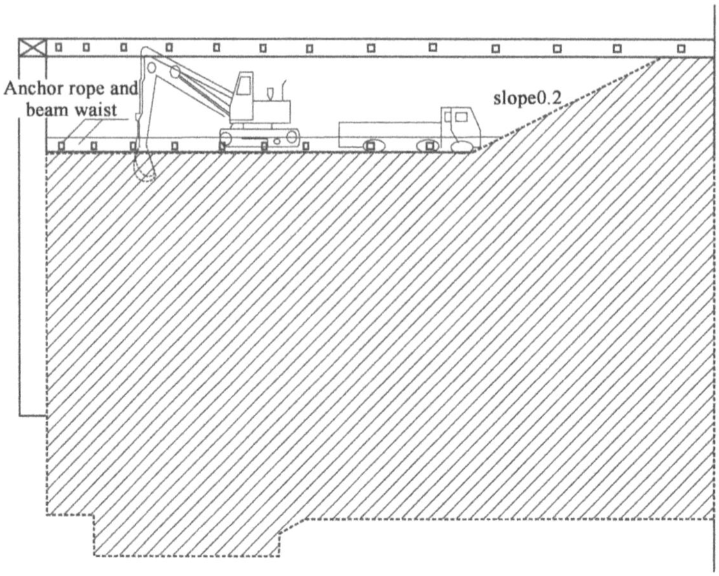

Fig. 3.8. Profile of step excavation of foundation pit.

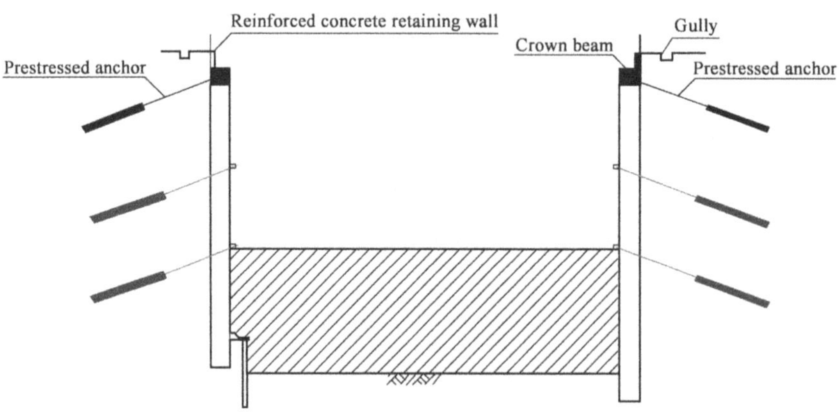

Fig. 3.9. Cross-section diagram of excavation of foundation pit.

- **Step 5:** The same construction method is utilized, according to the mortar bolt and the height of the anchor cable, excavating 0.5 m under the anchor and 0.5 m under the anchor cable, respectively. See Figs. 3.13 and 3.14.

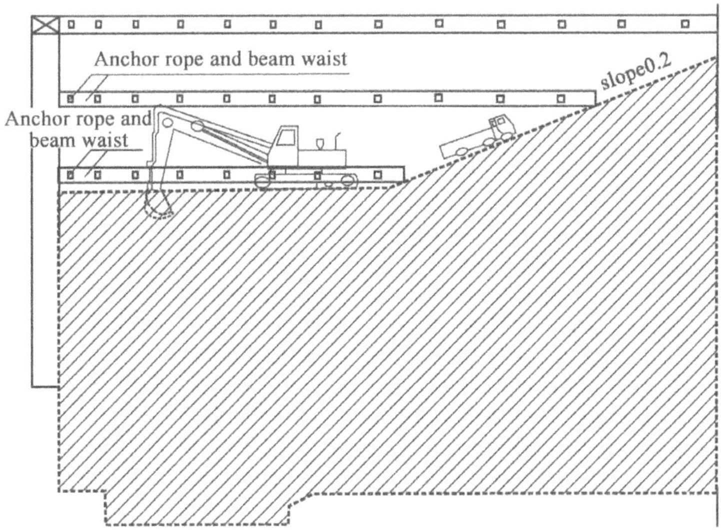

Fig. 3.10. Profile of excavation of foundation pit.

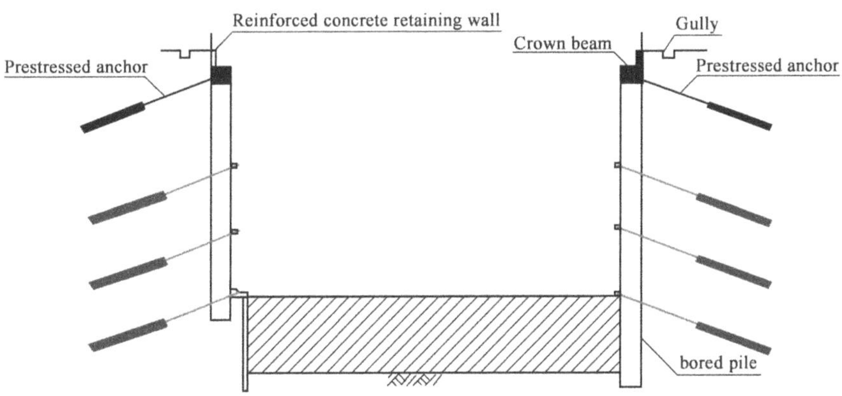

Fig. 3.11. Horizontal section of foundation pit.

- **Step 6:** Continue digging to the second mortar bolt and set the second mortar bolt. See Figs. 3.15 and 3.16.
- **Step 7:** The excavation is done to within 30 cm of the ground, and the remaining earth and stone is worked to the base, and concrete bedding is pouring. See Figs. 3.17 and 3.18.

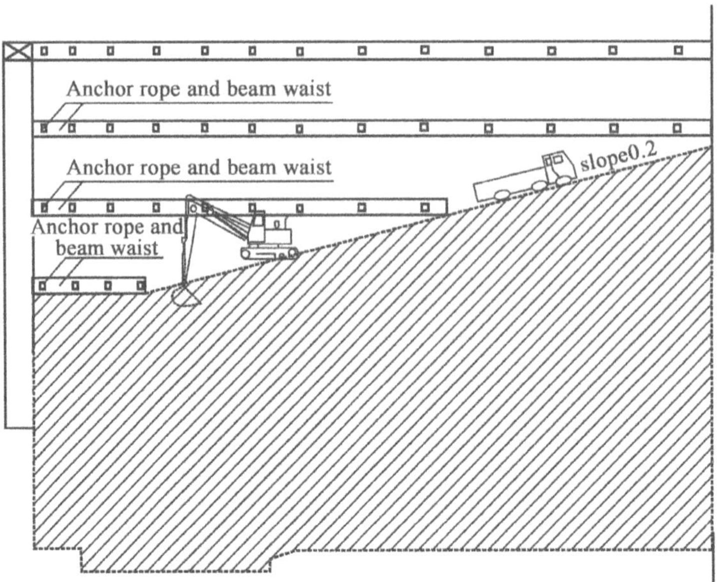

Fig. 3.12. Profile of excavation of foundation pit.

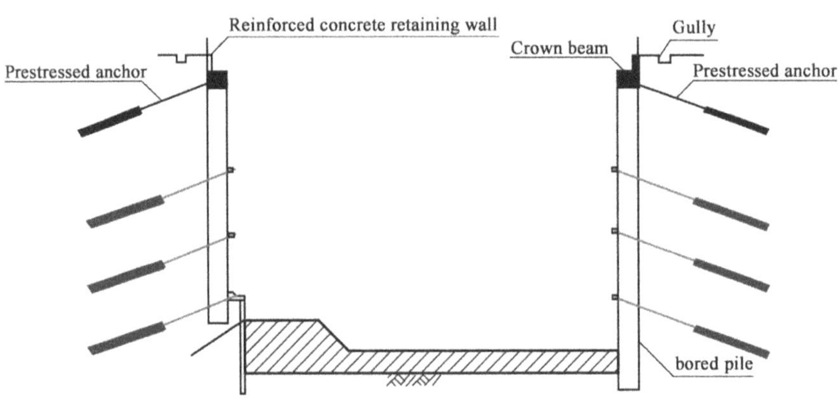

Fig. 3.13. Cross-section diagram of step excavation of the foundation pit.

3.2.4.5 *Excavation methods*

(1) The excavation of the foundation pit adopts the method of "longitudinal section and vertical stratification", and the anchor cable and anchor rod are designed for the excavation of foundation pit.
(2) The upper 8 m of earth is directly excavated and transported by excavator.

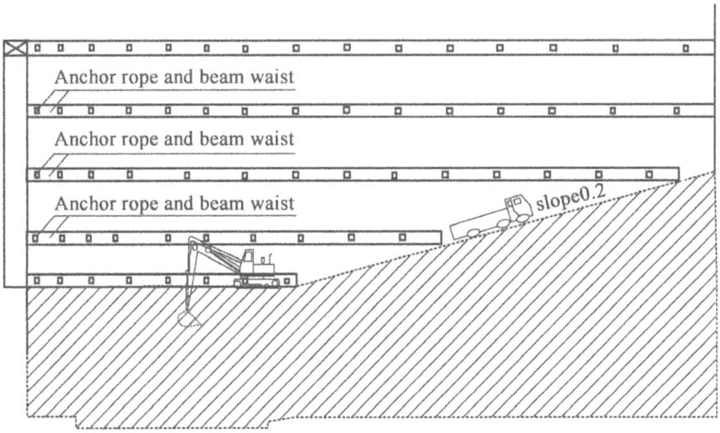

Fig. 3.14. Profile of excavation of foundation pit.

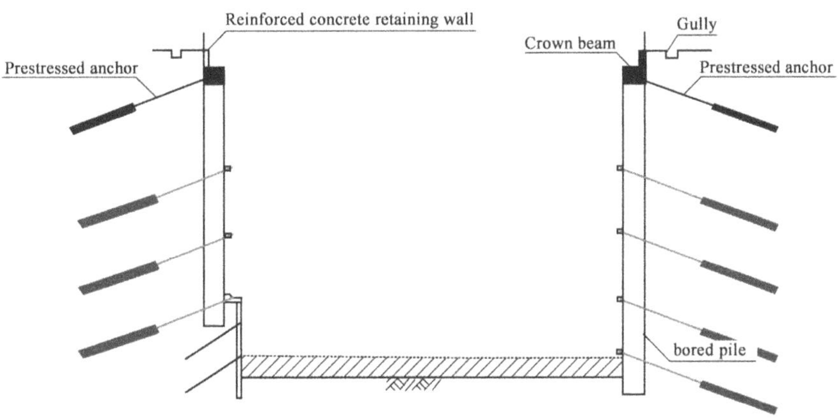

Fig. 3.15. Cross-section diagram of step excavation of the foundation pit.

(3) In the lower layer, two excavators are used, and the remaining earthwork is constructed with the long-arm excavators and gantry cranes.
(4) The soil slag is transported by the excavator to the temporary dumping ground.
(5) The upper part of the crown bean section of 1.1 m adopts a natural slope of 1:1. The slope of the refuse truck is always 0.2 and the slope of the working platform in the foundation pit is 1:1.5.

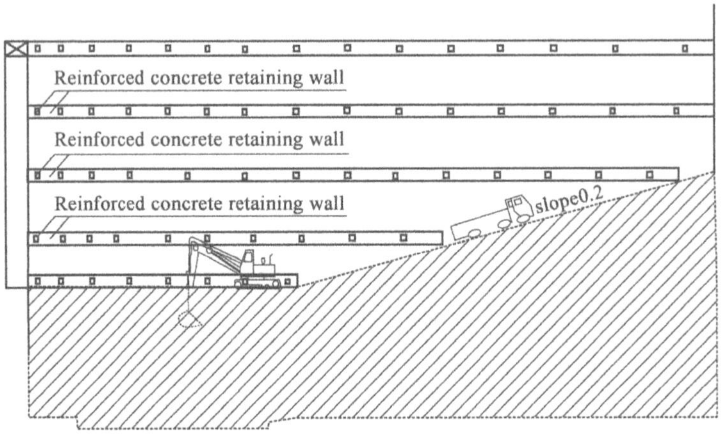

Fig. 3.16. Profile of excavation of foundation pit.

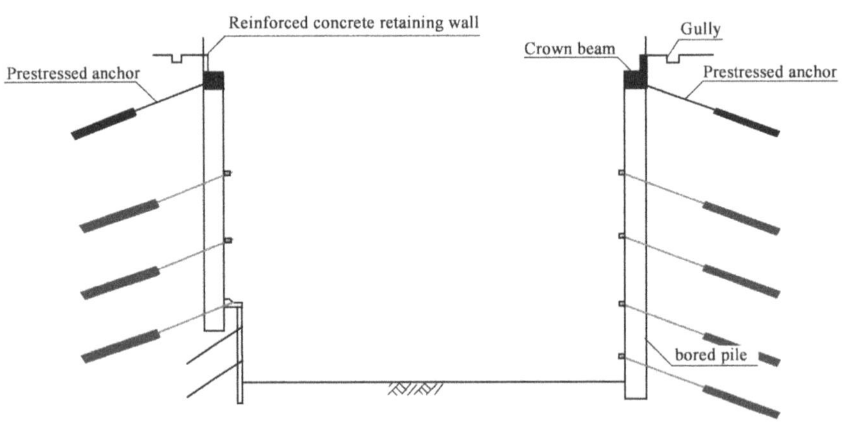

Fig. 3.17. Cross-section diagram of step excavation of the foundation pit.

(6) A small machine is used to manually excavate 30 cm above the base; the final part of the excavation is strictly controlled, so as to avoid over excavation.
(7) In order to ensure the stability of the foundation pit, the reinforced concrete floor will be laid as soon as possible after the bedding is finished.
(8) In the process of excavation, a skilled person will draw a geological sketch of the area. When the basement stratum is not in conformity

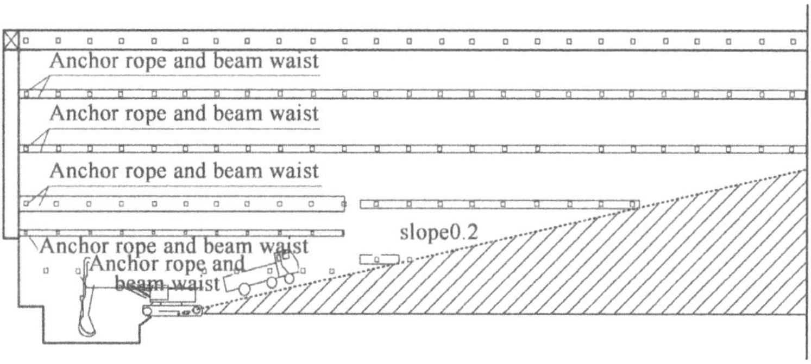

Fig. 3.18. Profile of excavation of foundation pit.

Table 3.1. Allowable deviation of excavation of foundation pit and the inspection method.

The serial number	Project	Allowable deviation (mm)	Inspection frequency		Test method
			The scope	points	
1	Bottom elevation	+10, −20	Each section of the foundation pit or 50 m.	5	With the level
2	Vertical and horizontal axis	50		2	With theodolite, crisscross the sides
3	Foundation pit size	Not less than design		4	Measure 1 point on each side
4	The foundation pit slope	5% of the design		4	Use a slope ruler

with the design, a timely report will be sent and the inconsistencies will be dealt with.

(9) During the excavation of the section, the intercepting ditch and drainage will be set, and the water seepage and rainwater will be drained in a timely manner.

(10) In the process of excavation, the foundation pit and surrounding environment should be monitored according to the established monitoring plan to provide feedback and guidance for the construction.

The allowable deviation and the inspection method for excavation of the foundation pit are shown in Table 3.1.

3.2.5 *Summary*

This section mainly introduces the construction points of the excavation of the open-cut station in hard rock, which mainly includes the following aspects:

(1) The construction transport ramp settings need to consider many factors, including the stability of the ramp itself and transportation convenience. The ramp set mode can be long along the foundation pit edge direction and can also be diagonal to the foundation pit.
(2) The excavation of the foundation pit in the hard rock area is generally conducted in the horizontal section for the supporting flow in different layers. For strong weathering hard rock and weathered soft rock, generally a kind of "chiseling method" of construction is adopted.
(3) If the conditions allow, traffic guide change is implemented as the requirements for moving the pipeline are not extensive. One should first choose the station construction scheme, because for the Ming method used for hard rock geological conditions, the risk is relatively small and the construction efficiency is higher. The optimal area can be chosen to put the slope, thereby increasing efficiency and saving investment.
(4) It is necessary to pay attention to the supporting procedures during the explosion part of excavation of the hard rock. Generally speaking, the first support should be reinforced concrete. At the same time, the blasting scale should be strictly controlled, and the cannon should be protected from flying stones.
(5) The depth of the buried section of the station should be fully considered, especially as the state law advocates that for comprehensive utility underground tunnel, the embedded depth of the roof should not be less than 3.5 m, reserved for the underground pipe gallery.

3.3 Excavation of supporting type of open-cut station

Deep foundation pit engineering involves the application of the slope support at the basement layer or buried depth, assessment of the geologic condition, understanding of the hydrogeological condition taking into account nearly enclosed environmental conditions, consideration of appropriate construction requirements, clearance from environmental protection, etc. Because Qingdao has a typical rock–soil combination, with a particularity of geological conditions, there is great difficulty in implementing construction mechanisms, and some traditional supporting structures often cannot be used. For example, the traditional concrete wall supporting form will lead to construction equipment getting embedded within the rock as the structure does not meet the fixed depth, and hence the equipment cannot be directly used. A similar situation will be encountered in the application of the pile or underground continuous wall in the rocky earth geological conditions. At present, the support system of the underground continuous wall is not visible under the soil–rock assemblage. In this case, the requirements of the supporting technology, construction technology and environmental protection are very high. At present, in Qingdao deep excavation and slope supporting application are commonly used for assessing the basic characteristics of the supporting materials.

The form of the foundation pit support for Qingdao Metro station is listed in Table 3.2.

3.3.1 *Slope excavation*

3.3.1.1 *Concept and characteristics of slope excavation*

During the excavation of the foundation pit, one can often choose and determine the slope based on safety considerations. The excavation of the soil mass should be done without reinforcement and support, relying on the strength of the soil itself, under the new state of equilibrium in the stability of the slope thereby maintaining the stability of the foundation pit. To build the foundation or the basement, a safe and reliable operation space should be provided, at the same time ensuring that the engineering environment surrounding the foundation pit is not affected by adverse conditions and meets predetermined requirements. The foundation pit excavation method under such unsupported measures is usually called slope excavation. Slope excavation is suitable depending on the foundation soil quality, if the excavation depth is not much and the construction

Table 3.2. Qingdao deep foundation pit supporting type and metro station foundation pit control summary.

S/N	Supporting type	Meaning and features	Foundation pit type	Adaptive requirements	An overview of joint applications with other forms of support	Notes
1	Perform the slope excavation	During the excavation of the foundation pit, often one can choose and determine the slope based on safety considerations. The foundation pit digging should be done without reinforcement and support, depending on the strength of the soil itself, under the new state of equilibrium achieved which maintains the whole foundation pit slope stability, thereby building the foundation or the basement and providing a safe and reliable operation space. At the same time, it can be ensured that the working environment around the foundation pit is not affected so as to meet the scheduled working environment requirements.	(1) Soil quality; (2) Soil rock assemblage.	Suitable for the following: (1) The perimeter of the foundation pit has sufficient space for the slope, and there are no adjacent structures, such as construction, underground pipelines, etc.; (2) No underground water or underground water development is detected; (3) There is no structural surface for production and slope failure in the H and field of the rock mass can occur; (4) It is possible to adapt to the second- and third-grade foundation pits.		The excavation of the slope is suitable for the project of good soil quality, great excavation depth, and sufficient slope area. When the foundation pit is deeper than 5m, w hen the slope excavation is adopted, the slope excavation shall be graded, the horizontal platform shall be set at the grading level and the width of the platform shall be 1–1.5 m. The grading platform of the rock slope is generally not less than 0.5 m.

2	Soil nail wall support	In the reinforced soil, soil nailing is set up and support is formed on the surface of the steel mesh, which is similar to the effect of the heavy force-retaining wall. It has the characteristics of a light structure, with simple, quick and low-cost construction and wide-ranging applications.	(1) Soil quality; (2) Rock mass; (3) Soil rock assemblage.	Suitable for the following: (1) The groundwater level above or the foundation pit after precipitation; (2) A good foundation pit for the self-stability of soil layer; (3) Not applicable; Not suitable for soft soil and water-rich gravel: (a) It is very rare to use the foundation pit with strict environmental conditions and strict requirements for variable shape. (b) Do not use as a water-retaining structure.	(1) The soil nail and the bolt are combined to form a composite soil nailing wall supporting form, which is suitable for the foundation pit with strict deformation requirements, and it is suitable to aid in increasing the excavation depth of the foundation pit. (2) The combination of soil nail wall support and other supporting forms is utilized .	(1) Soil nailing wall support is a kind of primary soil-strengthening technology; (2) The soil nail should be effectively connected with the surface layer, and structural measures such as pressure plate and reinforcing rib should be set. (3) The length of the soil nail should exceed the slip surface, and it should be 0.5–1.2 times the depth of the excavation.

(*Continued*)

Table 3.2. (*Continued*)

S/N	Supporting type	Meaning and features	Foundation pit type	Adaptive requirements	An overview of joint applications with other forms of support	Notes
3	Composite soil nail wall support	Based on the form of soil nailing wall support, add anchor rod, micro pile, cement soil pile or insert core material in cement soil pile respectively to form multiple forms of composite soil nailing wall support, so as to make up for some defects of the function of soil nailing wall support form; change the original stress state; appropriately increase the excavation depth of foundation pit, etc. It is appropriate to increase the excavation depth of the foundation pit.	Same as above	Suitable for the following: (1) The foundation pit of variable shape is stricter; (2) Improve the self-stability of the rock mass; (3) Function of stopping water and anti-seepage; (4) The development of "time–space efficiency" for soft soil.	(1) Combination of composite soil nailing wall support and soil nail wall support; the upper soil layer is limited by slope and soil nail wall support; lower weathered rock, composite soil nail wall support (soil nail + bolt + surface layer); (2) Combination of composite soil nailing wall support and pile anchor support; the upper soil layer, the pile anchor support, straight open excavation; the lower	

4	Compound anchor spray wall support	The support form is composed of soil (rock) anchor, shotcrete and prestressed anchor or micro pile or cement soil pile for advance support, so as to solve the problems of foundation pit deformation, soil self-supporting and water separation	(1) Soil quality; (2) Soil rock assemblage.	(1) Support form of "non prestressed anchor + prestressed anchor + prestressed anchor + shotcrete" is mainly applicable for large depth; soil layer thickness is relatively less, the groundwater is inadequate, the stability of rock mass is relatively good in the foundation pit; (2) The supporting type of "non-prestressed bolt + pre-support, micropile + shotcrete" is mainly applicable for the steep slope with loose soil, or breaking rock, where groundwater is inadequate and the foundation pit has poor self-reliance;	part of the wind-driven rock, retaining the slope excavation, composite soil nail wall support (soil nail + bolt + surface layer).

(Continued)

Table 3.2. (*Continued*)

S/N	Supporting type	Meaning and features	Foundation pit type	Adaptive requirements	An overview of joint applications with other forms of support	Notes
				(3) The supporting form of "non-prestressed anchor rod + pre-support. Miniature prestressed anchor pile + + injection coagulation soil" is mainly suitable for deep excavation, steep slope and displacement control to seek higher and loose soil or broken rock, where groundwater is not abundant, neither water-rich nor water-poor, independent of the foundation pit; (4) The supporting form of "prestressed anchor shot coagulation + + cement-soil pile and soil" is mainly to apply to the upper part in the case of water-bearing soft soil, where the cutoff curtain		

can be set up only in the upper soil layer, requiring a cutoff curtain in the relative water-resisting layer depth;

(5) The supporting form of "non-prestressed anchor rod + preying force bolt + soil-cement pile + shotcrete" is mainly applicable to the upper part. In the water soft soil layer, the excavation depth is greater, the slope is steeper and the displacement control is higher and the soil is loose and independent;

(6) The supporting form of "prestressed anchor + its should force reinforced cement-soil + + anchor spray shoot coagulation soil" is, mainly suitable for the upper part, where the anchor pole of a prestressing value is bigger, for aqueous soft

(Continued)

Table 3.2. (*Continued*)

S/N	Supporting type	Meaning and features	Foundation pit type	Adaptive requirements	An overview of joint applications with other forms of support	Notes
5	Pile anchor support	There is a rigid supporting type composed of pile body, prestressed anchor rod (cable) and anchor-bearing structure; the pile (supporting pile) is connected with the bolt (playing the back role), which has strong resistance to soil deformation. If there is an aquifer, it can be used to create a water-retaining curtain between the retaining piles, and the water-retaining pile can be arranged independently. The combination of supporting piles and water-retaining piles plays a dual role of retaining soil and stopping water.	(1) Soil quality; (2) Soil rock assemblage.	and weak soil layer, of certain excavation depth, slope steepness and displacement control, with high demand for the poor loose soil independence. Suitable for the following: (1) Foundation pit for vertical excavation; (2) The deformation of the foundation pit with relatively strict requirements; (3) There is a foundation pit with rich water, sandy soil layer and soft soil layer, which requires retaining of both soil and water.	(1) The combination of pile anchor support and composite soil nail wall support; (2) Combination of pile anchor support and soil nail wall support.	(1) Anchor-bearing structure, including anchor, gasket and section steel; (2) The supporting piles can adopt different piles, and the steel-reinforced concrete piles or steel pipe piles are not the same. (3) The water-retaining pile can be filled with a deep mixing pile or high spray pile.

6	Foot pile support	The pile bottom is strongly weathered or embedded in the weathering granite; considering the economic factors and the construction, the depth of the embedded rock is limited, but at the bottom bedrock of the foundation pit under several meters, when excavating to the basement, the supporting pile foot is hanging in the air, which is commonly known as the "out-seam pile".	Soil rock combination	It is required that the supporting pile should be fixed in the stable layer. Stopping the water pile (curtain) should satisfy the sealing requirements of the aquifer, preventing infiltration and water leakage. It is suitable for rock strata of shallow depth with soft ground conditions.	The supporting structure of "hanging foot pile" can be combined with inner support or bolt support.

site has enough space for the slope. When the depth of the foundation pit is more than 5 m, the slope excavation should be graded, and the platform shall be equipped with a transition area. The width of the platform is generally 1–1.5 m. The grading platform of the rocky slope is generally not less than 0.5 m. In the construction of the soft soil foundation pit, when the excavation is not deep (3–5 m) and the excavation of the foundation pit is not high or the surrounding engineering environment is not reliable, the slope excavation method shall be adopted. In general, the slope excavation method is the preferred method for iron station construction. This method is mainly applied for shallow buried underground tunnel engineering construction, because the impact on the surrounding environment is relatively low. As long as there is reasonable control in the construction of the slope rate, the construction costs and soil structure stability can be controlled.

3.3.1.2 *Application scope of slope excavation*

(1) There is enough space around the foundation pit for a slope, and there are no adjacent constructions, underground pipelines or any other displacement of sensitive facilities.
(2) Groundwater is not developed.
(3) The rock mass is in level II or III with good quality, and there is no occurrence of unfavorable structural surface or sloping.
(4) The second- and third-grade foundation pits can be adapted to the conditions.

3.3.1.3 *Application of slope excavation in Qingdao Metro*

In the pit of the Qingdao open-cut metro station foundation, the environment of the surrounding areas allows slope excavation engineering, for which the station workforce is often used. The soil slope ratio, usually 1:0.5, is now 1:0.2 for rocks, and may be adjusted according to actual conditions for different strata. In addition, due to the larger occupied slope of the excavation site, usually a combination of different forms of support is used, such as spray anchor bracing, composite soil nailing, wall supporting, forepoling using micro steel pipes, etc., and the slope supporting in the Qingdao Metro Line 3 or the application of Changsha road station is more representative. From the starting point of the station to the center, the right line is supported by a "stilted pile", and the pile is set with a high-pressure rotary jet pile with a pressure of 1200–1500, which is interrelated with the filling pile. The weathering strata of the bored piles is 1.5 m, and the slope is 1:0.3. The left line is made of a manual excavation pile and shotcrete support. The rest of the station adopts the form of slope + soil nail wall support, and the slope series is changed with the soil layer, with the slope rate of 1:0.2 and 1:0.5 as shown in Fig. 3.19.

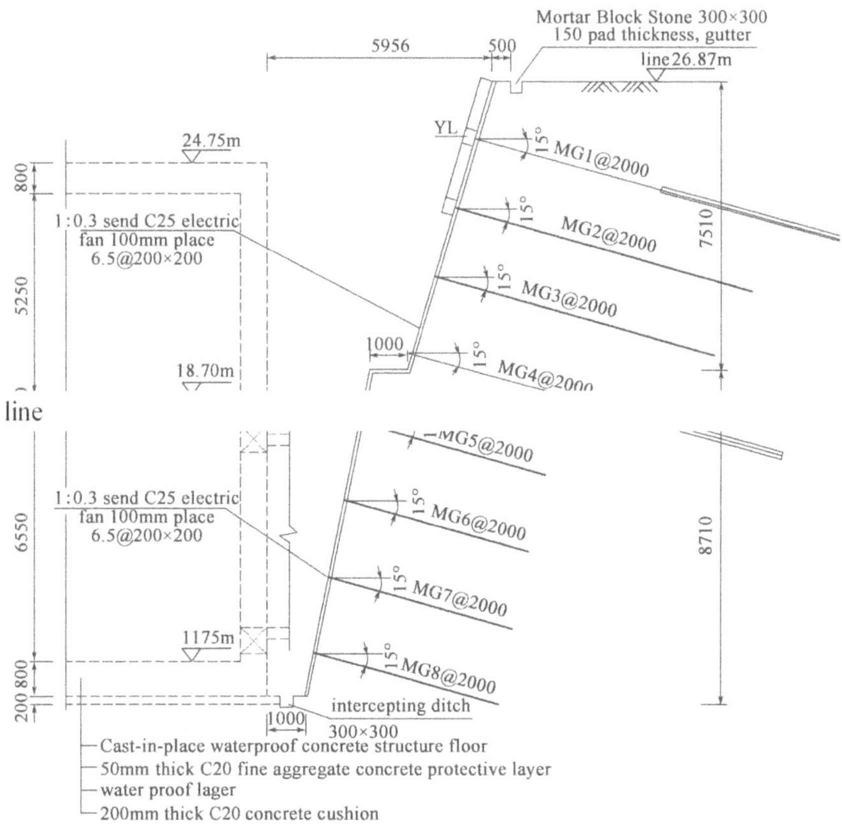

Fig. 3.19. Cross-sectional drawing of the main body of Changsha road station (size unit: mm).

3.3.2 *Soil nail wall support*

3.3.2.1 *Concept and principle of soil nailing wall*

Soil nailing wall (Soil Nail Wall) is a kind of in situ soil reinforcement technology; the foundation pit slope is reinforced by soil nails made of steel. A steel mesh is laid on the slope surface to spray a layer of concrete, and a combined-type slope reinforcement of earth side slope-supporting method is used. Its structure is set in the slope reinforcement bar (that is, the soil nail or bolt), its surrounding soil with a solid knot formation of the complex, as well as the surface layer of gravity-retaining wall-supporting structure.

The so-called "soil nailing" refers to the long and thin bar that restricts the soil, and the friction force of the soil surface can be used to better play the tensile strength of the soil nail. The soil nail can also be used in the steel tube, angle steel and so on straight into the soil without the need for grouting. It is mainly subjected to tension, the frictional force between the grouting body and the soil is provided for the resistance of the soil nail and the development of soil deformation is arranged.

The soil's low shear strength and tensile strength can almost be ignored, but the soil has certain structural integrity, in the foundation pit excavation, requiring the critical height of the slope to be in an upright position, but should there be greater depth, the ground will be overloaded, leading to sudden overall damage. Generally, the slope protection measures are based on the passive restraint mechanism of the retaining slope, which can withstand the lateral compressive force of the soil and maintain the overall stability of the soil. The soil nail wall technique involves a soil nail body with a certain length and distribution density in the soil, which is combined with the soil to make up for the deficiency of soil strength. So by enhancing the stability of the slope soil itself, an active restriction mechanism on the basis of composite foundation not only effects the improvement of the whole stiffness of the soil, making up for the low tensile and shear strength at the weak point, but through the interaction, the soil structural strength also potentially comes into full play, changing the slope deformation and failure characteristics, improving the overall stability; more importantly in the soil nail wall process under load a sudden collapse or slope sliding is not going to occur, as the soil nailing wall not only delays the plastic deformation stage of development but also ensures that obvious progressive deformation and cracking of the whole sliding wall will not occur.

3.3.2.2 *Construction process of soil nail wall*

The main construction process of the soil nail wall is as follows:

(1) **Digging and drainage:** Earthwork excavation is carried out from the top down by an excavator, with soil nailing performed for each layer. In the 5-m range, mechanical excavation shall be made up of a 200-mm protective layer according to the design slope. In addition, work should be coordinated with earthwork excavation of the foundation pit drainage, completing drainage measures at the top and bottom, the top to set up 1% of the slope, and within a certain range with a C15 concrete plaster, allowing the foundation pit effluent water to flow away freely. Slope drainage is achieved with the tube wall with a phi 50 perforated plastic drainage pipe placed at the soil level near the side wall. A plum flower-form horizontal layout with vertical and horizontal spacing fixed at 2.0 m helps in the flow of water from the bound geotextile filter material.

    The drainage ditch and water pit are set at the bottom of the slope to drain the water from the drain pit.
(2) **Early shotcrete:** Immediately after the slope dressing, a layer of 50-mm thick C20 concrete is sprayed to make the exposed soil close in time.
(3) **Drill hole:** According to the geological conditions, the drill hole is selected and the diameter is 120 mm, and the hole position should be determined according to the design requirements before drilling.

    In the areas with poor geological conditions, water drilling is adopted to avoid the use of bentonite or other slurry as drilling pads. After the hole is opened, the hole should be cleared, and there shall be no dirt, impurities and mud in the hole.
(4) **Setting the soil nail:** According to the design requirements, the length of the steel bar should be 150 mm longer than that of the design hole. The hole is rechecked for the presence of debris and mud

before pushing the reinforcement, and then the steel bar is pushed along the hole to the bottom of the manhole.

The reinforcement should not be rotated in the push process to prevent damage of the hole wall of the positioning bracket, and at the same time to prevent a forced push, in order to avoid the reinforcement of the head of the soil. If the steel bar cannot reach the bottom of the hole, it should first be pulled out and then pushed. After the push is completed, the exhaust pipe and the grouting pipe should be checked to ensure that they are clear.

(5) **Grouting:** Before adopting the bottom grouting method, grouting is performed first at the orifice to set the pulp plug and tighten the flush with the hole in the wall. The mouth of the slurry grouting pipe is set at a depth of 200–500 mm from the bottom of the hole. When grouting, the slurry should be mixed evenly with the mixer, the grouting pressure is 0.4–0.6 MPa, the grouting pipe is connected to the grouting pump, and the side of the grouting is drawn in the direction of the orifice until the filling is stopped. In addition, in order to increase the early strength of the cement slurry, a moderate amount of early strength agent was added to the slurry.

(6) **Steel mesh:** Take steel mesh and weld it into the soil nail head. Lay eight pieces of 150 × 150 steel mesh on the slope and, according to the design requirements in the soil nail head in the vertical and horizontal direction using an 18 Wells glyph stiffener, weld along the length of the soil nail with 225 (100 mm) long locking reinforcement, with the soil nail and the stiffener taken together as a whole. Welding is done strictly according to the specific requirements.

(7) **The end shotcrete:** Spraying should be conducted before setting the control and should be buried deeply in coagulation soil. Spraying should be done from the bottom of the gathering in a straight line. Jet flow is also set in the same direction as the tags. Injection is done gradually from the bottom of the jet flow. This should be done in a direction perpendicular to the jet surface, with the shooting commonly done from 0.8 to 1.5 m distance. The rebar should be sprayed, and then the front of the reinforcement should be sprayed to prevent any gap in the back of the reinforcement. The segmented spray face should be at a 45° incline, so that the shotcrete can be firmly held together as whole.

3.3.3 *Composite soil nail wall support*

3.3.3.1 *The concept and classification of composite soil nailing*

Wall composite soil nailing uses one or several supporting technologies, or a water technology unit synthetic composite supporting system, with its main constituent elements of soil nailing, prestressed anchor, cutoff curtain, micropile, hanging net chute shot concrete pavement, in situ soil, etc. The composite soil nail wall is composed of soil nailing, surface layers and auxiliary support.

Soil nailing is the main force component of the composite soil nailing wall, and there are two types of soil nail, qigong injector and grouted soil nail. It is usually made of angle steel, round steel or steel pipe, and its general length is not more than 6 m, or it is 0.5–0.7 times larger. This kind of soil nail is usually not used for drilling, grouting and construction speed, but it does not apply to gravel soil, hard cement soil and loose sandy soil layers. The qigong projectile is usually made of a straight steel rod of a length of 25–38 mm and a width of 3–6 m. This kind of soil nail is developed in the UK and has not been reported in domestic applications of this type of soil nail. The grouting soil nail is the most commonly used type of soil nail, and the whole soil nail is composed of the nail body and the surrounding grouting body. The nail body generally adopts the phi 16–32 mm III level or II level of hot-rolled deformed steel materials. The grouting material is mortar or cement mortar. The pore diameter of the grouting soil nail is generally 75–150 mm, its length is not limited and the length of general engineering is 6–10 m or 0.5–1.2. Compared with other types of soil nails, qigong has unique advantages in construction in gravel soil, hard cementing sites and loose sandy soil.

The surface layer includes the supporting surface and the waterproof floor. The surface layer of temporary soil nailing is usually made of netted concrete with a thickness of 50–100 mm. The surface layer is equipped with a steel mesh, the grid is square and the side length is 150–300 mm. The steel mesh spraying concrete surface of the permanent soil nailing support is 150–250 mm thick, and it is generally equipped with two layers of steel mesh. In recent years, for unsaturated soil layers, sometimes precast concrete slabs or geotextiles are used as surface layers. The width of the concrete sheath at the top of the soil slopes is generally 1–2 m. The width of the waterproof ground is generally taken as the staple length.

The common structure of the composite soil nailing wall auxiliary support is the mixing pile and micropile. Using the auxiliary cement mixing pile, usually the watery mud acts as the main curing agent, and the cement commonly used is label number 425 or 525 ordinary Portland cement. A cement mixing ratio greater than 10% ensures the right amount of leakage protection.

The micropile is generally composed of a seamless steel tube and welded pipe. The diameter of the steel tube is 48–108 mm, usually 48 mm. Generally, the spacing is less than 1 m along the edge of the foundation pit or the side slope, and the depth of the soil is usually 1–3 m at the bottom of the foundation pit.

Apart from the slope of the foundation or the environment of the pit, different factors such as geological conditions, soil nailing wall with prestressed anchor, cutoff curtain, miniature pile of one or several kinds are also considered important factors. Here are some common types of composite soil nailing.

3.3.3.1.1 Combination of soil nail wall and prestressed bolt

This kind of supporting type is often used in the foundation pit which has certain requirements for deformation, and can effectively increase the stability of the foundation pit slope. Among them, the prestressed anchor often uses a steel bar or has a rod body. In general, grouting material for cement paste or mortar is placed at the end of the anchor cable. The waist beam connections and the pressure plates are set depending on the intensity of stress. The soil nailing and anchor cable are two different components, and the distribution is the main work force. In the real work process, the control system is set such that it is not too big and the anti-surface layer is broken as shown in Fig. 3.20.

3.3.3.1.2 The combination of soil nail wall and cutoff curtain

In areas with higher groundwater levels, when the foundation pit is surrounded by existing buildings, roads and important lines, in order to prevent the precipitation from causing subsidence damage, there is a great need to set up a water curtain. It is one of the main functions of a

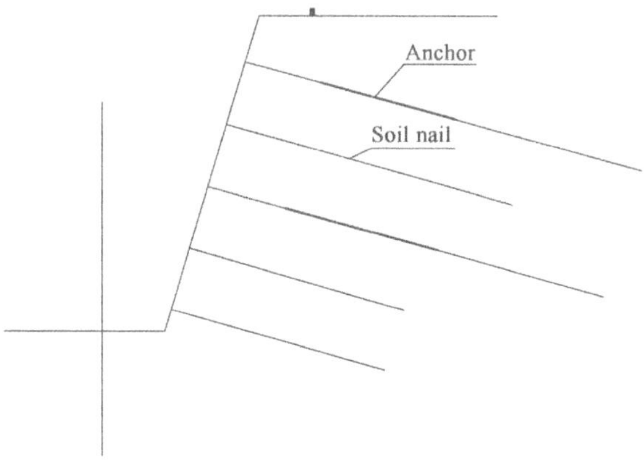

Fig. 3.20. Schematic diagram of combination of soil nailing wall and prestressed bolt.

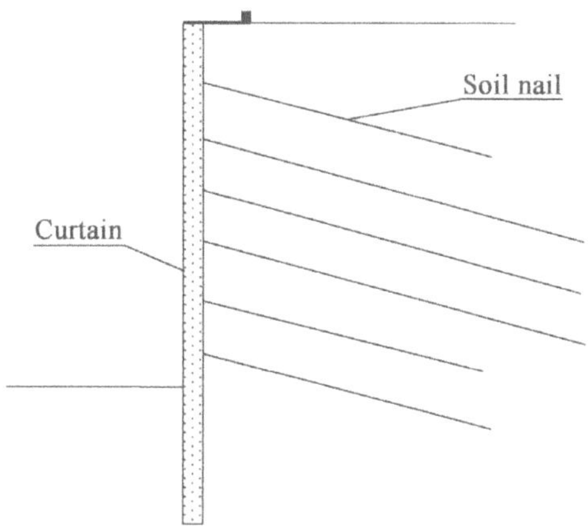

Fig. 3.21. Schematic diagram of the combination of soil nailing wall and cutoff curtain.

water-cutting curtain which also has other roles: the curtain can advance surface excavation and, at the same time, its role in the stability of the foundation pit slope is to be reckoned with, as it can also prevent pit surges where groundwater is abundant. To achieve the above effect, the strength of the curtain needs to be guaranteed. The construction of the curtain usually adopts a mixing pile, which has a certain length of overlap. It is also possible to use a spinning pile, which can be drilled into the gravel and pebble beds as shown in Fig. 3.21.

3.3.3.1.3 Combination of soil nailing wall, cutoff curtain and prestressed stroke

There is a need to control the deformation and increase stability, using a prestressed anchor under complex geological conditions, thus forming the composite form of that which is more commonly used. It is suitable for filling soil, mud soil and thick sand layers, under non-slope conditions and in deep foundation pits as shown in Fig. 3.22.

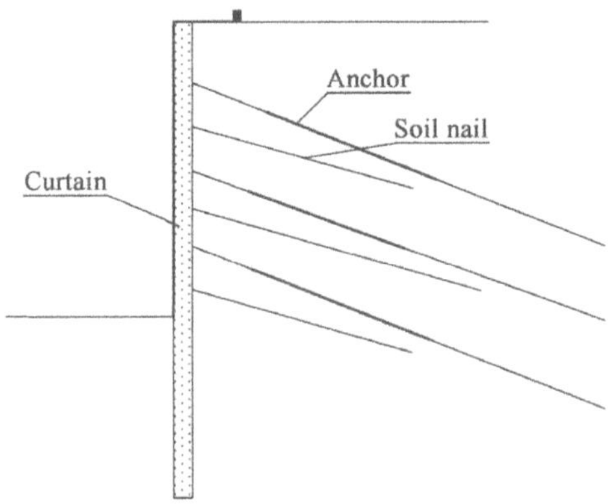

Fig. 3.22. Schematic diagram of the combination of soil nailing wall and water cutoff curtain.

3.3.3.1.4 Combination of soil nailing wall, micropile and prestressed pavement

The micro-column in the composite soil nailing wall support as the reinforcement to enhance the supporting stiffness is used for the vertical slope excavation of the foundation pit, as this kind of supporting form can increase uprightness to aid the excavation slope stability and provide deformation control, which is a very important role. The diameter of the micropile is less than the diameter of the drill hole, and the steel pipe or section steel is placed and the grouting is formed. After the strength reaches a certain extent, the excavation is allowed as shown in Fig. 3.23.

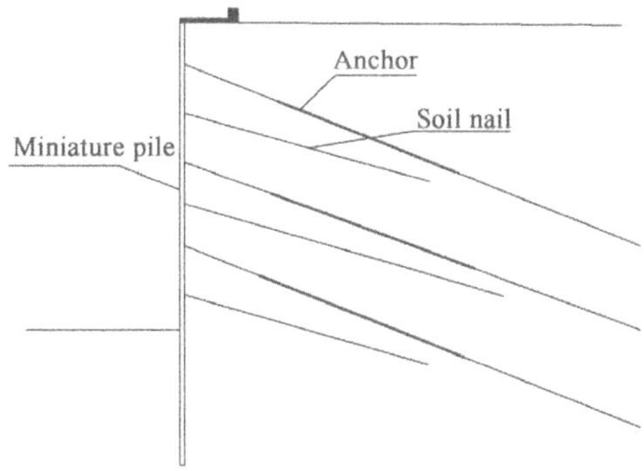

Fig. 3.23. Schematic diagram of combination support of soil nailing wall, cutoff curtain and prestressed bolt.

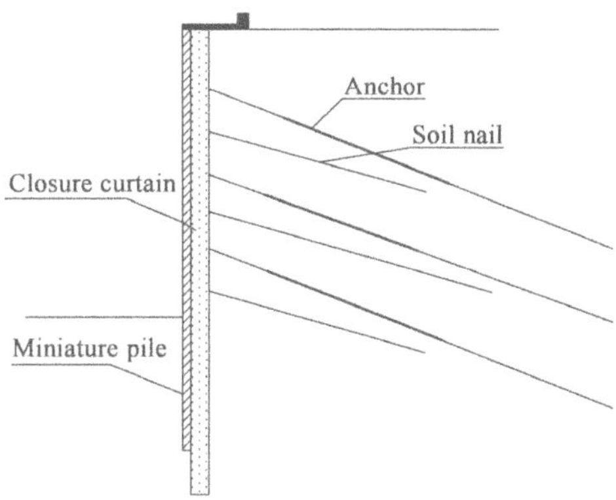

Fig. 3.24. Schematic diagram of soil nailing wall and cutoff curtain, prestressed anchor and micropile composite support.

3.3.3.1.5 Combination of soil nailing wall and prestressed bolt, cutoff curtain and micropile

This kind of support type is generally used in the following situations: deep foundation pits, poor formation conditions and strict environments. This kind of support type is more complicated than other composite soil nailing wall support forms due to the combination of many components. However, the supporting structure is stable and the deformation is small, which is suitable for supporting under complicated conditions as shown in Fig. 3.24.

3.3.3.2 *Construction process of composite soil nail wall*

The composite soil nailing wall working sequence is as follows: The work surface is opened to dig a bore along the clear-cut slope to install the soil nail reinforcement by spreading a grouting steel fabric shotcrete layer in the next step of excavation. According to the different soil characteristics and the supporting structure method, the sequence can be changed.

3.3.3.2.1 Excavation of working face

The soil nailing support shall be constructed according to the design requirements of the layered excavation depth according to the operation sequence, and the soil nails and shotcrete shall not be excavated at the next level before the completion of the upper working surface of soil nails and shotcrete. When the foundation pit surface is larger, one is allowed to excavate freely in the middle of the foundation pit from 8 to 10 m from the surrounding slope, but it should be in coordination with the excavation of the stratified work area. The support layer excavation depth and the construction of the work order should be ensured within the prescribed time after finishing the bare slopes, thereby maintaining the time table of the different stages of construction, namely setting soil nails and sprayed concrete being done in a timely manner. The excavation of the foundation pit in the horizontal direction should also be carried out in sections, which can be divided into 10–20-m sections.

3.3.3.2.2 Clear the slope

After the excavation of the foundation pit, the side wall of the foundation pit is suitable for cutting into a slope with a small machine tool or shovel to achieve the designed slope.

3.3.3.2.3 Hole position

The holes shall be determined, marked and numbered according to the design requirements before the soil is nailed into holes.

3.3.3.2.4 Hole specifications

According to the design requirements of the plane position, hole depth, lower dip angle and aperture selection of the drilling equipment, a manual hole is often dug using a Luoyang shovel. The hole diameter, hole depth, hole pitch and inclination must meet the design requirements, and the deviation value cannot be greater than that mentioned in the provisions in the technical planning of the foundation pit as per the soil nail support technology (CECS 96.97). In the process of forming holes, the holes should be recorded and the soil characteristics, hole quality and accident treatment should be recorded according to the soil nail number. The design parameters of soil nail should be modified in time when the soil body and the initial design are compared.

3.3.3.2.5 Hole cleaning

After drilling, the hole should be cleared, and the local seepage hole or loose soil in the hole should be treated immediately. Soil nail reinforcement and grouting should also be set up in time.

3.3.3.2.6 Installation of soil nails

The steel bar should be straightened, de-rusted and coated before use. In order to ensure the position of the steel bar in the center of the borehole, soil nail reinforcement of the manhole is done first by establishing stents and stent nails along a distance of 2–3 m, and using metal or plastic parts; the construction should not impede the free flow of grout when grouting.

3.3.3.2.7 Grouting

After the soil nail rebar is inserted into the manhole, it can be filled with heavy force and low pressure (0.4–0.6 MPa) or high pressure (1–2 MPa).

For a horizontal hole one should use a low-voltage or high-pressure grouting method. During pressure grouting one should ensure that the borehole mouth is stopped with a pulp (for piecewise slurry injection, check the plug in the middle position of the specified borehole), and pressure is maintained after 3–5 min.

For the inclined hole, as enforced by gravity or low-pressure grouting at the end of the grouting method, the grouting tube should be inserted in the lower end of the manhole; during the time of grouting, the tube should be slowly withdrawn at a constant speed, with catheter pulp flowing out of the mouth under the surface of the slurry, confirming the escape of gas from the hole.

An early strength and swelling agent should be added to the grouting, and the friction force of soil nail and pore wall soil can be improved by high-pressure injection.

3.3.3.2.8 Laying steel mesh

The steel mesh in the surface layer should be firmly fixed to the side wall to meet the requirements of the protective layer thickness. The steel mesh can be fixed with reinforcing bars in the earth, and no vibration should be seen in the concrete ejection.

The mesh can be welded or tied up, and the grid's allowable deviation is plus or minus 10 mm. The lap length of each side should not be less than one grid's side length or 200 mm; for example, the length of welding is not less than 10 times the diameter of the mesh.

3.3.3.2.9 Ejection concrete surface layer

When the shot concrete beam is ordered from bottom to top, the surface of the tube head should be controlled within the range of 0.8–1.5-m distance; the jet flow direction should be perpendicular to the chute surface, first in the reinforced area behind the bar, then filling the reinforced front, to prevent the occurrence of voids in the steel back. It can also be used before laying the steel mesh, and then after laying the mesh, it will not be less than 40 mm. It is necessary to pour water on the side wall before ejecting the concrete. During ejection, we should add accelerator to improve the concrete's coagulation speed and prevent concrete collapse.

In order to ensure the thickness of the shotcrete, the steel bar used for fixing the steel mesh in the soil can be controlled. When the thickness of the face is over 100 mm, it should be ejected twice, and the thickness should be 50–70 mm.

After the final coagulation of shotcrete, 2 h should be allotted for 5–7 days, for use of a tube curing agent, according to local conditions.

The next step of the concrete layer should be below the bottom of the foundation pit, and the depth should not be less than 0.2 m. At the top of the foundation pit, it is appropriate to set the tube concrete roof with a width of 1–2 m.

For the soil nailing wall support of appropriate construction, drainage measures should be taken to include supporting the internal water and foundation pit drainage, ensuring groundwater discharge and surface water drainage, to avoid soil saturation and ease of hydrostatic pressure acting on the surface. The water and drainage ditch can be made around the top of the foundation pit, and the drainage ditch and sump pit should be set in the pit, at a distance of 0.5–1 m that is reserved in the side wall, and the water in the sump pit should be pumped out in time.

If the water pressure on the side wall of the foundation pit is great, the length of the back plug of the supporting surface can be 400–600 mm while the diameter cannot be less than 40 mm. A horizontal guide pipe is used at the outer end of the support surface, at a spacing of 1.5–2 m, so that the concrete face layer is set after the water discharge.

3.3.4 *Compound anchor wall support*

3.3.4.1 *Concept of composite anchor tube wall support*

Anchor tube support is widely used in rock slope engineering. In the soil–rock combination found in the foundation pit, the anchor chute supporting the body has the limitation of the supporting type which cannot be fixed in the upper soil layer composed of loose, soft soil, with a plastic flow station under the conditions of rich underground water. In order to expand the application range of the anchor tube support, the conventional anchor tube support is modified, and the concept of generalized composite anchor tube support is proposed. A generalized composite anchor chute wall is composed of the soil (rock) anchor rod, shotcrete and prestressed anchor or forepole miniature pile and cement–soil pile combination, to solve the problem of deformation, self-reliance of soil, water and form of the supporting type.

3.3.4.2 *Type of composite anchor tube wall support*

The commonly used composite anchors are composed of the following combinations: non-prestressed bolt + prestressed bolt + tube; non-prestressed bolt + pre-supported micropile + tube concrete. Non-prestressed anchor + prestressed micropile + prestressed bolt + shotcrete; non-prestressed bolt + soil–cement pile + tube concrete; non-prestressed bolt + prestressed bolt + cement soil pile + tube concrete; non-prestressed bolt + prestressed bolt + reinforced water soil + tube concrete. In order to make full use of the slope space, the soil and water pressure in the supporting structure is reduced, and the soil pile and the micropile are often designed to be tilted at a small angle. Figure 3.25 is the type of composite anchor tube support.

The characteristics and conditions of the six composite shotcrete walls are discussed.

3.3.4.2.1 Non-prestressed bolt + prestressed bolt + shotcrete

When the horizontal displacement and settlement of the foundation pit are strictly required, the prestressed bolt can be used in the bolting support. By means of a non-prestressed bolt, a prestressed bolt and a reinforced concrete surface layer, the solid structure of the foundation pit is constructed. The prestressed anchor increases the potential slip surface of the soil foundation pit under the normal stress, improving the shear resistance and greatly reducing the slide force of the soil, effectively controlling the deformation; therefore, the supporting patterns can be applied to great depth, where the thickness of the soil layer is relatively small, the groundwater is not adequate and the rock mass stability is relatively good for the foundation pit.

3.3.4.2.2 Non-prestressed bolt + pre-supported micropile + shotcrete

The micropile is set in the back of the shotcrete surface layer, which is usually made of a grouting steel pipe with a leading vertical or inclined

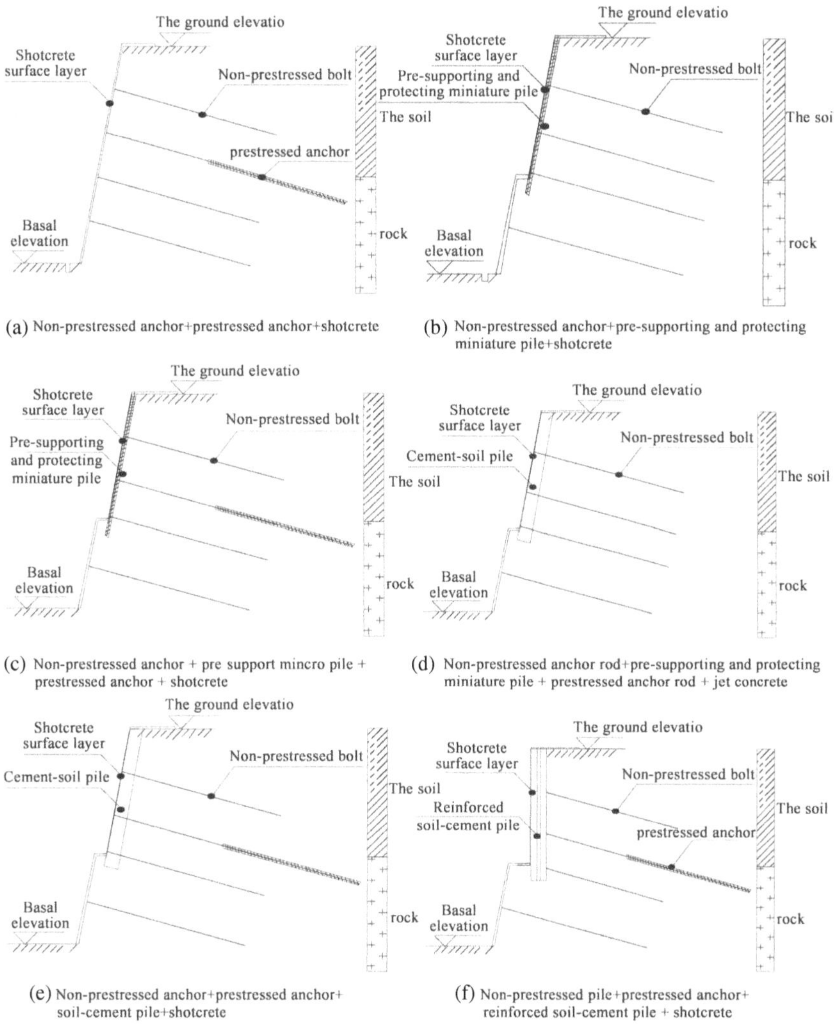

(a) Non-prestressed anchor+prestressed anchor+shotcrete

(b) Non-prestressed anchor+pre-supporting and protecting miniature pile+shotcrete

(c) Non-prestressed anchor + pre support mincro pile + prestressed anchor + shotcrete

(d) Non-prestressed anchor rod+pre-supporting and protecting miniature pile + prestressed anchor rod + jet concrete

(e) Non-prestressed anchor+prestressed anchor+ soil-cement pile+shotcrete

(f) Non-prestressed pile+prestressed anchor+ reinforced soil-cement pile + shotcrete

Fig. 3.25. Schematic diagram of compound anchor tube support type.

setting. The steel pipe has a small diameter and is convenient for fast construction. The function of the micropile is to solve the problem of self-reliance of the stratified rock mass before the excavation of the foundation pit. The large area of the micropile can support the weight of the finished shotcrete surface. This kind of support type is mainly applicable to a foundation pit where the slope is steep, the soil is loose or the rock is broken, the groundwater is not adequate and the self-reliance is poor.

3.3.4.2.3 Non-prestressed anchor + prestressed micropile + prestressed bolt + shotcrete

On the basis of the second type of support, the prestressed bolt is added. Widespread distribution of the root pile can not only support the weight of the upper shotcrete layer that has been completed but also strengthen the surface stiffness, which is conducive to the prestressed anchor rod or the spread of the force on the surface that is laid on the foundation for applying prestress. This kind of support form is mainly applied to the foundation pit with a great excavation depth, a steep slope and high displacement control, and loose soil mass or breaking rock, inadequate groundwater and poor self-supporting capacity.

3.3.4.2.4 Non-prestressed anchor + soil–cement pile + shotcrete

This kind of support type is firstly carried out before the excavation of the foundation pit; the construction of water soil pile is carried out before the curtain is completed and then the earth excavation is carried out along the curtain. The water soil pile is used to cut the well and make the advance supporting structure, and the soil pile can play the role of the above micro-type pile. This kind of support method solves the problem of self-reliance and anti-seepage of the layered rock mass before the excavation of the foundation pit. This support method is mainly applicable for the upper part of the water-bearing soft soil layer, as it can only be set in the upper soil layer, which requires the cutoff curtain to enter a certain depth of the relative aquifer. If the mixing pile cannot be realized, the spout pile is used. If the soil is not embedded in the foundation pit, a transitional platform should be set at the foot of the pile, and then the lower slope excavation shall be supported by the foundation pit.

3.3.4.2.5 Non-prestressed bolt + prestressed bolt + soil–cement pile + shotcrete

On the basis of the fourth type of support, increasing the prestressed bolt can effectively control the deformation of the foundation pit. Likewise, overlapping the cement–soil curtain can not only support the weight of the upper shotcrete layer that has been completed but also strengthen the surface stiffness, which is conducive to the counterforce role of a rock (soil) or prestressed anchor bolt on the surface after further proliferation that is laid on the foundation for applying prestress. This kind of support form is mainly applied where the upper part is the weak soil layer with high water content, the excavation depth is greater, the slope is steep, the displacement control is higher and the soil is loose and independent.

3.3.4.2.6 Non-prestressed bolt + prestressed bolt + reinforced cement soil + shotcrete

On the basis of five types of support, insert a worker steel core material such as a cement–soil pile in order to further improve the panel after prestressed anchor stiffness and prevent loss of prestress which might be too excessive to further reduce the displacement. This support form is mainly suitable for soft soil layer anchor prestress of a greater value, with water cutting, where excavation depth, slope steepness and displacement control are higher and soil is loose and independent.

These kinds of composite shotcrete wall support types are part of the concept of generalized composite shotcrete wall. In the narrow sense, the composite shotcrete wall is supported by non-prestressed bolts and prestressed bolts.

Advanced protection of the miniature pile and cement–soil pile is solved in the process of foundation pit excavation by step-by-step self-reliance of the layered soil, solving the water problems and the existence of the prestressed anchor, changed by the force of support by the rock bolt and shotcrete, thereby reducing the deformation. Of course, the influence of the prestressed bolt on the composite shotcrete is determined by the number of prestressed anchors and the prestressed value.

3.3.5 *The pile anchor support*

3.3.5.1 *Concept and structure of pile anchor support*

Pile anchor bracing refers to the slope protection with one or more anchor supporting forms, as it has a statically indeterminate structure, characteristics of good stability and high safety performance, it is an important support to deep foundation pits, and it combines the pile (anti-slide pile) support and bolt support methods.

The deep foundation pit pile anchor support structure is a kind of comprehensive slope protection technique which combines the construction technology of the drill pipe and the construction technology of the prestressed bolt (cable).

The pile anchor structure is composed of a retaining structure (pile) and an external pull system. Its soil-retaining structure with a cantilever-retaining structure is the same for the underground continuous wall and reinforced concrete pile, while the pile anchor-retaining structure mainly has the pile and anchor system in common. The fully mobilized formation has a steady ability at the same time, making the palisade structure as the load source. This has become a part of the retaining structure, thus satisfying the design of "new Austrian method" construction requirements. But because of the pull system, most of the weight placed by the anchor-retaining pile on the pull system is passed on to the anchor solid in the stable region, spreading from the load to the surrounding stable rock and soil layers, so as to give full play to the formation of the bearing capacity.

3.3.5.1.1 Pile supporting structure

The pile support structure is the supporting structure of the soil-retaining slope of the foundation pit. The design scheme of the pile row relates to the type of pile row, which mainly includes the artificial digging hole filling pile, drilling (punching) hole filling pile and precast pile. The detailed parameters of the system include the pile margin, the pile diameter, the silicon strength grade of the pile, the embedded depth of the pile, the type of steel reinforcement and the square measurements of the pile.

The pile design utilizes the following principles: The pile diameter should not be less than 350 mm. The distance between piles (center distance) shall be 2–3 times the pile diameter, and the length of the pile head embedded in the crown beam shall not be less than 50 mm. The crown beam should be connected with reinforced concrete. The width of the crown beam should not be less than the pile diameter, and the height should not be less than 400 mm. When the crown beam is connected with the beam, it can be reinforced by the structure. The concrete strength of piles and crowns is not lower than C20. In the foundation pit with a great excavation depth and poor soil stability, the pile diameter should be larger, the pile spacing should be smaller, but a smaller pile diameter and larger spacing could be selected. The connection point of the waist girder should be set near the support point, and should not exceed 1/30 of the support interval. The soil protection between piles can be composed of steel wire mesh shotcrete, bricks, wood and precast concrete slabs. In the case of low-level support, it is convenient to use steel wire mesh shotcrete and board construction, but shotcrete with wire mesh is most widely used. When water seepage occurs between piles, drain holes should be arranged on the protected surface. The main force member of pile anchor support is the structure of the pile and prestressed bolt.

3.3.5.1.2 Anchoring structure of bolt (cable)

The anchor (cable) structure refers to the structure of the support pile which improves the flexural capacity of the supporting pile. The design scheme is the layer number of the bolt, which is generally designed to be 1–3 layers when combined with the pile-supporting structure. The details of the structure parameters include anchor rod and anchor bolt angle setting depth of horizontal free length, anchor length, anchoring section diameter and anchor bar type.

A prestressed anchor rod (rope) is divided into a free section and anchor section; in the free section the scope of internal forces is the same, with the pressure on the anchor of the potential sliding body side, and part of the pressure on the beam waist, then passed to the anchor. Therefore, it is necessary to guarantee the structure of the anchorage as well as the anchor. The anchor-bearing structure is composed of the anchor head (tool), the gasket and the waist girder. One can transfer the prestress from the prestressed tendon to the pile surface. In the prestressed area, the prestress is transferred from the tendon to the supporting steel beam permanently. But often the variety of prestressed reinforcement determines the form of the anchor head, as anchor head is fixed in the anchorage by a tension lock. The anchorage of the anchor head basically has the following types: a locking cone-type anchorage of prestressed steel wire, extruded anchors for locking prestressed steel wires, such as XM anchor and OVA anchor, screw anchors for locking steel bar, etc. In the support engineering of the foundation pit, a prestressed wire conical anchor is generally used.

3.3.5.2 *The principle of pile anchor-supporting slope*

The pile anchor-supporting mechanism of soil pressure around the deep foundation pit, with groundwater pressure and additional loads such as the row of buildings surrounding the deep foundation pit, has a tendency to cause a collapse inside the deep foundation pit, producing relative lateral displacement. Deep foundation pit bottom pile is embedded within the scope of the soil depth due to lateral displacement of the pile in order to bear the influence of passive earth pressure against the pile.

In addition, the effect of the top of pile on the deep foundation pit due to the prestressing anchor (because the effect of lateral displacement of the pile will produce resistance) also prevents the pile from moving against part of the active earth pressure. Therefore, the active soil pressure of the supporting pile is borne by the passive earth pressure and the anchorage force. When the active earth pressure is less than or equal to the passive earth pressure and the anchorage force of the bolt, there is no lateral displacement of the enclosure pile, i.e., the supporting system is effective. When the active earth pressure is greater than the passive earth pressure and the bolt limit anchorage force, the enclosure pile will produce lateral displacement, and when the displacement exceeds the allowable limit, the supporting system will fail. The pile anchor supporting principle (anti-sliding pile) is a combination of the row pile and anchor rod (rope) supporting principles, wherein the sliding resistance in the block foundation pit slope slide mainly comes from the sliding force and anchor anti-slide pile with anchoring force.

3.3.5.2.1 Principle of slope protection of piles (anti-slide piles)

Pile-supporting structure mechanical analysis: For the vertical built-in pile in the soil of the foundation pit bottom, the lateral load is outside the foundation pit. For the lateral soil pressure, the critical state of the whole system depends on its self-balancing capability, solving the issue of maintaining the stability of a pile body at great depths and a pile body's internal force as the ultimate bearing capacity of pile support structure. Further, because the lateral load and elastic embedment reaction (passive earth pressure) both cause soil pressure, and they are related with continuous distribution, their variation law is more complicated.

Basic principle of piling support: The supporting structure of the pile is opposite to the gravity rigid-retaining structure, which can be considered as a kind of flexible retaining wall-supporting structure, bearing the huge load of the soil. The supporting surface layer generally includes the filling pile, the crown girder, the waist girder and the protective surface layer of the pile. The main effect of the girder is to enhance the overall stability of the perfusion pile and improve the stiffness of the supporting structure. The other part of the girder (crown girder) is used as the support point, and the pile part is divided into the prestressed anchor rod (cable) in the soil layer. The main function of the pile row is to bear the lateral pressure of the rock soil and transfer the lateral pressure of the rock to the pile body, and the partial pressure is passed to the prestressed anchor rod through the girder to limit the collapse of the rock mass. The joint action of the row pile, crown beam, waist girder and anchor is formed to support the whole and maintain the stability of the foundation pit.

3.3.5.2.2 Anchoring mechanism of bolt (cable)

In foundation pit engineering, the bolt (cable) as an anchorage system of foundation pit-supporting structure provides protection force (i.e., anchoring force), including for the bolts and grouting, the surrounding rock mass and supporting four elements on the surface. The system has three interfaces: the first is the interface between the anchor rod body and the grouting body, the second is the interface between the grouting body and the surrounding rock body, and the third is the interface between the anchor part and the supporting surface layer. The interface between the grouting body and the surrounding rock mass is the weakest link in the system. The anchor is effective and can offer much protection force for the supporting structure, generally depending on the bolt grouting body, the surrounding rock and soil body leaching junction of the friction (shear force) size and function.

The effect of prestressed anchor rod (rope) is exerting force in advance to cause a free period of elastic deformation of the reinforced anchor rod body; the slope protection pile (or even the beam, crown beam) is passed on to the soil, to prevent the excavation slope from sliding. In addition, the tension generated by the action fulcrum maintains the stability of the rear pile, which has to support the row piles and improve the stress state of the pile. Due to the use of the bolt, it can provide support to the pile bearing horizontal resistance, bringing the pile and the stratum chain together, forming a complex work, bringing its effectively under shear and increasing the sliding resistance of pile and slope-reinforced body. Each bolt is strictly distinguished; in order to free the anchorage segment, the anchoring is set outside the active critical slip surface of the soil, using pressure grouting, and for the anchor rod body a steel strand or fine rolling thread steel is generally chosen. For anchor prestress, the pile surface is set with enough stiffness of beam waist yu steel, channel steel or time to transfer the bolt-pulling force.

3.3.5.3 *The application scope of pile anchor support*

3.3.5.3.1 Applicability to pile anchor support technology

Pile anchor support technology is applicable to the I, II and III safety levels of the deep foundation pit excavation due to space limitation, and is also applicable to the miscellaneous the fill soil, residual soil, powdery soil, sand and weathered rock, pebble bed and other geological conditions, including foundation pit which is weak near the top and is surrounded with overload or residential buildings, and so on and so forth. The concrete filling pile is used as the foundation pit supporting structure, which can be selected according to the diameter, pile length, pile spacing and hole-forming method. The pile anchor support itself has spacing, which is simpler than the construction of the dense pile. Its advantages are good stiffness, a variable shape control system, excavation of foundation pit and anchoring of anchor rod, small impact on the working period and wide adaptability.

3.3.5.3.2 It is not applicable to pile anchor support technology

The use of the pile anchor supporting structure also has certain constraints; for example, the prestressed anchor rod or anchor shall be available for the soil layer, the prestressed anchor will be near the building foundation, the influence of the basement and underground pipe network will be great and the prestressed anchor should not be used when the liquid limit is greater than 50% of the soil and sandy soil relative density is less than 0.3. Compared with the soil nail and bolt support, the pile anchor structure uses a large amount of cement, and the underwater casting concrete is not easy to control. If the water level is higher than the foundation and the water volume is larger, the water stop effect is not as good as the continuous wall, which is not suitable for the pile anchor-supporting structure. The features of the pile anchorage method are that the cost is relatively high, the excavation depth of the foundation pit is relatively shallow and the stratum is stable; this method of support is not suitable from the economic aspect. For extremely complex geological conditions, especially in soft soil areas, the excavation depth is 12–15 m or more, the required security level is high and the deformation of the foundation pit engineering, in terms of safety and quality, is not suitable for the pile anchor support method.

3.3.6 *Pile-braced anchor combination support*

3.3.6.1 *The characteristics of anchor combination support*

In the design of the foundation pit support system, one or more supporting structures can be used in the same foundation pit. Due to the dual structure of the "upper earth lower rock" in the Qingdao area, the lower weathering and breeze rock can provide greater anchor tension. In the excavation of the deep foundation pit, in order to control the deformation and facilitate the construction, the anchor and inner support can be combined, and the supporting form of the anchor is adopted, as shown in Fig. 3.26.

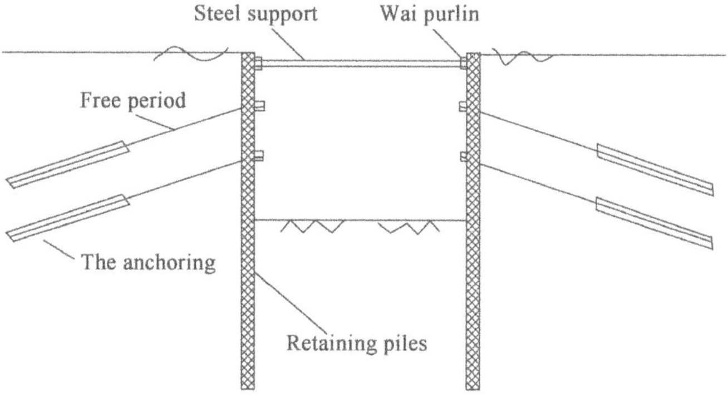

Fig. 3.26. Schematic diagram of anchor and inner combined support.

3.3.6.2 *The application scope of the pile one anchor combined-type support*

(1) It is suitable for the long and narrow foundation pit with strict deformation control and wide-open width.
(2) It is suitable for a deep foundation pit support project with few underground pipelines and no unknown underground features.
(3) Because the excavation requires large machines and makes special demands of the soil, such support can provide a large operation surface for easy construction. This kind of support type is especially suitable for the geological conditions of complex soil and rock stratum in the Qingdao area, because the lower rock mass can provide great pretension for the anchor cable (rod).

3.3.7 *"Foot pile" support*

3.3.7.1 *The concept and characteristics of "stilted pile"*

At present, due to the deep excavation depth of the foundation pit, it includes two different layers of rock–soil medium, namely, the upper layer is a certain depth of soil and the lower layer is the rock mass. The geotechnical binary structure of the foundation pit is special, because the embedded pile bottom is strong or composed of weathering granite. When the economic factors of the construction are taken into consideration, with the limited embedded rock depth and the foundation pit bottom being several meters under the rock, when the excavation to the basement is carried out, the supporting pile foot is hanging in the air, which is commonly known as the "out-seam pile". During design and construction, a certain width is often set aside under pile-embedded rock face or rock shoulder to support the pile foot, but because of the limitation of the architectural space, the reserved rock shoulder width cannot be too wide, and therefore, its build-in force on the pile foot is limited. In general design, a bolt is added at the foot of the pile to compensate for the insufficiency of the rock shoulder.

The "out-seam" pile-supporting structure can be combined with a brace or bolt, on the top with a pile anchor bracing for the foundation pit, and the department adopts advanced support micro steel pipe piles, so as to achieve the aim of saving lives and ensuring safety; this kind of support form on multiple cuts is used in the station.

3.3.7.2 *Application of "stilted pile" support in Qingdao Metro*

The "stilted pile" is a kind of foundation pit support type of the Qingdao Metro. This section describes the application of the "stilted pile" in combination with the envelope structure of the Qingdao Metro Line 3, Shuangshan station.

3.3.7.2.1 Project overview

Qingdao Metro Line 3 Shuangshan station is located at the intersection of Hefei road and Heilongjiang road, west of Heilongjiang road. The structure of the station is a two-storied two-stringed frame structure system, the total length of which is 251 m, the width is 18.8–20.8 m and the height is 13.2–14.4 m. The base plate is buried deep at a depth of 15.82–18.50 m, and the roof of the station covers the southern depth, the deepest part being at 4.10 m with the shallowest place at 2.62 m. The main structure of the station adopts the construction of Ming excavation, and the supporting structure adopts three forms as follows: bored cast-in-place pile + steel tube internal support + anchor cable mixed support system; bored cast-in-place pile + anchor cable system; and hammock pile + advanced micro steel pipe pile system. In this chapter, a new hybrid support system is summarized based on the geological conditions of Qingdao and the design experience of the local foundation pit.

3.3.7.2.2 Engineering geology

The station geomorphic type for the pediment accumulation slope is relatively flat terrain, with an existing ground elevation of 27.56–31.97 m and local depression height of 25.69 m. The upper soil layer in the fourth system of the site is as follows: one layer of artificial filling and alluvium, five to seven layers of silty exclusion soil and 12 layers of sandy soil (gravel sand). The underlying bedrock is Yanshanian granite, with strong and great weathering depth, and the burial depth of the breeze rock is

4.8–28.80 m. The soil layer in the fourth part of the site is relatively developed, and the local area is distributed with coarse gravel sand. The rocks are in a strong weathering zone, with poor water performance and groundwater is not adequate.

3.3.7.2.3 Foundation pit support scheme

After several demonstrations and trial calculations, the economic and engineering progress factors were taken into consideration, and finally different supporting combinations were adopted according to the different burial depths of the weathered rock. Weathering in the basement at a depth of 3 m above the out-seam pile is prevented by using advanced micro steel tube pile supporting form, upper out-seam pile of bored piles and steel support (or anchor cable). Piles in the lower part of weathered rock layer at a depth of 1.5 m are corrected using advanced micro steel pipe piles. The upper soil layer anchor force is small, and the surrounding environment does not allow long anchor cable construction, so the out-seam pile with steel support is closer from the rock at the bottom inclined anchor cable; in order to increase the construction space and for the sake of convenience, the out-seam pile bottom adopts the anchor cable.

3.3.7.2.4 Design of the supporting system of stilted pile + advanced micro-steel pipe pile

(1) Design principles
The geological structure of this section is the hard soil layer of the upper soft layer, and the weathered rock layer is about 3 m higher than the base, and the physical parameters of each layer are shown in Table 3.3. In this chapter, two independent foundation pit support segments are available for the supporting system of the hoisting pile and the advanced micro steel pipe pile. The upper out-seam pile is supported by piles + steel (or anchor cable) at the bottom of the weathered rock drilled pile of 1 m, and it uses the lock foot waist Liang Wending out-seam pile bottom, which can be thought of as an out-seam pile bottom anchor for the stability of the rocks, to aid their analysis and calculation according to the elastic fulcrum method. Advanced micro steel pipe piles at the bottom of the rock slope are

Table 3.3. Physical formation parameters.

Layer No.	Soil name	Depth (m)	Severe (kN/m³)	Floating weight (kN/m³)	Internal friction angle (°)	Friction with anchor body obstruction (k Pa)	Drilling cohesion force (underwater, k Pa)
1	Miscellaneous	3.90	17.5	10.0	15.00	15.0	0.00
2	Drilled soil	0.80	19.7	9.7	13.00	30.0	21.00
3	Drilled soil	3.6υ	19.7	9.7	10.00	20.0	17.00
4	Drilled soil	1.7	20.0	10.0	22.00	80.0	23.00
5	Spall	3.80	20.0	10.0	30.00	80.0	0.00
6	Strongly weathered rock	2.8	23.5	13.5	45.00	200.0	0.00
7	Intermediate weathered rock	1.80	24.5	14.5	55.00	350.0	0.00
8	Micro-weathered rock	20.00	24.5	14.5	65.00	650.0	0.00

due to hardness and good stability, the analysis and calculation of which are done using the limit equilibrium method. Rock bolt and shotcrete are used for support in micro steel pipe piles as stable construction measures of the lower strata, preventing damage to the lower strata from blasting excavation and mechanical drilling. The supporting cross-section of the suspension pile and the advanced micro steel tube is shown in Fig. 3.27.

According to the requirements of the relevant regulations and the actual conditions of this project, the design standards of the foundation pit are determined as follows:

(a) The deformation control protection level of the foundation pit is level 1, and the stability and deformation of the foundation pit are checked according to this grade. The important J insurance coefficient of the side wall of the foundation pit is 1.1.

(b) The envelope structure has to meet the stability requirements of the foundation pit without causing overturning, slippage and local instability. The supporting system does not lose its stability;

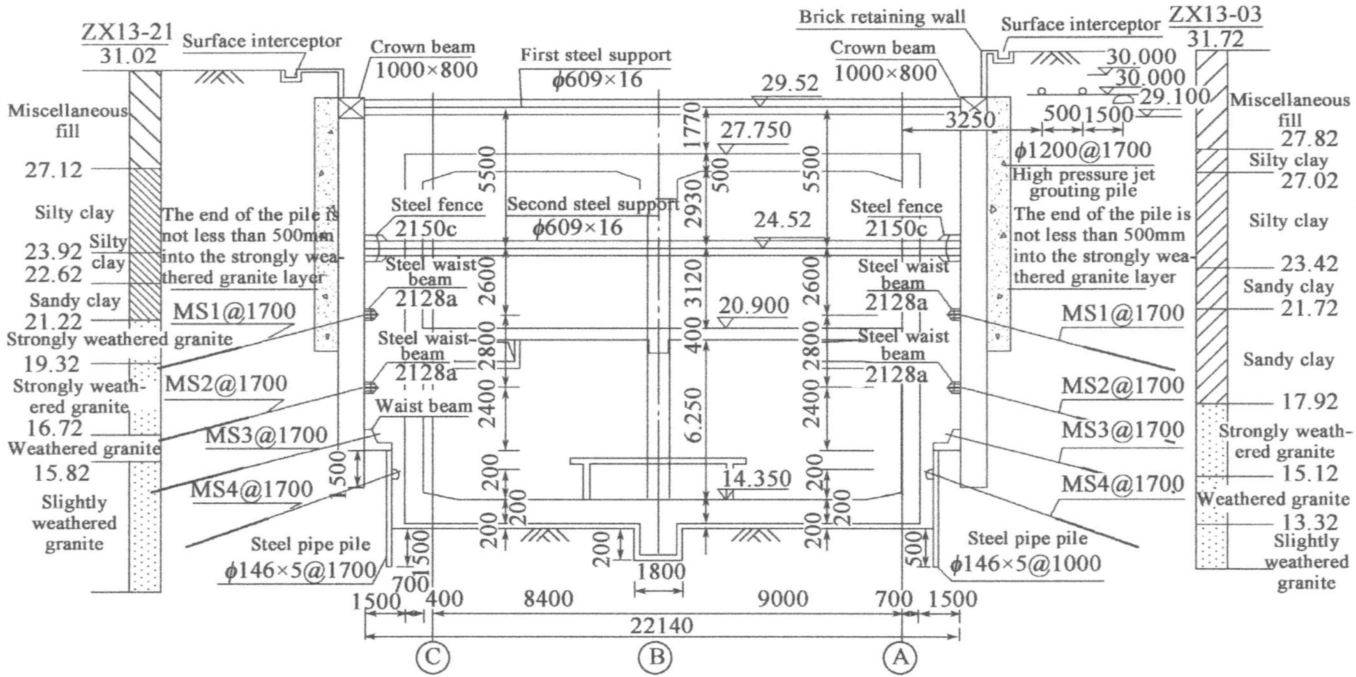

Fig. 3.27. The lateral cross-section of the hoisting pile and the leading micro steel pipe pile (size unit: mm; elevation unit: m).

the anchor cable and the girder are not damaged by the strength. The preloaded axial force in the steel tube is 40–60% of the axial force of the support. Anchor cable tension should be drawn to 105–110% of the design prestressing force, and then locked according to the regulation value.

(c) Land subsidence caused by construction should be controlled within the allowable limits of environmental conditions. According to the sensitivity of the surrounding environment and the distance of the underground pipeline to the transformer, safe and reliable measures should be taken. During the construction period, the subsidence of the ground around the foundation pit is 0.15%H (H is the excavation depth of the foundation pit), and the maximum horizontal displacement of the surrounding structure is 0.2%H, and both are less than 30 mm.

(d) As a foundation pit supporting structure during the construction period, the bored cast-in-place pile is considered to bear all the external earth pressure during the construction period. The bored piles are designed according to the strength control, and the crack width is no longer checked. However, the pile participating in the anti-floating action needs to be calculated by the crack, and the crack width is not greater than 0.2 mm.

(e) The standard section of the ground overload is calculated by 20 kPa, and no load is allowed within 2 m around the foundation pit.

(f) As per the foundation pit design, it can be used for 20 months.

(2) Analysis and calculation

(a) Calculation of upper suspension pile

The foundation pit depth is 15.50 m, and the supporting form of bored pile + steel support (anchor cable) is adopted. The upper part of the foundation pit is supported by two steel supports and the lower part by three anchor cables. In this chapter, analysis of the foundation pit is carried out, and the pile support model is used to simulate the stability and deformation of the foundation pit in the whole construction process, as shown in Table 3.4 and Figs. 3.28–3.30.

After calculation and analysis, the maximum horizontal displacement of the foundation pit is 19.39 mm, smaller than 0.15% H, the

Table 3.4. Material specifications.

Construction part	Gauge lattice	Species and classes	Limit strength standard value (M Pa)	Design value of steel compression strength (M Pa)
Steel shotcrete	Φ609 × 16	Q235	—	215
Anchor wire	1 × 7φ815.2	Steel strand	1860	—

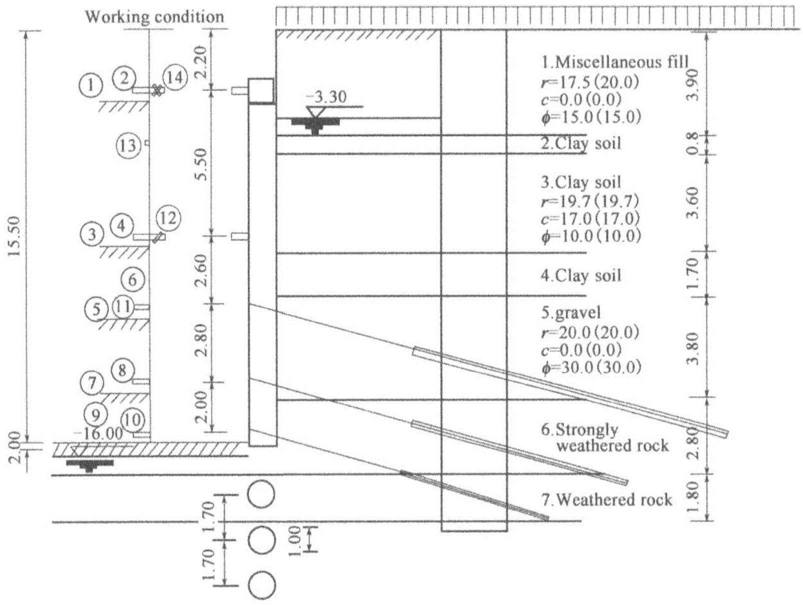

Fig. 3.28. Calculation model of suspension pile support (size unit: m).

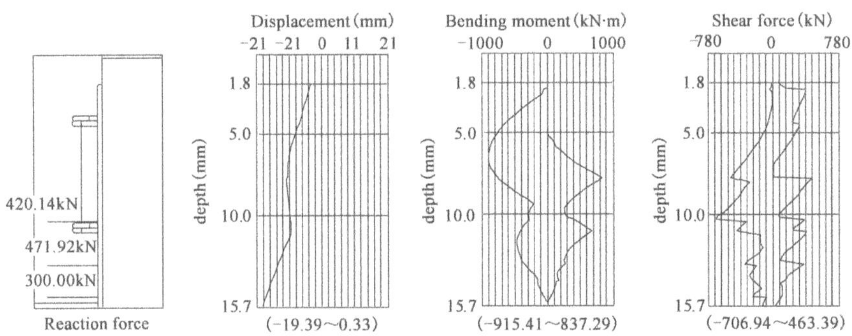

Fig. 3.29. Diagram of internal force and displacement envelope.

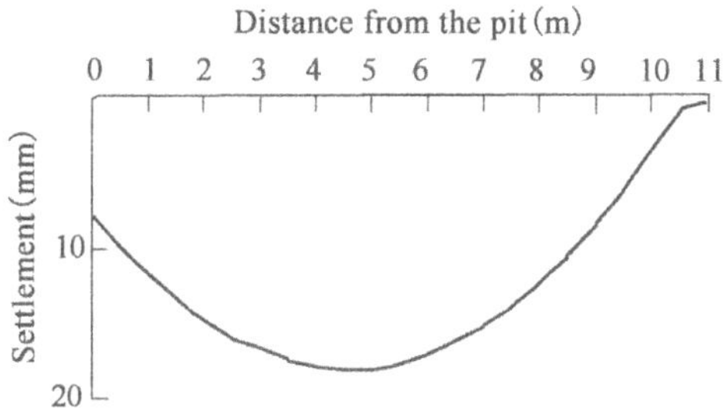

Fig. 3.30. Surface subsidence.

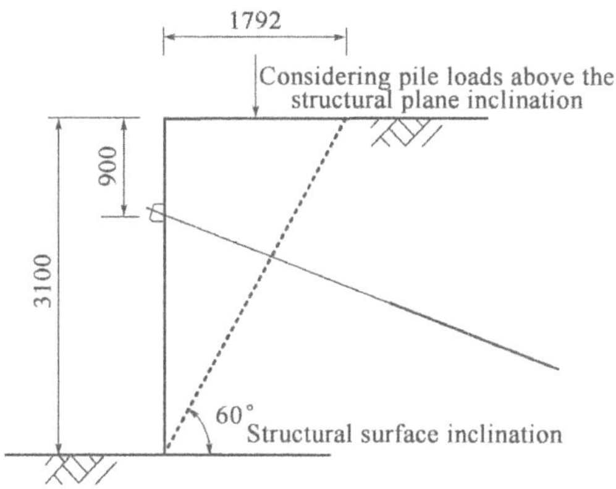

Fig. 3.31. Stability calculation model of rock slope (size unit: mm).

maximum sink value of the side surface of the foundation pit is 18 mm, less than 0.2% H, and the deformation value is less than 30 mm, meeting the deformation control requirements. The design values of the steel support-bearing capacity, the anchor cable tension and the shear force, and the bending moment of the filling pile are all controlled within the specified range.

(b) Calculation of lower rock slope

The calculated height of the lower rock slope is 3.1 m, and stability analysis software is used to analyze and calculate the plane sliding stability of the rock slope. In the calculation, the gravity of the suspended pile on the slope of the rock is applied as the external load on the side slope, and the calculation model is consistent with the actual situation as shown in Fig. 3.31.

Due to the low slope, and most of the strata being open to the breeze, the slope stability is better, so on the upper set of the vertical slope the anchor cable should be used to stabilize the outseam pile bottom soil and increase the anchoring force of the rock mass at the same time; through calculation, the slope stability coefficient is found to be 1.395, at the very edge of slope safety.

Through the above calculation, it can be seen that the results are safe and reasonable, and the foundation pit design meets the relevant requirements, which can be used as the design basis.

(c) Foundation pit support parameters

The support parameters of the foundation pit are as shown in Tables 3.5 and 3.6.

Table 3.5. Steel brace axial force.

Steel brace	Steel pipe specifications (mm)	Level interval (m)	Prestress (kN)	Axial force design value (kN)
The first steel bracing	φ609 × 16	6	550	1979
The second steel bracing	φ609 × 16	3	650	6238

Table 3.6. Anchor cable count parameter.

Number	Length of free segment (m)	Anchorage length (m)	Anchorage diameter (mm)	Lever pitch (°)	Steel strand specifications	Strength design value of steel strand (M Pa)	Pre-applied force (kN)	Pullout resistance design value (kN)
Ms1	6.55	12.5	Φ150	15	4φ′15.2 mm	1860	300	577
Ms2	6.55	8.5	Φ150	15	4φ′15.2 mm	1860	300	648
Ms3	6.55	5	Φ150	15	4φ′15.2 mm	1860	300	412
Ms4	6.55	5	Φ150	15	4φ′15.2 mm	1860	300	400

Fig. 3.32. Construction site of stilted pile + advanced micro-steel pipe pile support section.

3.3.7.2.5 Project implementation

Third-party testing data showed that out-seam pile + advanced micro steel pipe piles supported the 5.2-mm maximum horizontal displacement of the retaining pile top, the maximum value of settlement of the pile top was 4.0 mm, the maximum deflection displacement of the retaining pile body was 8.6 mm and the maximum ground surface settlement was 4.0 mm. In the process of surface subsidence, deformation data were monitored to ensure foundation pit supporting structure safety, as shown in Fig. 3.32. It can be seen from the monitoring data that the actual deformation of the foundation pit is very slight, as the foundation pit supporting design security coefficient is too large, and the existing formation physical and mechanical parameters of the old-fashioned supporting structure should be optimized.

3.3.8 *Summary*

This section focuses on the concept, type, characteristics, function principle and scope of application of the common supporting types of excavation foundation pit in the hard rock area. It mainly includes the following types:

(1) Slope excavation is suitable for the foundation soil, the excavation depth is not great and the construction site has enough place for slope engineering, as long as reasonable control in the work slope rate is shown, which can effectively control the construction into the soil and the structure stability.
(2) The soil nail wall support is formed by setting soil nails on the basis of the strength of the soil in situ. In the construction, it is necessary to follow the principle of stratified section excavation and edge excavation, with no over-excavation.
(3) Composite soil nailing uses one or several single supporting technologies or a water technology unit synthetic composite supporting system and, by a variety of combinations, forms a composite foundation pit supporting technology, which can make forepoles and support section water effects.
(4) The characteristic of pile anchor support is to replace the internal support of foundation pit with anchor bolt, to provide anchor tension for supporting row piles, so as to reduce the displacement and internal force of supporting row piles, and to control the deformation of foundation pit within the allowable range.
(5) Characteristics of the pile anchor support are the anchor rod instead of the foundation pit, supporting pile anchor tension, reducing the displacement and internal force of supporting pile and to the extent permitted to control the deformation of the foundation pit.
(6) The anchor support structure is combined with the advantages of the supporting structure of the pile, which can achieve the effect of a "rigid and flexible" structure.
(7) The hard rock geological conditions are given priority, the scheme of the out-seam pile is a typical form of construction and the upper part of the pile soft soil part adopts a supporting type, meeting the safety requirements for the lower part of the hard rock + steel pipe pile, in the form of anchor construction. With regard to investment and construction, the use of large force should typically be promoted.

(8) The construction quality of the prestressed anchor bolt (cable) is very important for structural safety. Construction should be strictly according to the plan and specifications, and a drawing test shall be conducted to check the quality of construction (double control).

3.4 Main structure construction of open-cut station

This section takes Qingdao Metro Line 13 Chaoyang Mountain CBD station as an example to elaborate on the main structure of the station. The CBD station of Chaoyang Mountain is the second floor of the metro line 13, the main body of the station is 225.05 m, the standard length is 22.7 m and the effective platform is 80 m. At the station roof, the depth of the two floors is 16.41–17.94 m.

The depth of the foundation pit is about 25.4 m. The station has to use the method of the smooth excavation. The roof thickness of the main structure of the station is 800 mm, the thickness of the medium plate is 400 mm, the thickness of the bottom plate is 900 mm and the lateral wall thickness is 700 mm. The station is the main body structure. According to the principle of construction, "longitudinal section, established to delamination, piecewise, parallel running water", the main structure's construction progresses alongside the excavation support, which in turn corresponds with resources for the construction. Construction shall be carried out on the first floor, rear wall and then middle board (roof). The standard section of the main structure of the station is shown in Fig. 3.33.

3.4.1 *Construction of integrated collision*

3.4.1.1 *Basement cleaning*

According to the survey lofting elevation control point in the process of controlling the excavation depth, excavation should be done to 300 mm from the pit design elevation, then mechanical excavation must be stopped and artificial stripping and leveling should be done, lest one overbreaks or disturbs the basal undisturbed soil. After excavating to the base, one can use the broken hammer to repair the individual under-excavated drum bag, so as to make sure that the basement slag is cleaned up manually after

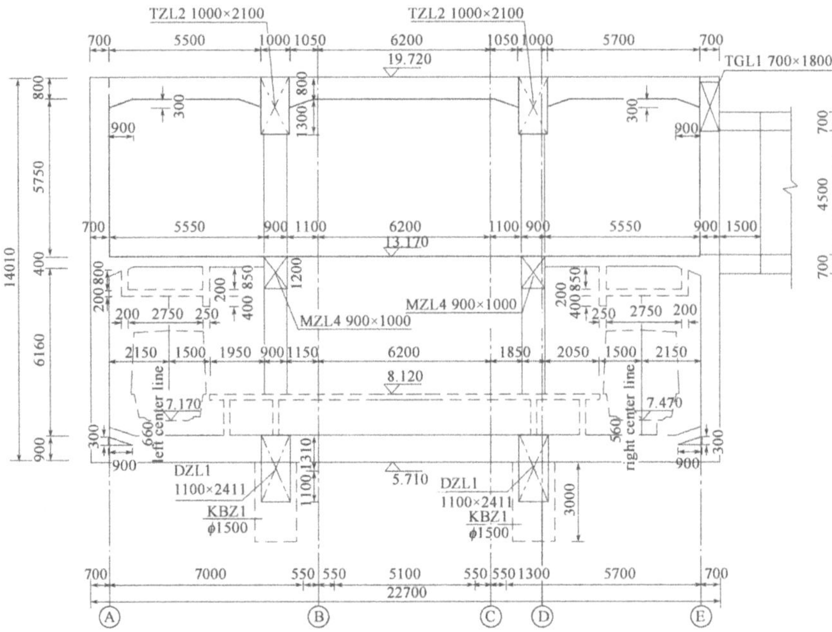

Fig. 3.33. Section of the standard station (size unit: mm; elevation unit: m).

excavating. It is strictly forbidden to level the base concave and convex parts with the empty slag. After cleaning up the slag, the supervisor, owner, designer and geological prospecting unit shall carry out the basement inspection, and the construction of the next part can be carried out through the rear. At the same time, the construction of the bottom layer is carried out, and the groove is dug under the longitudinal beam, and the external mold is repaired. Drilling in the vertical ground hole, excavating the horizontal earth groove and completing the other aspects of comprehensive grounding are done. When the ultradeep phenomenon occurs, the gravel pad is used. After the excavation is finished and the experience is qualified, the bedding construction should be carried out in time to prevent the base from softening.

3.4.1.2 *Grounding construction*

The station grounding construction technological process is as follows: measurement positioning is used to dig a trench to check a groove in the vertical grounding body or horizontal grounding body, laying an exothermic welding connection in the backfill. The construction process of station grounding is as follows: Measurement and positioning, trenching, trench inspection, driving into vertical grounding body, laying horizontal grounding body, exothermic welding connection, exothermic welding, grounding lead-in line, backfilling and compaction, detection of grounding resistance, protective treatment, waterproof treatment of water stop flange and intermediate water stop plate when lead-in line passes through the structural bottom plate, and setting of grounding terminal fixings.

3.4.1.2.1 The integrated grounding measurement is released

According to the ground floor plan of the integrated grounding grid, the comprehensive ground is measured and the line is measured. After the excavation of the soil and the base of the main structure of each section, the line is measured and a self-gray mark is used.

3.4.1.2.2 Trench excavation

After the completion of the measurement and positioning of the integrated grounding network, mechanical cooperation is carried out according to the design requirements.

3.4.1.2.3 Vertical grounding body

According to the measurement positioning mark, the vertical grounding body (50, L =5 m copper tube) is entered and drilled through a hole diameter of 120 mm. In order to prevent the slurry from being diluted, after the construction water has been pumped out, it can be put into the ground. Water is drilled in and pumped out.

3.4.1.2.4 Laying of horizontal grounding body and horizontal pressure belt

The horizontal grounding body is flat copper of 50 mm × 5 mm. The horizontal grounding body is 0.8 m under the backing layer and the longitudinal girder of the bottom plate. If the elevation of the bottom plate is changed, the relationship of 0.5 m should be maintained.

3.4.1.2.5 Welding between ground system components

The grounding system is welded by hot welding. The bonding length is not less than 2 times the width of the ground body and six times the diameter. Hot melt solder joints should not be naturally loosened. The welding points should have the same electrical conductivity as the welded conductor, the welding joints can withstand repeated surges of fault current without degradation and the resistance of the welding points is not affected by the time change. The copper content of welding points should be above 97%. After welding, the solder should be coated with asphalt and other anticorrosive materials.

3.4.1.2.6 Comprehensive grounding backfill

For the integrated earthing system section after laying connections, using the low resistivity of leaching to backfill poor soil compaction, each layer of the backfilled soil thickness is controlled to be less than 300 mm, using a small machine combined with poor artificial compaction.

3.4.1.2.7 Grounding wire installation

The grounding lead is not less than 0.5 m above the structural plate, and the waterproof concrete should be filled around the water stop ring (especially the lower part). The grounding line upward point (leads to the station floor) is located as follows: Derivation leads to a point inside the platform under the plate sandwich cable pit; near the platform layer is the strong I cable interlayer within the weak current equipment room, which avoids the rail bottom air duct, structure and orbital wall and so on. The substation is set on the platform layer (underground floor 2), and the outlet point of the power grounding wire is located in the cable interlayer within the range of the transformer. The outlet point of the weak current grounding wire should be located near the weak current cable well.

3.4.1.2.8 Connect the grounding wire to the mother row

The grounding wire can be applied to all the main completed structures after the platform board and the grounding bus; according to the requirements of the drawings on the platform near the plate girder side plane of the support wall or floor, the final wire will be grounded on a line fixed on the grounding bus, once the comprehensive grounding is completed.

3.4.1.2.9 Grounding resistance test

Comprehensive grounding is conducted in accordance with the division section on construction of a body structure. To assess the stage of completion, the integrated earthing system grounding resistance measurement is done, which is not greater than 0.9 Ω. If the electrical resistance value of the measurement is not desirable, the method of applying the drag reduction agent is adopted in the system.

3.4.1.2.10 Method of application of reducing agent

(1) Vertical grounding: First, excavate the trench and drill a hole with a diameter of about 120 mm.

Second, drain the hole with a water pump or a tube with a valve at the bottom to prevent the slurry from being diluted, and the vertical grounding body and the horizontal grounding body are welded. In the end, the slurry will be pressed from the pipe mouth until it fills the whole tube body and the reducing agent filling area. The dosage of the reducer is about 30 kg/m, and the vertical body should be kept in the central part of the filling area of the reducer. If mechanical slurry pump grouting is used, the pipe wall in the pipe wall of the lower part of the copper tube shall be staggered with a diameter of 10–15 mm.

(2) Horizontal grounding body: Only the horizontal grounding body of the ground network is applied to the dropping inhibitor. First, dig the trench and drain the internal water. Next, lay the horizontal grounding body and connect to the adjacent ground body as required. Finally, ensure that the level of the grounding body should be in the filling area of the drag reduction agent near the center (on level ground before grouting at the bottom of the body to make the necessary support), and the drag reduction agent quantity should be about 20 kg/m.

3.4.1.2.11 Construction requirements for reducing agent

(1) The filling material should be fully exposed to the grounding body when the material is filled with the reducer or the plain soil in the groove and hole of the grounding body.

(2) It is strictly controlled according to the requirement of manufacturer's product specifications.
(3) When the grounding network construction is carried out in a timely manner, the grounding resistance will be subject to strict specifications. For example, if applying a drag reduction agent does not meet the design requirements, afterward further drag reduction measures should be performed, such as the deep earthing, expanding networks, deep dozen vertical grounding electrodes and the backfill method, to increase the resistivity of the soil to reduce the grounding resistance. After the final grounding is completed, the joint test shall be carried out before the construction of the concrete pad. If the design requirements are not reached, the remedial measures should be taken in a timely manner.

3.4.2 *Construction of masonry structure*

3.4.2.1 *Construction process of masonry structure*

The masonry structure of this project is mainly the lateral brick wall of the main structure. The construction process flowchart is shown in Fig. 3.34.

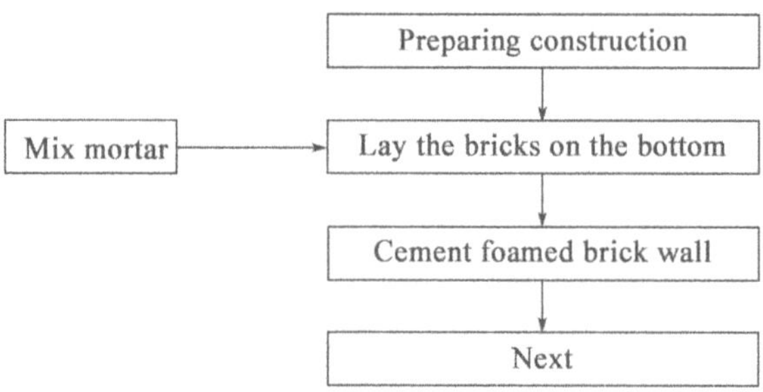

Fig. 3.34. The Construction process flowchart

3.4.2.2 *Construction method of masonry structure*

(1) **Constructing a wall:** Before the masonry building, the wall should be erected at the masonry position.
(2) **The bottom of the row of bricks:** According to the method of group building, starting from the corner or positioning to the side of the row of bricks, the inner wall should be lined with bricks, and as the

vertical side is the wrong lap joint, the upper and lower skins are staggered by 1/2 bricks. Where there are not enough half brick rows, ordinary bricks or stones are laid. Where there are more than required half bricks a saw tooth machining is used. To make whole bricks, the hewing method will be utilized using broken bricks, which shall not be used for the first skin brick masonry as it must be set.

(3) **Selecting bricks:** Bricks of 600 mm × 240 mm × 240 mm made of cement and gas are used. The appearance of the bricks is checked and they should have the edge drop angle and crack phenomenon, and must not be crisp. They are required to have a uniform appearance with no pressure on the surface.

(4) **Angle:** The corner should be set before the brickwork starts, and no more than 3 bricks should be used in each corner. If there is a deviation in the large angle of the new disc, in time to be suspended, it should be corrected in a timely manner. At the corner of the plate, the brick layer and elevation of the skin should be carefully controlled, as should the size of the mortar, so that the horizontal gray seam is uniform. After the large angle plate is rechecked once again, the horizontal and vertical angles should be completely in line with the requirements, and then the line wall can be hung.

(5) **Hang line:** For building such a wall hanging line, a tube line should be used, along with the setting up of several branch points, with middle string cable tension, and each layer of bricks is threaded flat, so that for the uniform flat straight horizontal seam on both sides of the roof takes care of the brick wall, laying a foundation for the next process to control the thickness of the plaster.

(6) **Cement and masonry:** In addition, it is suitable to use the slurry method. Vertical joints should be first fixed using batch mortar and then masonry. Bricks should be laid flat. The width of horizontal gray seam and vertical gray seam is generally 10 mm, but should not be less than 8 mm and should not be greater than 12 mm. Along the height of a one-step frame, it is appropriate to have two vertical lines in the t-brick position. Masonry mortar should be used as follows. General cement mortar shall be used within 3h, and cement mortar shall be used within 4h and shall not be used overnight. The wall should be plastered, and the residue from the masonry should be blown away by spraying water. The height should not exceed 1.8 m per day.

3.4.3 *Cushion construction*

(1) In foundation pit construction, it is key to pour out the concrete bedding. Construction should follow the excavation process block operation, that is, after the digging comes the pouring part. At the same time during the stage of foundation pit excavation, a small machine with artificial excavation is used for the basement above 30 cm, but mechanical digging should be prevented to avoid basal soil disturbance and destruction, and it is required that the water tank bottom surface should not be hard.

(2) Before pouring, the grounding wire should be carefully checked. The commodity concrete pump should be adopted to set the mold, along with vibrating compaction and continuous pouring.

(3) Because the base plate is directly constructed on the finished cushion, it is necessary to create conditions for the construction of the bottom plate.
   (a) Mechanical excavation shall be conducted as soon as possible, so as to avoid the disturbance of the original foundation without increasing the quantity of backfill and construction difficulty.
   (b) The constructed floor slab is more than 2.0 m, which can be used to facilitate the construction of the bottom plate.

(4) According to the elevation of the beforehand buried pile cushion thickness which is controlled, the design requirements are met with construction control maintained in a timely manner, thereby preserving the health and safety of the construction crew, and ensuring that cushion layers are without honeycombs, pits and cracks, with an allowable deviation, as shown in Table 3.7.

Table 3.7. Cushion allowable deviation.

S/N	Project	Allowable deviation (mm)	Frequency detection rate		Checking square method
			Scope	Number	
1	Thickness	−20– 30	Each construction section	≥4	Dipstick metering
2	Elevation thickness	−10 – 5	10 m	≥4	Level quantity

3.4.4 *Base plate construction*

3.4.4.1 *Construction sequence of floor board*

The cushion layer concreting is complete and after reaching a certain intensity, the lofting is measured, the bottom pad leveling and elevation are checked, and then tie plate steel is qualified after "three inspection" by casting slab concrete. According to the construction of the joints, the main project is divided into 10 sections along the plane. The overall arrangement is carried out from east to west. The construction process flowchart is shown in Fig. 3.35.

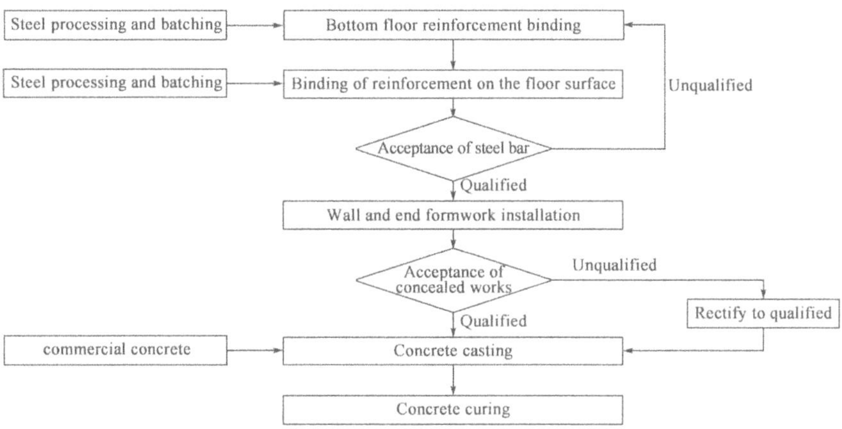

Fig. 3.35. Construction process flowchart of reinforced concrete construction.

3.4.4.2 *Construction method of floor boards*

(1) According to the design requirements, the bottom plate of the station is poured according to the construction joints.

(2) The surface is concreted with commercial concrete. Each time the floor is poured, two car pumps are arranged directly and the trap vibrator is adapted. According to the flowing direction of the concrete, in each of the pouring belts before and after arrangement of the two vibrators, all the way vibration actuators are used to cast the concrete work, in order to ensure the high quality of concrete. During the pouring period, the continuous supply of site water and electricity

and access to the road should be ensured. The on-site construction test personnel should be responsible for testing the concrete and the test blocks, and also for the collection of various technical data.

(3) Construction measures for the bottom plate

(a) For cushion concrete after reaching a certain intensity, measurement of internal structure lofting is conducted, elevation and smoothness of the cushion layer are reviewed, then the base plate reinforcement is installed and the internal structure of the stairs and the platform board are reinforced strictly according to the size of the loft. The following features are utilized: dislocation, reinforcement of construction, other specifications, quantities, requirement of design drawings, requirement of the welding plates, side wall and stray current steel medium plates, weld line drainage, pouring concrete, self-check reports, and supervision after the acceptance of the qualified report.

(b) The concrete shall be controlled in accordance with the design requirements and tested in the current setting at the site. The anti-pressure and anti-permeability test blocks shall be completed and the standard curing shall be carried out.

(c) A comprehensive review is prepared before concrete pouring and technical clarification, with a clear construction order for each team. The debris should be removed before the water and moisture affect it. During the construction, the layer difference shall be strictly controlled to prevent cold joints. The concrete for the bottom slab shall be pumped with commercial concrete, and the electric immersion vibrator shall be used for vibration. The vibrator shall be inserted into the lower concrete not less than 5 cm during the concrete vibration, and attention shall be paid to avoid missing vibration and over vibration.

(d) The reinforcement is strengthened, and the junction of the subdivision extends the vibration to 1.5 m, ensuring that the concrete is solid outside and controlling the relative settlement.

(e) Rebar and carpentry workers are encouraged to perform their duty and discover any problems in time so that they can be dealt with, ensuring normal construction conditions; when changing shifts, it is important that there is clarity and with no circumstances when a post is left empty.

(f) When the concrete work is finished, it will be covered with a plastic film, and when necessary, a straw bag or sack will be placed on it with water added; the curing time is not less than 14d.

(4) The floor longitudinal beam is constructed with the base plate.

3.4.5 *Construction of wall and end wall*

(1) Before the side wall is constructed, the waterproofing construction shall be completed according to the design. When leakage occurs at the joints, it is required by design and specification to stop the leakage, and a waterproof layer should be constructed according to the design requirements.

(2) The steel bar type as per the size and specification given by the design shall be welded with the reinforcement of the finished side wall and the end wall.

(3) The end wall structure, column structure and a small number of side walls adopt a wooden template, with a large steel standard section of the side wall, triangular steel truss support, and templates and scaffolds of sufficient strength, stiffness and lateral stability.

(4) Construction joints and deformation joints should be set according to the design requirements, ensuring that the template is stable and reliable, with no deformation and no leakage.

(5) After the reinforced installation is complete, welding should be checked by a qualified person before the supervision engineer is notified to check the junction structure for waterproofing and reinforcement, and inspect the embedded parts and stray current engineering. After passing inspection, the next procedure on the construction schedule can be carried out.

(6) The side wall and column concrete are layered and watered, and the height of each layer is not more than 30 cm. The concrete is poured continuously and is checked for leakage or over-vibration, so as to ensure that it is airtight and watertight.

(7) The well-constructed side wall should be reserved long enough to be firmly anchored to the side wall of the next construction. The reserved joint of rebar should meet the requirements of the construction. At the same time, the upper and lower sides of the wall between the construction joints should be installed with a water belt, constituting waterproofing treatment.

3.4.6 *Construction of floor, sheet and longitudinal beam*

3.4.6.1 *Building, roof and beam formwork and support system*

(1) The plate beam model adopts a wooden template, which uses the support of the mandrel steel tube bracket; the template and the support system are checked and calculated to ensure that the strength, rigidity and stability meet the construction requirements. In order to ensure a proper structure for the headroom, the plate and beam heights are increased by 2 cm. Along the longitudinal and transverse camber setting, the bulging height is set approximately to a degree of 1/1000–3/1000, which is tentatively 3 cm. The concrete arch camber values according to different span and site conditions can be adjusted a little.
(2) The model is set to the designed arch degree and the bracket is removed after the roof reaches the design and specification strength, so as to avoid sagging and cracking. During construction, the steel pipe, wood and bracket used in the supporting system are often inspected so that any hidden danger is eliminated in time.
(3) The steel bar is machined into the on-site rebar, and the steel bar is installed in the foundation pit.
(4) In this process, the concrete is used for symmetrical pouring. Cement is poured on the top floor for proper compaction, slurry and surface work.
(5) For the floor slab after pouring, the bracket can be removed after the floor reaches the designed strength due to the large span.
(6) In the construction of the steel bar, the air duct and the rail roof are preserved.

3.4.7 *Structural column construction*

(1) Construction of the structural column is done after construction of the floor.
(2) The column template and the support system use a wooden formwork, which is further supported by a column and steel pipes. The formwork and the support system are checked to ensure that their strength, rigidity and stability meet the construction requirements.
(3) The steel reinforcement is processed on the ground and installed in the foundation pit.

(4) The concrete model is poured, and the layered section is symmetrically poured according to the design elevation. In order to ensure concrete compactness, when necessary an attachment type vibrator can be used.

3.4.8 *Internal construction*

3.4.8.1 *Platform plate construction*

After the construction of the main structure is completed, the construction of the platform plate shall be carried out. The structure of the platform plate is constructed subsection by subsection, in accordance with the main body structure. The first part is the supporting wall construction and the second part is the plate construction. The construction method of the platform plate is as follows:

(1) A wooden template is used, the supporting system adopts a portal frame and the wall is reinforced with a wall bolt.
(2) A concrete-conveying pump is used to pour into a mold, with artificial hand-holding, and the vibratory rod performs the compaction.
(3) The steel bar is designed and processed in the processing shed to be transported to the site, and the embedded steel bar connecting the supporting wall and platform plate is welded.
(4) After the initial coagulation of the concrete slabs, the compaction is done, and it is covered with a plastic film after the final coagulation.

3.4.8.2 *The winding road and building*

A rail escalator is set up against the wind during the construction of the station. The stairs and escalator are arranged in the main structure after the construction using segments, utilizing the same construction method as the main body. Attention should be paid to the construction of the stairs: the concrete should be constructed from the bottom to the top.

3.4.9 *Summary*

(1) This section mainly introduced in great detail the main structure of the hard rock area and the main engineering points of station

construction technology for the Qingdao Metro Line 13 Chaoyang CBD mountains station, including comprehensive grounding construction, construction of the masonry structure, cushion construction, base construction, construction of the side wall and floor, roof and girder construction, construction of the column structure, internal structure and different structures, the construction methods and applied work processes. Positioning of the pile outside the water stop construction is critical. For the spin chute water stop curtain, two rows are suggested and coring inspection is also essential. The conditional field is chosen at the TRD curtain to stop water; alternatively, we can also choose occlusal pile water stop. In general, it is necessary to ensure the quality of the water stop, otherwise the leakage will be serious after excavation, and the safety of the foundation pit shall be highly uncertain.

(2) After the excavation of the foundation pit is completed, the supporting structure should be treated with water, and the basic principle should be to eliminate the bright water and create conditions for the construction of the waterproof plate.

(3) Using the "out-seam pile", a brace or pile anchor bracing structure in the foundation pit, one should control the quality while constructing the main structure, using plain concrete for its waterproof function, good flatness and solid basis.

3.5 Waterproofing drainage construction of open-cut station

3.5.1 *Mining and drainage construction technology*

The Ming excavation of the foundation pit during construction of the station is done to avoid sand flow and piping, prevent soil pit wall collapse, to ensure the construction safety and reduce the influence of the excavation on the surrounding environment. Further, when the excavation depth is greater and conducted within saturated soft soil layers near aquifers and there is a confined aquifer under the area, there is a great need to choose a suitable method for foundation pit dewatering and drainage. This section takes the Lingshanwei station of Qingdao Metro Line 13 as the background, and elaborates on the drainage construction technology of the open-cut station foundation pit. Lingshanwei station is about 191 m long and 45 m wide, and the excavation method is proposed for it. The main procedures involved in excavation of the station are the following: The 1st artificial fill layer, the 4th layer containing silt medium and coarse sand, the 6th layer containing silty clay, the 7th layer containing silty clay, the 9th layer containing medium and coarse sand and the 12th layer containing cohesive soil gravelly sand, locally strong and moderately weathered breccia tuff can be seen. The ground floor of the station was buried deep at about 17 m. During the survey, the depth of the water level was 1.7 mm, give or take ~3.8, and the thickness of the fourth layer was about 17.9 m.

3.5.1.1 *Precipitation well setting*

The foundation pit of Lingshanwei station is designed with 9 drainage wells at a well depth of 20.6 m.

The large length of the excavation section is designed with 21 precipitation wells, and a well depth of 20.6–29 m. In order to prevent the peripheral buildings and the ground, pipeline, sewer settlement and so on from flooding or sinking, the water level view log is set inside and outside the foundation pit. Figure 3.36 shows the layout of the foundation pit of Lingshanwei Station.

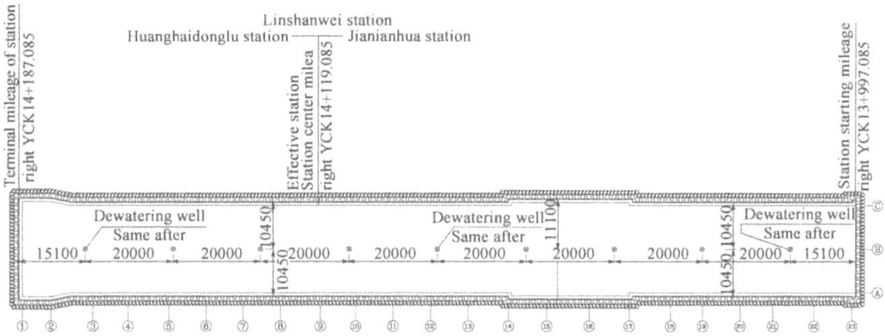

Fig. 3.36. Layout plan of the foundation pit dewatering well in Lingshanwei Station (size unit: mm).

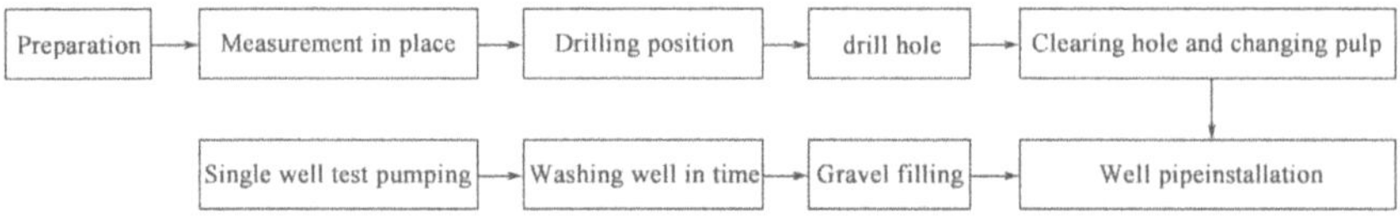

Fig. 3.37. Drainage construction process flow.

The dewatering well is divided into two stages: tube well construction and tube well pumping. Tube well construction involves socket installation and wash well drilling; casing well pumping involves pumping before earthwork excavation and after the excavation process. The drainage construction process is shown in Fig. 3.37.

The detailed map of the station pipe well is shown in Fig. 3.38.

Cross-sectional drawing of the foundation pit and dewatering well (1:20)

3.5.1.1.1 Preparation

(1) The construction site shall implement the "three-way one level" process and meet the requirements of equipment, facilities and entry and exit.
(2) Before drilling, preparations must be made for the following: Ensure good water supply (equipped with generator) and that the drilling machines are in good condition; the filter material, pipe and filter are required in good quality and quantity, according to the order of the supervision inspection. Pumping equipment (deep well submersible pumps or deep well pumps) and drainage system are required. Mud pits should be set up near the rig to prevent the flow of mud. Informational materials (precipitation records, construction records, etc.) are required as well.

3.5.1.1.2 Test and release well position

When the location coordinates and elevation measurement are taken into consideration, plane deviation should be less than 50 mm. If the well cannot be positioned due to obstacles and other factors, it can be properly displaced, but the maximum displacement is within 1 m, and the height of the bottom hole must meet the design depth requirement.

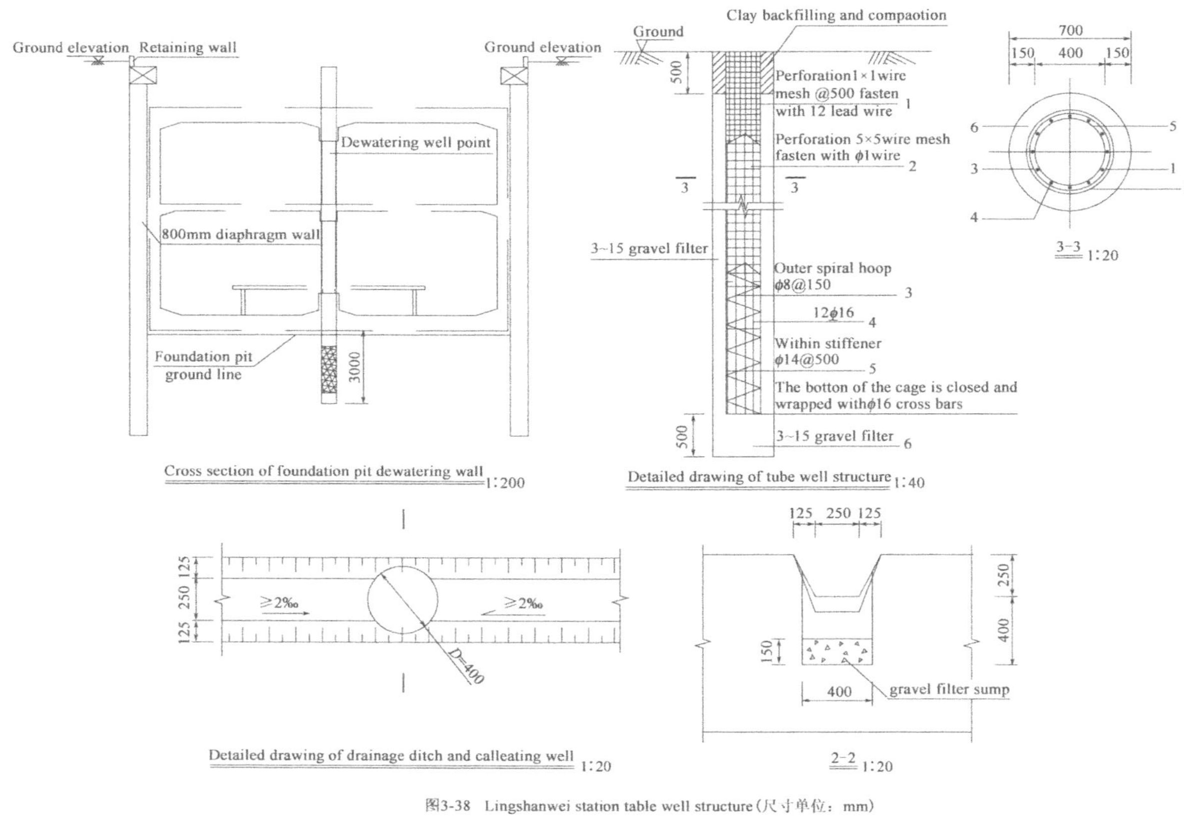

Fig. 3.38. Construction of Lingshanwei station pipe well (size unit: mm).

3.5.1.1.3 Drill position

The drilling rig must be stable and firm, with the hook, grinding disc and hole position in pairs. The horizontal error is calibrated with the flat ruler, and the median error is less than 5 cm. The site should be flat, and the foundation strong enough to prevent the drill from sinking. The quality inspector should fill out the opening inspection certificate and inspect the drilling tools before work begins.

3.5.1.1.4 Pipe well drilling

(1) The diameter of the drill shall not be less than 700 mm.
(2) The pump pressure must not be lower than 2 MPa and the pump flow cannot be less than 20 m^3/h.
(3) The diameter of the drill pipe should be adjusted to the diameter of the hole, so as to avoid damaging the drilling hole of the fine drill pipe.
(4) The drill pipe upright and do not swing during drilling.
(5) Mud reaming is adopted in the drilling, and the relative density of the mud is controlled within 1.15–1.2.
(6) The drilling depth is at least 1 m deeper than the design well.
(7) All landowners control the relative mud density to 1.10, control verticality within 1% and wellbore apex angle of tilt limit to 10°, after drilling to the designed depth and conducting an inspection of the end hole.
(8) In the process of drilling, one should observe and record water level, water temperature, the presence of washing liquid, signs of borehole wall collapse, sand flow and escape of gas.

3.5.1.1.5 Clear hole change pulp

(1) In order to prevent the mud from settling in the hole and ensure the depth of the pipe, the hole should be cleared in time after the final drilling, and the sand content should not be more than 5%.
(2) The mud in the hole should be diluted to relative density not greater than 1.08.
(3) No more than 30 cm of sediment can be deposited in the hole, and the sand content cannot be more than 5%.
(4) Only after the inspection is qualified can the well tube can be lowered.

3.5.1.1.6 Well tube installation

(1) The quality of the well tube should be inspected in advance, and any unqualified well pipe cannot be lowered.
(2) The bottom of the well is sealed with an iron plate. The upper part of the well tube should be at least 50 mm above the ground.
(3) The guide device should be used for drilling holes before installing the well.
(4) The bottom of the pipe is sealed with an iron plate, and the centering device of the tube is set up every four meters, which is required to be welded vertically. The steel wire mesh, steel cage and nylon cloth should be set according to the design requirements.
(5) It is strictly forbidden to force the well into the hole.

3.5.1.1.7 Packing

(1) The upper pipe of the well tube should be sealed before the filling.
(2) If the water source is put into the well, the filter material should meet the quality requirements.
(3) When filling the gravel, apply the spade evenly around the well to ensure that the gravel is even, compacted and filled to the top.
(4) The amount of filter material should not be less than 95% of the calculated value.

3.5.1.1.8 Wash well

(1) In order to prevent the mud from aging, the well should be cleansed immediately.
(2) Anti-stamping and piston joint should be used alternately to wash the well, to destroy the mud lining of the wall and to wash the permeation layer of the well.
(3) After completion of the well washing, the water content of the well water should be less than 1/10,000 (volume ratio).
(4) The sediment in the well shall not exceed 2 PI (2) Cm, and the speed of the hoisting and falling of the piston shall be controlled at 0.6–1.2 mI8, and it shall not be hard drawn.
(5) It is forbidden to drop a plunger into the well.

3.5.1.1.9 Pump pumping

(1) After completion of the well washing, the water should be pumped immediately, the water source should be connected and the water should be replenished until cleared.
(2) When dewatering a deep well pump using a submersible motor, the insulation of the cable connector must be safe and reliable, and it must be equipped with a protection switch control to ensure the safe operation; at the same time one should pay attention to the motor.
(3) A pumping test should be done after cleaning; the water level after the pumping test should be checked and the delay time is 8–16 h for water stability. A single well's water yield requirements shall not be less than 85% of the single-well calculation values after the overall pumping.

3.5.1.1.10 Pumping

(1) In both stable and unsteady flows, the pumping must be maintained in a continuous manner.
(2) If the water source has not broken into the well pipe, the well pipe filter shall not be blocked as a precondition, and a pump with suitable specifications (displacement and pipe diameter) shall be provided according to this principle.
(3) After the completion of construction of each pipe well, the pumping should be carried out separately.
(4) Deep well pumping equipment such as deep well pumps and deep well submersible pumps are often used, according to the different water yields and depths. After selecting a suitable pump for each well, one cannot arbitrarily change the pump.
(5) During normal pumping, adequate spare pumps should be provided to ensure continuous pumping.

The water should be pumped within 4–8 h, and then the water level measured.

3.5.1.2. *Design and precautions of drainage system*

(1) At the top of the foundation pit at Lingshanwei station, a drainage ditch is arranged along the surrounding structure, which is used to

carry water from the foundation pit and the ground rainwater. After the sedimentation tank, the sewer is drained. A water-retaining wall is built on the edge of the surrounding structure to prevent the surface water from filling the foundation pit. The precipitation near the foundation pit in the excavation section of Jialing interval is collected into the pipe well, and the drainage ditch is auxiliary. The vertical spacing of the four-wire large section is about 15 m, and the longitudinal spacing of the line is 20 m.

(2) Around the perimeter of the excavated area of the foundation pit, there are open drains at intervals of about 20 m, set near one set of wells, into which water seepage and the waste water of the construction area flow; there are also pump sewage treatment facilities from which waste water is discharged after standard treatment. The drainage ditch and the catchment well are deepened by digging, the bottom of the ditch is not less than 0.5 m below the foundation pit and the collecting well is less than 0.5 m below the trench.

(3) In order to stop the surface water flow from the foundation pit, a drainage ditch is set outside the excavation contour line of the foundation pit, a water well is set twice every 30 m and the water is discharged into the sedimentation tank after precipitation.

(4) Each collecting well is equipped with a water pump, which can be arranged in a row with the set, and the backflow of water is strictly prohibited. No less than 2 spare pumps are retained by the construction units for adequate drainage facilities during the rainy season.

(5) There is no sand concrete pipe well in the precipitation well. The hole diameter is 0.7 m, the diameter of the well tube is 0.4 m and the well depth is about 27.0 m. The thickness of the filter layer is 0.15 m, and the material selection of the filter layer should comply with the relevant regulations to prevent the removal of mud or sand.

(6) The bottom plate drain hole of the foundation pit of should be sealed after the backfill of the roof to meet the anti-floating requirements of the structural construction stage.

(7) Precautions must be taken for the use of well point systems.

    (a) The location of the precipitation well should be 1 m below the excavation surface of the foundation pit.

    (b) After the removal of the well point, the remaining water outlets shall be retained and applied as a cushion.

The drainage ditch in the foundation pit should be subdivided in front of the bedding layer to prevent the flow of groundwater into the trench which will destroy the foundation soil.

The well spacing should be adjusted according to the actual amount of water in the construction site. When the location of the precipitation well is in conflict with the position of the supporting temporary column, the position of the precipitation well should be adjusted.

3.5.1.3. *Construction requirements for precipitation wells*

3.5.1.3.1 Structural error of well

(1) The well diameter error is plus or minus 2.
(2) The vertical error is less than 1%.
(3) The well depth should meet the requirements mentioned in the text as well as in the well structure diagram.

3.5.1.3.2 Packing requirements

(1) In the aquifer section, the filter material should have certain psephicity, hydraulic gravel ≤3%, and flakes and needle stone chips larger than size 2–4 mm are strictly prohibited.
(2) To avoid uneven packing due to speed or deviation caused by filter tubes and filter material in the hole bridge, filter material should be promptly supplied after washing the well, with the actual packing quantity not being less than 95% of the theoretical calculation values.

3.5.1.3.3 Washing well requirements

(1) The washing well requires "clean water".
(2) After the completion of the pipe and packing of the equipment, the well shall be cleaned immediately and the interval between the completion and cleaning of the well shall not exceed 4h.
(3) It is recommended to use the isolation plug section to wash the well. If the well contains a large amount of mud, dredging can be carried out first and then the well is cleaned.

(4) When the conventional well washing is not good, the well can be used to soaked and rewashed.
(5) After the completion of the well, the mud in the hole is cleared with an air compressor, and the water in the well is completely cleared. The well washing should be carried out within 4h of the shaft completion, so as to avoid excessive time, and the protection provided by the mud skin gradually ages and can be damaged, affecting the seepage. After washing the well, experimental pumping can be carried out to determine whether the water level of a single well can meet the design requirements.

3.5.1.3.4 Pumping requirements

(1) It takes no less than 15d of advanced pumping time before the elevation of the underground water level is affected.
(2) Pumping of sand content is controlled to prevent the formation of fine particulate matter in groundwater causing land subsidence; hence, sand content must be pulled out of water: pumping sand content early $< 1/10{,}000$, the normal pumping sand content $< 1/50{,}000$.
(3) For the first time (before) after washing the well, the pumped sand content is tested and qualified, and time interval is not more than three months between the pumping sand content tests; there should be encrypted testing numbers for abnormal cases.

3.5.1.3.5 Maintenance of groundwater observation during precipitation period

(1) During the precipitation period, the groundwater should be observed and analyzed in a timely manner.
(2) If the water level changes rapidly in the local area, it is necessary to analyze the causes (such as the water pump damage, a sudden burst of water in the underground water structure or a rise of the water level in the region) and take corresponding measures.
(3) After the dewatering well construction, official measures should be taken before pumping static water levels and precipitation levels, which fall within the scope of design before drawdown has been

achieved. The frequency of observation shall be not less than three times a day; if the water level reaches the designated level or if the changes after drawdown are not much, it can be observed once a day.

3.5.2 *Excavation using the vehicle with waterproofing construction technology*

Waterproof engineering design and construction quality are not only directly dependent on the cost of construction and the ease (success) or difficulty (failure) of maintenance but also affect the project itself, its sturdiness and durability, which affects the normal operation of the metro engineering camp. This section takes the Qingdao Metro Line 13 two-river station as the background, and the open-cut station's waterproofing construction technology is elaborated.

Two-river station is an open-cut station, adopting the Ming excavation method of construction. The main body of the station utilizes the cast-in-place reinforced concrete box frame structure, which has a waterproof layer on the outside. The site has space for two groups of wind wells, a fire evacuation area and an outside waterproof layer; in addition to the local practices, suitable construction practices are adopted, including Ming construction. The roof of the main structure is thick with a soil thickness of 3.33–2.94 m, and the bottom plate is buried deep at 16.74–16.15 m; the roof of the wind tunnel is at a depth of 1.0–4.44 m; and the thickness of the roof is 0–4.48 m. A defensive water system of reinforced concrete structure is established, which is based on self-waterproofing of the structure, taking measures to control the cracking of structural concrete and increasing the anti-seepage performance of concrete. For deformation of joints, ensuring that construction joint seams are waterproof is the key point.

3.5.2.1 *Full package flexible waterproof scheme*

3.5.2.1.1 The waterproof scheme of the excavation method

The structure roof can adopt a 2.5-mm thick single-component polyurethane coating or any other waterproof coating. A waterproof coating layer of 4-mm thick linoleum insulation and a concrete protective layer of 70-mm thick fine stone are added. If there is a planting requirement for the structure roof, it should be added to the waterproof layer. An anti-root puncture layer of a 1.5-mm thick plastic waterproof panel is used.

The side wall of the envelope structure is made of non-black asphalt polymer self-taught film with waterproof rolling material (yps1.5 mm, GBIT23457 2009), and cement mortar is used for leveling.

The base plate is made of non-black asphalt polymer self-taught film with waterproof rolling material (yps1.5 mm, GBIT23457 2009), and a 50-mm thick concrete film is used as the protective layer.

The film thickness requirements of the waterproofing materials are as follows: the material thickness of the main body is not less than 0.7 mm, the thickness of the rubber is not less than 0.8 mm and coil thickness is not less than 1.5 mm.

3.5.2.1.2 Construction technology of flexible waterproof layer

(1) The construction process requirements of a single-component polyurethane-coated waterproof layer

The one-component polyurethane coating method is used for waterproof coating construction. Polyurethane waterproof coating products include polyurethane sealant vinegar and enhancement layer material. The reinforced layer can use 30– 40 g/m^2 of vinegar in a non-woven or fiberglass mesh cloth (16–20 mesh).

(A) Basic processing requirements

(a) After the concrete for the roof structure is poured, water compaction should be repeatedly collected.

Concrete surface should be at the base level (it is flat and is of 2 m and the clearance between the grassroots and ruler was less than 5 mm, allowing only a gradual change), solid, and free of leather, sand and oil pollution.

(b) The protrusion of the base surface is chipped from the root, and a squeegee is used for compaction with the poly-ammonia vinegar sealant at the cutting part. When there are pits on the base surface, it is then washed with high-pressure water. After drying in the tank, the compaction is filled with polyurethane sealant. When there is a crack of more than 0.3 mm on the base, a 10-cm joint should be coated with a 1-mm-thick polyurethane coating waterproof reinforcement layer and then immediately pasted as the reinforcement layer; finally the waterproof layer is brushed.

(c) All internal corner parts follow a 1:2.5 ratio of the cement mortar. These corner parts should make an obtuse angle corner that is of dimension 5 cm × 5 cm or make one or more rounded corners. All Yang angles should be made into 1 cm × 1 cm obtuse angles or should have one or more rounded corners. Angle range at the grassroots level should be smooth or be levelled off.

(B) Construction sequence and method of waterproofing

(a) After the primary treatment has been completed and inspection has been passed, a special base coat of ammonia vinegar will be applied first (a special thinner is used to dilute the polyurethane waterproof coating with a dosage of 0.15–0.2 kg/m^2). After the bottom coating is done, the waterproof coating is strengthened in the important parts such as the yin–yang angle and the construction joints, and the thickness of the layer is 1 mm and the width is 500 mm. After laying the waterproof coating to strengthen the layer, the layer coating on the surface is immediately strengthened to paste the enhancement layer, and finally the surface layer is brushed. It is strictly forbidden to apply the film waterproof layer to strengthen the surface and paste the reinforcement layer. The deformation joint is first applied to the reinforcing layer of 1-m wide butyl rubber rolling material, and then the coating waterproof reinforcement layer is applied according to the construction seam.

(b) The waterproof layer coats a large surface. The waterproof layer is coated with multi-channels (generally 3–5), and the upper and lower two coating brushes should be perpendicular. After each coating is dry, it can be used for the construction of the lower road coating.

(c) Polyurethane coating for waterproof layer construction and application of vinegar after acceptance should be done in a timely manner for the covering of the waterproof layer; the surface protection layer uses 7-cm thick fine stone concrete on the waterproof layer, covering a layer of III isolation linoleum or 4-mm thick PE foam plastic sheet isolation layer. The facade waterproof layer shall be protected by a PE foam sheet of a thickness not less than 6 mm or a masonry 120 brick wall. All plastic sheets are foamed 25–30 times.

(C) The construction requirements of roof waterproofing construction for the open-cut structure

(a) There is a green requirement on the surface of the local roof covering at the northwest corner of the station, and the root puncture layer should be set on the surface of the coating waterproof layer. An EVA plastic waterproof panel with a thickness of 1.5 mm is used to construct the root puncture layer.

(b) The setting range of the resistance of the plastic waterproof panel shall be over 5 m beyond the edge of the planting roof, and the insulation layer shall not be set up in the scope of the plastic waterproof plate resistant to the root system. However, the isolation layer between the plastic waterproof board and the ceiling is required to be at least 1 m.

(c) For grade I roof waterproofing, the structure needs to be on a slope of 1.5%.

(d) The width of the plastic waterproof board should not be less than 2 m, and the roll joints shall be welded with double weld joints, and then the gas will be tested after welding.

(e) The specific practice requires the testing of the EVA waterproof panel.

(2) The construction process requirements of the waterproof rolling material for the pre-shop are as follows.

When using the "outer protective paste" method to lay the pre-laid waterproof rolling material, the plane part is laid with a hollow paving method, and the facade is laid by a mechanical fixation method.

(A) Base processing requirements

The base surface of the waterproof layer should be cleaned up, and the smoothness should be satisfied: D/L is less than 1/50, D is the maximum depth between two adjacent convex surfaces and L is the minimum distance between adjacent convex surfaces. The concave and convex parts should be smooth. All the protruding parts that do not meet the above-mentioned requirements shall be chipped and used with 1: 2.5 cement mortar for leveling. The pit area is filled with 1: 2.5 cement mortar. The base surface should be clean, flat and firm, not loose, peeling and filled with sand.

(a) Any uneven parts shall be covered with a 1:2.5 cement mortar round and smoothened. When the base surface conditions are poor, a 400g/m^2 geotextile buffer layer can be laid.

(b) The base surface may be slightly damp, but there should be no clear water flow, otherwise it should be used for water blocking treatment or temporary drainage.

(c) For all the negative angles 1: 2.5 cement slurry is used at 5 cm x 5 cm of the obtuse angle or R greater than 5 cm, and the positive angle is 2 cm by 2 cm, or the round angle of R is greater than 2 cm.

(B) Waterproof layer construction technology

(a) First, the waterproof rolling material should be laid on the Yin and Yang corner as per the design requirement.

The reinforcement layer is used as a single-layer pre-laid rolling material with a width of 50 cm and the thickness of a single waterproof layer. Along 25 cm on both sides of the corner, strengthen the adhesive or stick method on the base surface. The large surface waterproof layer should be fixed on the reinforced layer surface.

(b) When there is a pipe through the waterproof layer, first the strengthening layer of the coil should be laid with the glue method to strengthen the layer fixed on the base, then the bedding face waterproof layer should also be stuck on the surface strengthening layer.

(c) When the waterproof layer is pre-laid, the insulating film near the side of the pile will not be torn off. The insulating film on the surface of the cast-in-place concrete structure should be torn off before the concrete is poured.

(d) The mechanical fixation method is adopted to fix the waterproof layer on the pile or cushion, and the fixed point is 2 cm from the edge of the coil, and the nail distance is not greater than 50 cm. The nail length shall not be less than 27 mm, and the waterproof layer shall be firmly fixed on the base surface with a gasket, and the diameter of the gasket shall not be less than 2 cm. Care should be taken to ensure that nothing falls off when the concrete is poured.

(e) The effective lapping width of the adjacent two coils is 10 cm (not including nail holes). The upper pressure is overlapped. In the lap joints, the diaphragm must be torn off. Short side lap joints should be staggered by more than 1 m.

(f) When the waterproof layer is laid, the reinforcement layer shall be laid on all construction joints and deformation joints. The width of the construction joints shall be 50 cm and the width of the deformation joints shall be 1 m. The strengthened layer and waterproof layer of the construction joints are stuck with adhesive, and any deformation cracks within 10 cm on either side of the isolation membrane on the surface of the waterproof layer should not be torn off (that is, the scope of the waterproof layer and strengthening layer paste).

(g) Once the floor waterproof layer is laid, one must get rid of the isolation membrane and coil and immediately pour a 50-mm thick C20 fine stone concrete protection layer; for the side wall waterproof layer, temporary protection measures should be taken to avoid damaging the waterproofing.

(h) The damaged part of the waterproof layer shall be repaired with the same material; a patch shall cover the damaged part, and the minimum distance between the patch and the damaged edge shall not be less than 10 cm. The patch adhesive layer should be oriented to the cast-in-place concrete.

(3) The method of overall waterproof at the entrance and exit includes the method of protecting the waterproof layer of the bottom plate at the entrance and exit and the method of protecting the waterproof layer of the top plate and the side wall at the entrance and exit.

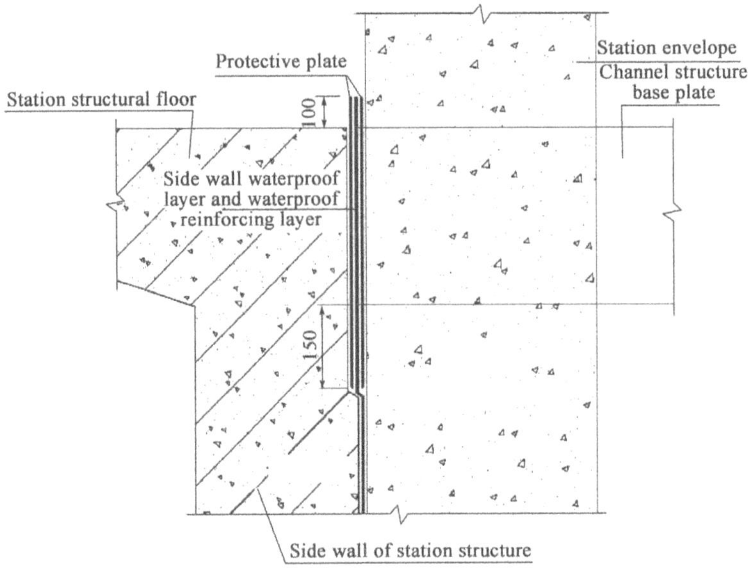

Fig. 3.39. Protection method of anti-water layer of population base plate (size unit: mm).

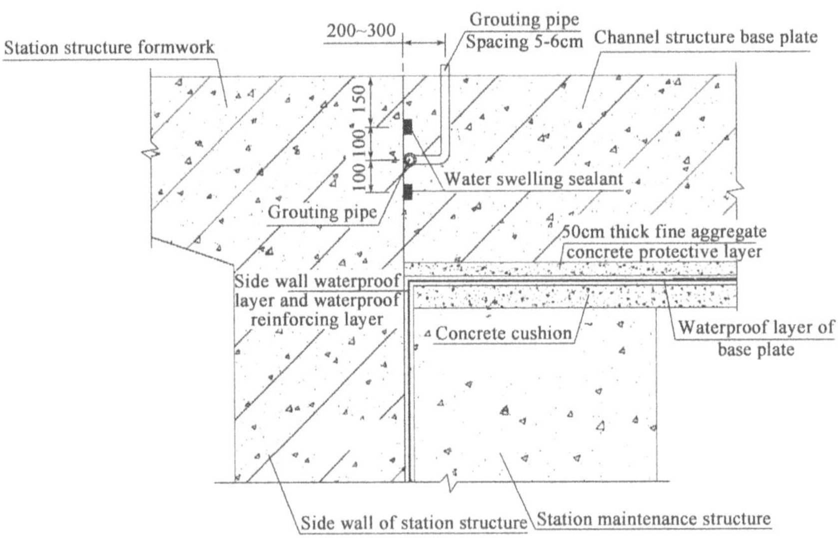

Fig. 3.40. Main body of the station and the waterproofing construction of the joint of the population base plate (size unit: mm).

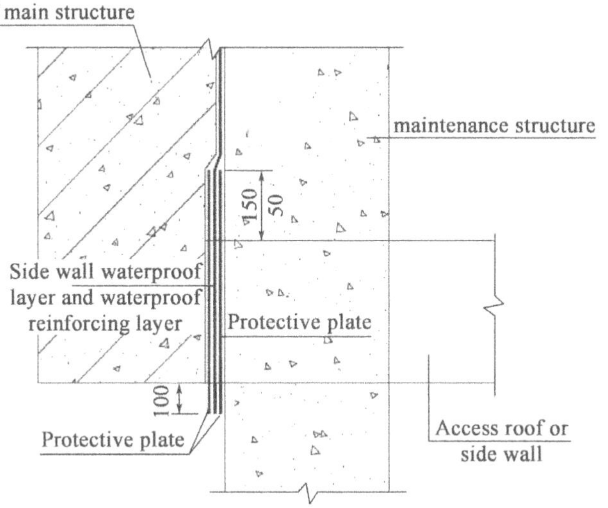

Fig. 3.41. The protection method of anti-water layer in the roof and side wall of the population base plate (size unit: mm).

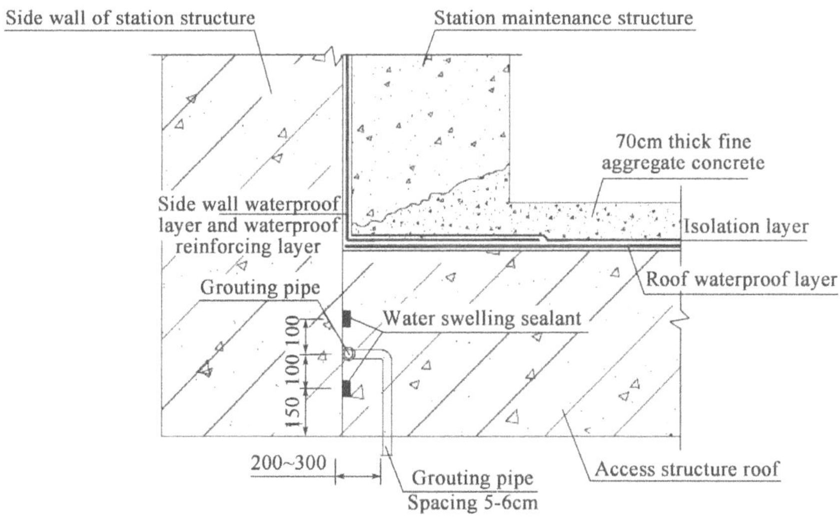

Fig. 3.42. Main body of the station and the waterproofing construction of the roof joint of the population base plate (dimension: mm).

3.5.2.1.3 Waterproof measures in important areas

Waterproofing of construction joints, deformation joints, post-pouring belts, wall pipes and pile heads.

3.5.2.1.4 Waterproof technical requirements for construction joints

(1) General technical requirements

The setting of the structure of the ring to the construction joints shall not be greater than 18 m. The longitudinal construction joints of the wall shall not remain on the wall with maximum shear force or the bottom plate and the side wall, which shall not be less than 300 mm on the surface. When the wall has a reserved hole, the hole edge of the construction gap should not be less than 300 mm.

(a) The surface of construction joints shall be firm and flat, and shall not be covered with slurry, oil, loose gravel, empty or crushed stone, etc. Otherwise, it shall be cleared.
(b) The ring should avoid the groundwater and fissure water, and should be combined with the deformation joint.
(c) The horizontal construction joints should be secured before pouring the concrete on the surface to avoid laitance and the surface should be sun-dried. Then the laid cement pulp is cleared using besmear brushes. Following this, the concrete interface is treated using suitable agents. Cement-based permeable crystallization type waterproof coating materials are spread to a thickness of about 30–50 mm with a 1:1 cement–concrete ratio (Fig. 3.42).
(d) The surface of vertical construction shall be cleaned before the concrete surface is laid and then coated with concrete interface primer or cement-based waterproof coating, and then concrete shall be poured in time.

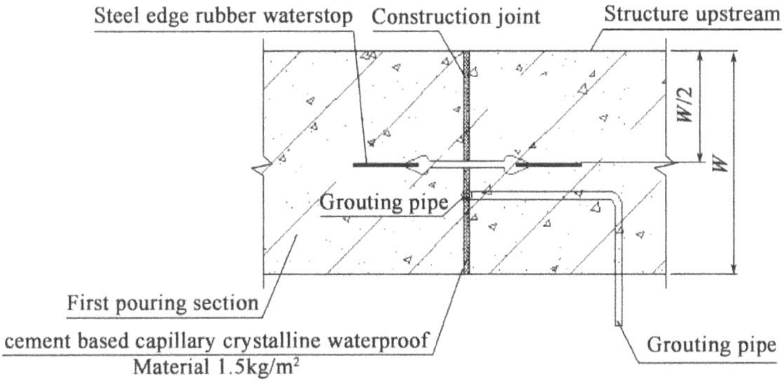

(a)

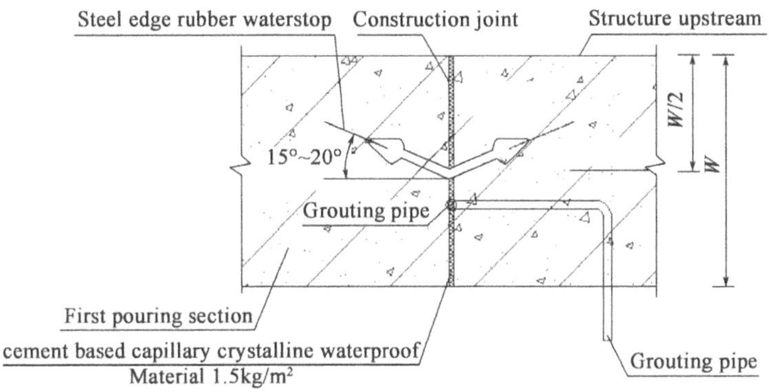

(b)

Fig. 3.43. (a) Lateral wall (ring) vertical construction joint waterproofing. (b) Roof (ring) vertical construction joint waterproofing.

When the design requirements are applied to the surface of the construction joints, the cement base penetrates the crystalline waterproofing materials without the other kinds of interface agents as shown in Fig. 3.43.

(2) The selection of a waterproof scheme for construction joints shall be determined according to the structure waterproof level and different parts, and the waterproof scheme shown in Table 3.8 is adopted.

Table 3.8. Waterproof scheme of construction joint.

Method	Waterproof grade	Circumferential construction joint	Longitudinal construction joint
Open excavation structure	First-level fortification	Copper edge rubber waterproofing tape, watertight adhesive	Copper edge rubber waterproofing tape, watertight adhesive
Special site construction joint	—	Double watertight grouting pipe	—
Floor construction joint	—	Single-stop strip	—

3.5.2.1.5 Waterproof technical requirements for deformation joints

(1) General technical requirements
   (a) The deformation joints shall be arranged in combination with different structural interfaces and intervals; the station, the location of the air duct interface and the changes in the geological conditions shall be considered, and the width of the deformation joints shall generally be 20 mm.
   (b) The thickness of the concrete structure at the deformation joints shall not be less than 300 mm, and the structure within 500 mm on both sides of the joints shall be treated with equal thickness when a variable section is encountered.
   (c) The water stop at the deformation joints shall be of a medium pass with a width of 350 mm.
   (d) When the structure is constructed, the concrete surface on both sides of the deformation joint shall be reserved in the roof and side wall, and the stainless steel plate shall be arranged in the groove to facilitate the timely drainage of the seepage. When the structural roof exists, the concrete should be pre-arranged in advance to avoid the formation of water on the back beam.
   (e) The embedding location of the buried water belt shall be accurate, and the central line of the hollow ring and the deformation joint shall coincide.

Table 3.9. Deformed joint waterproof scheme.

Method	Roof deformation joint	Side wall
Open excavation structure	Facing water surface joint	External water stop belt
	Medium buried steel edge rubber water stop belt	Medium buried steel edge rubber water stop belt
		Back surface joint

(2) The selection of the waterproof scheme of the deformation joints shall be determined according to the grading and different parts of the waterproof structure, and the waterproof scheme as shown in Table 3.9 should be adopted.

3.5.2.1.6 Waterproof technical requirements of wall tube

(1) The wall tube (box) shall be embedded before pouring concrete.
(2) The distance between the wall tube and the inner wall angle and the concave and convex part should be greater than 250 mm.
(3) The wall tube can be used according to the deformation amount, the fixed waterproof method or the casing-type waterproof method.
(4) If the wall tube protrudes out of the outer wall, measures should be taken to ensure that it does not damage the pipe body when backfilling.

3.5.2.1.7 Waterproof technical requirements of post-pouring belt

- The post-pouring belt should be used for the construction site where deformation joints are not allowed.
- The post-pouring belt shall be constructed after the concrete age of both sides reaches 42d.
- The post-pouring belt should be made of compensating shrinkage concrete, and its anti-seepage strength should not be lower than the concrete on both sides.
- The post-pouring belt should be located in an area with less stress and deformation. The spacing should be 30–60 m and the width should be 700–1000 mm.

The casting requirements: (a) The casting plan should be formulated before pouring and whether the setting of the post-pouring belt conforms to the design requirements, and the pouring site should be cleaned up. (b) In case of rain, snow or hail during construction, the new concrete part shall be covered with plastic film immediately. When the concrete has hardened, it should be covered with a 30–50-mm thick cement mortar with no coarse aggregate and then concrete. (c) Before pouring, the concrete surface should be cleaned and thoroughly wetted. (d) Horizontal pressure shall be applied by mechanical or artificial means before the final coagulation, and the concrete surface shall be subjected to repeated pressure.

3.5.3 *Summary*

This section mainly introduces the main points of the construction in the hard rock area, including the waterproofing construction technique and the drainage construction technology.

(1) Qingdao Metro Station No. 13 uses waterproofing construction technology for the foundation pit, mainly introducing the full package flexibility for the open-cut station, waterproof schemes, waterproof material and waterproof measures for important parts. For the waterproof project construction, the location is the key to control the waterproofing process as waterproofing is very difficult, so one has to formulate feasible technical guarantee measures and supervise implementation according to the requirements, while ensuring the construction quality of the waterproof layers and effectively improving the stability and security of the metro operation.
(2) Qingdao Metro Line 13 Lingshanwei guard station uses draining technology from the construction of the dewatering well set, the draining process to the drainage system design and considerations, construction requirements and other aspects of dewatering including the Ming excavation drainage. The foundation pit draining control work is not only confined to controlling the dewatering well construction quality but also many other aspects such as adhering to the process, while adopting other flexible measures for auxiliary

draining. At the same time, it is necessary to comprehensively analyze the situation by means of monitoring data and field tests, so as to ensure the safety and construction quality of the foundation pit.

(3) Waterproof board operations should be performed under the conditions of no water in order to guarantee the construction quality. Further, proper management of operating and management personnel should be adopted such as utilizing a real-name system, setting up a traceability quality system, etc. A waterproof board of polymer from coil should also be chosen.

(4) It is also very important to control the quality of the roof and the outer part of the structure, especially the main body and the settlement joints of the subsidiary structure, and all the joints should be strengthened.

(5) In case of seepage, the principle of curing and the system of the grouting should be adopted; the space between the main body and the supporting structure should not have any cracks.

(6) Cracks in the structure are a common quality fault in open-cut stations, and system research should be strengthened in terms of concrete material selection, section length, time control and maintenance to ensure quality.

Bibliography

[1] Liu Guibin, Wang Wei dong. *Foundation Pit Engineering Manual*. Beijing: China Building Industry Press, 2009.

[2] Zhang Yuanlin. A study on the construction technology of metro Ming excavation. *Engineering Construction and Design*, 2017 (03): 140–142.

[3] Tang Zhongxin. Research on the waterproofing construction technology of metro mining station. *Jiangxi Building Materials*, 2017 (13): 156–157.

[4] Li Yuanxun. Analysis method and experimental study on the internal force and deformation of deep foundation pit pile anchor support. *Lanzhou University of Technology*, 2016.

[5] Yuan Hai. Research on the envelope structure of the metro station in Qingdao. *Qingdao University of Technology*, 2012.

[6] Zhang Shizhen. Waterproof technology for metro station construction. *Science and Technology and Enterprise*, 2016 (08): 141–142.

[7] Zhu Danhui. The use of the steel pipe pile in the ground iron foundation pit with the anchor pile and the leading micro-type steel pipe pile is used. *The Iron Track Standard is Set*, 2014 58 (05): 90–94.
[8] Zhang Ping. Design and application of anchor support structure for deep foundation pit construction. Guizhou University, 2015.
[9] Ren Yanjun. Construction technology of soil nailing wall support and its application in foundation pit construction. *China Construction*, 2015 (03): 146–147.
[10] Song Zhengping: The current situation and development of deep foundation pit support technology. *Technology Innovation and Application*, 2014 (30): 263.
[11] Duan Jian. Construction technology of foundation pit pile anchor support. *Chinese Residence* (Next Issue), 2014 (10): 308.
[12] Liu Guilin. Application of pile one anchor compound support technique in deep foundation pit construction. *Jiangxi Building Materials*, 2014 (16): 46.
[13] Yu Xuebin. Design and construction of waterproof engineering in metro stations. *China Building Materials Technology*, 2014 (04): 96–97.
[14] Xu Junxiang. Application and control of pile anchor support structure in deep foundation pit engineering. Tianjin University, 2014.
[15] Yang Xu. Construction technology of water proofing construction in metro station. *Science and Technology Innovation Guide*, 2014 (05): 41–42.
[16] Li Guilin. Numerical analysis of anchor cable support system in deep foundation pit in the metro station of tuyan composite formation. Ocean University of China, 2013.

Chapter 4

Subsurface-Excavated Station Construction Technology

4.1 Overview of subsurface-excavated station construction technology

4.1.1 *The development of subsurface excavation*

The station is an important part of the metro. For the advancement of metro construction, the choice of and innovation in the excavation method become important research topics. The construction methods of metro stations at home and abroad have mainly included several different types such as the Ming excavation method, the dark excavation method and a light and dark combination. The method is most commonly used, which has the advantages of convenient construction and good work quantity, but due to the immense site requirements and various other factors, its application has been gradually restricted. In recent times, in order to reduce the influence of metro construction on urban traffic, on the basis of the Ming excavation method, the subsurface-excavated method was developed. The subsurface-excavated method mainly covers uniaxial digs. In this method, after completion of the structure of the roof transportation is back to normal, the station construction is through shaft or slope excavation which supports the main body, safety is given a high priority, traffic impact is low, field time is comparatively short and nearby buildings are not impacted as much of the construction and allied activities are conducted under the surface. With the increasing development and maturation of shallow tunnel digging construction technology, metro builders have developed the concealed excavation method, with improved construction technology, less land usage, less moving of equipment and changes in the surrounding area, and the remarkable merit of low impact on the surrounding environment, thereby solving the difficult problem of metro construction in urban China. With the fast development of urban metros, the design of metro stations and construction technology have improved; in cities the construction of metro stations needs to take into consideration pipelines, transportation, buildings and the influence of various factors on the environment, leading to the wide usage of the concealed excavation method. For metro car parks that cover a large area, the use of underground construction technology can greatly reduce the excavated volume, decrease the need to move buildings, avoid the requirement of moving underground pipelines and directly reduce the impact on ground transportation.

The excavation section of the open-cut station is more complicated than the tunneling section, and the excavation section is larger than the interval tunnel, so the stress is more complex and the construction risk is increased correspondingly. The structure commonly used in the open-cut stations is the single-arch double-deck structure, which is a standard island platform station of about 20 m. Because of the influence of the structure, the underground open-cut station is buried deep within the tunnel at 5–6 m, which is often used for shallow excavation. In recent years, with the development of underground spaces, the construction of metro stations at home and abroad shows the development trend of large spans and shallow depths. The increasing density of intensive ground construction and city construction at the same time as greater urban underground engineering has led to the need for increasingly strict environmental protection, with construction methods required to not only ensure the safety of personnel and the underground structure itself but also limit the impact on the surrounding environment.

In early domestic metro construction, all basic stations were constructed in the soil or sand, including Beijing, Shanghai and other places in China, where the quaternary layer is overburdened and thicker. The Beijing CRD method was the commonly used method, followed by the traditional shallow tunnel excavation method and when combined with the characteristics of the subsurface-excavated method, it formed the basis for the PBA excavation method, which improved Beijing construction technology industry by leaps and bounds. Later, in Dalian and other places that required the construction of the metro, there was soft and hard rock formation under the chosen metro station area leading to a problem with the excavation method. The excavation methods for the soft soil layers and for hard rock formation are completely different. Then, an innovation of the arch cover method was proposed, which was matched with a reasonable excavation method.

Many other domestic cities have less quaternary soil layer thickness than Qingdao. A lot of underground stations are made up of whole-weathered or breeze granite, with high rock strength in the hard rock strata, where innovation is a big problem and hence the long-span Beijing excavation method is a good choice. Through the conclusion of several lines in Qingdao Metro, a new method for the construction of the hard rock stratum has been utilized. The arch cover method is optimized for the initial and surrounding rock arch cover, which is suitable for the construction in the hard rock stratum. This book combining the construction of Qingdao area in hard rock geological conditions, introduces arch cover construction key technology in Beijng, analysing the arch cover method, the second lining arch cover, arch of second lining arch covers a variety of methods, such as construction technology key points. The underground open-cut station in the Qingdao area is built in strong weathered granite, of which the upper part is the artificially filled coarse gravel sand layer, the formation is weak and the permeability is strong. At the same time, the site is mostly located at the bottom of the road, with busy traffic, a dense system of underground pipelines and controlled surface subsidence, which along with construction safety are the determining factors of the choice of construction method.

4.1.2 *The main construction method*

In the process of the current urban metro construction, the shallow tunnel excavation method is used for the main construction of the full face, after which the CD method, cross the door (CRD) method, the double-side heading method, the hole pile method (PBA), the side hole method, the Zhu Dong method, etc., are used. However, the selection of appropriate construction methods should be combined with the appropriate structural engineering, geological and hydrogeological conditions of the location, city planning requirements, the situation of the surrounding buildings and roads, traffic and site conditions, planned depth of the construction, form of the structure, the time limit of the project, the construction cost and other factors that influence the decision.

4.1.2.1 *Lateral hole method*

The lateral hole method is a commonly used method for building long-span tunnels. Based on the steps involved in the method, a two-sided or one-sided guide hole is first constructed, and then the middle part is excavated. Because the first construction is the two-sided hole, which should be synchronized, the surface disturbance is immense and the safety precautions cannot be properly maintained. When using this method in the underground open-cut station, the tunnel sometimes requires three steps or even four steps to excavate. The construction process is extensive and the temporary support consumption is immense. The construction process is shown in Figs. 4.1–4.4.

4.1.2.2 *Middle hole method*

The middle hole method is the core of the CRD method, based on the step method, which is performed according to small blocks or quick short steps,

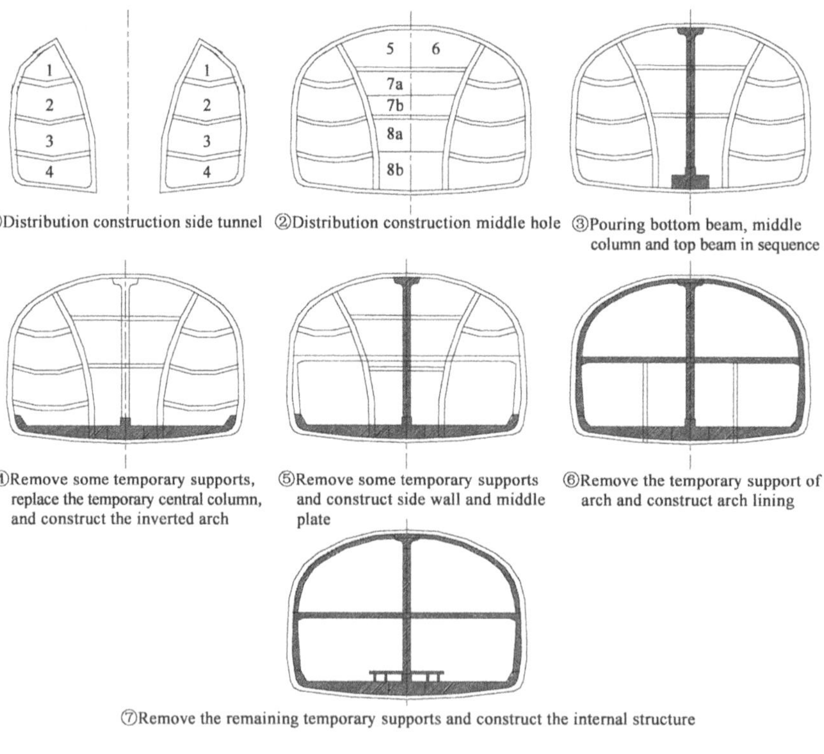

Fig. 4.1. The construction flowchart of the side hole method in the underground station.

based on the principle of closed, step-by-step construction, where the safety degree is high, the ground settlement and influence scope are smaller as compared with the side hole method. However, it is difficult to limit the additional displacement generated in the transformation process. It is the same as the side hole method, with immense temporary support and excessive engineering waste. The construction process is shown in Fig. 4.2.

4.1.2.3 *Cylinder method*

On the basis of the steps method, the lateral wall and the guide hole of the beam column are constructed first, then the structure of the beam column is constructed in the guide hole of the beam column, and then the arch is excavated and the initial supporting structure is applied to it.

When the bottom layer does not meet the requirements, a pile foundation can also be constructed in the side wall and the guide hole of the beam column, and then the side wall and the structure of the beam column are fixed. It is often used in soft soil areas for double-span single columns, three-span double columns and other structural forms of underground open-cut station construction. The construction process is shown in Fig. 4.3.

4.1.2.4 *PB method*

The principle of the PBA method is to cut and frame the structure construction to carry on an organic combination, namely, when the conditions are not suitable on the ground for the construction of the foundation pit retaining structure, the change can be applied underground first underground for the retaining pile and pile top girder, making the retaining piles, pile top longitudinal beam, roof arch, beams and arch support framework (PBA Pile) bear the external load of the construction process. Then under the support of the arch top and side pile, and a step-by-step excavation pretension horizontal support (if necessary), construction of the internal structure is finally done using the outer edge and top arch primary support and a combination of the secondary lining layer of the permanent bearing system. Its characteristics are as follows:

(1) In a strong permeable layer, there will be water formation during construction in an anhydrous or low-water area, where long-term ground

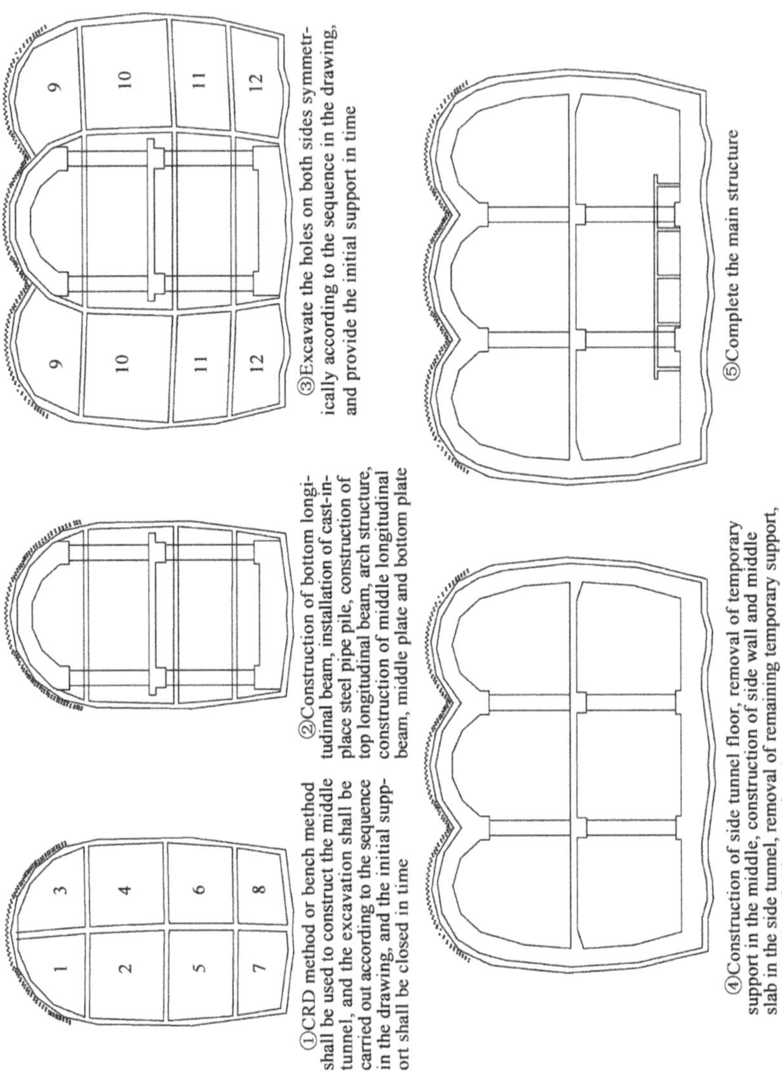

Fig. 4.2. Process flowchart of tunnel construction in the underground station.

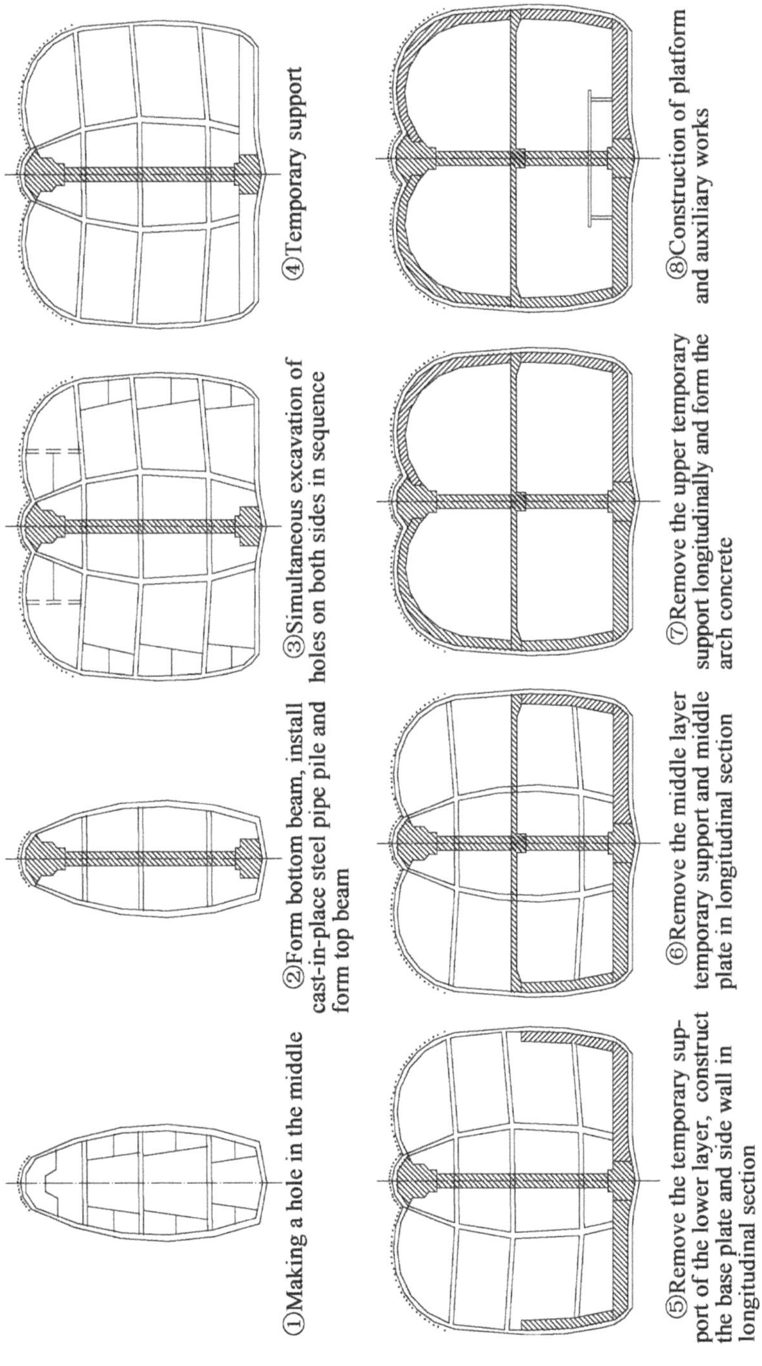

Fig. 4.3. The construction flowchart of the pillar hole method in the underground station.

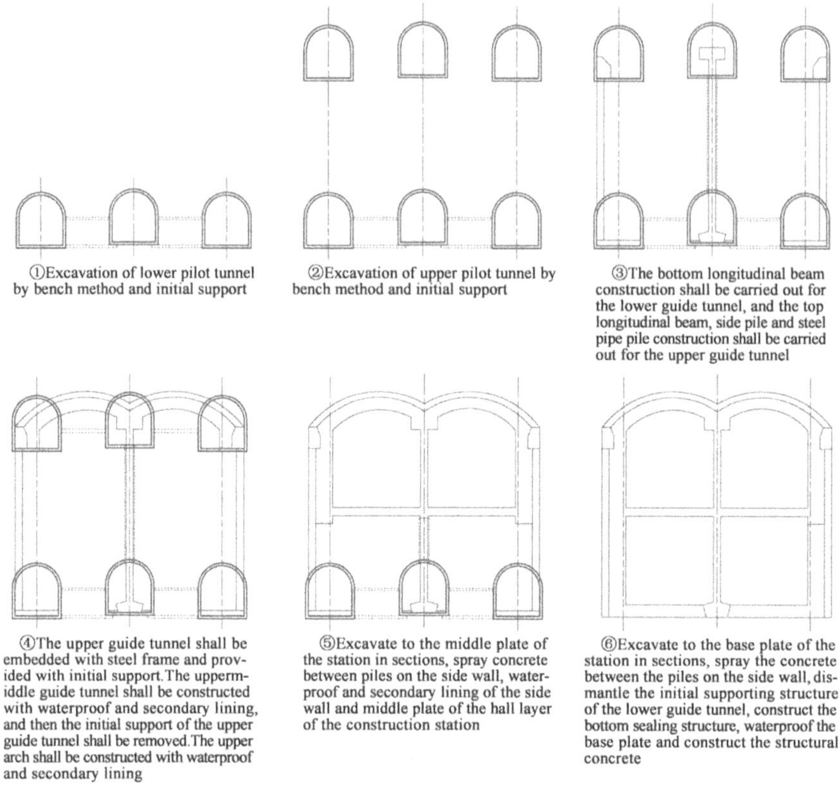

Fig. 4.4. The construction flowchart of the pillar hole method in the underground station.

subsidence caused by large amounts of precipitation must be avoided as it would increase cost; such a situation is advantageous to the protection of groundwater resources and reduction of construction cost.

(2) To make supporting piles safe, and also control strata subsidence, the hole method must be avoided; the CDS, CRD and double-wall pilot tunnel methods and repeated excavation cause defects leading to ground subsidence and rigidity of the primary support leading to its weakening.
(3) For the CRD and double-side wall methods, the amount of temporary work is relatively less, and the structural stress condition is good, which is relatively economical and reasonable.
(4) The number of structural layers is limited, which is beneficial to the safety of underground structures and the surrounding buildings.

(5) After the formation of the piles, beams and arch-bearing system, a large construction space becomes available, which is convenient for mechanized operation, thus speeding up the progress.
(6) Within the pilot tunnel above the formation of open hole piles, the "pile effect" plays a role in retaining the wall on both sides of the soil, which can crumble because of the hidden danger during construction such as quicksand.

4.1.2.5 *New tube method*

A new pipe curtain method (also called New Tubular Roof (NTR) method) was adopted in 2008 in the construction of air ducts for Shenyang Metro Line 2 using the full Beijing method; in the quaternary strata the construction of a large underground station was realized, and the shield construction of the station was synchronously implemented. This method was used on mainland China for the first time. Construction technology on the basis of reference to both foreign technology and domestic standards and specifications leads to a concrete engineering environment and conditions for further development, thereby enriching the content and means of domestic underground excavation in underground engineering.

With the development of Beijing domestic construction technology, especially in recent years, the PBA method is used widely in quaternary soft soil layer, leading to an increase in bearing model research and construction technology of soft soil layers.

4.2 Evolution of subsurface-excavated station construction technology in Qingdao

4.2.1 *Construction characteristics of subsurface-excavated station in Qingdao area*

Qingdao urban traffic conditions are congested especially on the main roads with a high density of buildings, mostly concentrated in the urban metropolitan area, but quickly multiplying to other areas. Hence, the station covers a broad area, with easy, flexible access, so as to encourage wide usage.

The Qingdao area has a vast coastline and river system, and all river flow is obviously controlled by precipitation with some seasonal variation. As for groundwater conditions, the main types of fissure water in the Qingdao area are unconsolidated rock pore water, clastic rock pore fissure water, bedrock fissure water, littoral lateral surface water and underground water. The overall flow direction of underground water is consistent with the slope. The groundwater level is 1.5–2.0 m with seasonal variation. In the area of Linhai, the groundwater is affected by tidal sand and changes regularly with the rising and falling tide.

According to a geological survey, Qingdao city has the typical structure of two-dimensional composite stratum. In general, the fourth layer of soil with different thickness is covered by granite rock with strong, medium and micro weathering. The fourth system has a soil thickness of 0–20 m, cutting through all kinds of weathered rock and soil layers, which is very uneven lengthwise, with characteristics that are "soft under the hard".

Based on the above geological conditions of Qingdao, the Qingdao Metro underground station has the following characteristics:

(1) The holding layer is of a hard rock formation, and has the characteristics of "soft and hard" rock.
The "soft hard" formation is the representative characteristic of the underground excavation project of Qingdao Metro. It is very difficult for underground excavation due to the "upper soft hard" formation, which is mainly reflected in the following:
    (a) There is a large span of underground excavation, and there is considerable risk of instability of the surrounding rock of the arch due to structural weakness, the presence of mainly silty soil, strong weathering strata, sandy strata, etc., and improper handling can easily cause a collapse.
    (b) Although only mechanical or engineering excavation can be done on the weak surrounding rock, strength is higher in the lower part of the tunnel rock mass and hence it can be excavated with the drilling and blasting method, and blasting vibration can be withstood to a certain degree by the surrounding rock without arch disturbance.

(2) There is a presence of some marine strata.
Due to the adjacent sea, the construction environment of some metro lines is characterized by marine conditions, and the chloride ion content in the groundwater is higher. The marine condition is a great challenge to the durability and mechanical maintenance of the construction of the underground open-cut station.

(a) The condition of the surrounding rock changes suddenly.
The geological conditions are especially complicated for the "land reclamation" projects in some offshore areas of Qingdao. Offshore rocks are affected by ongoing geological processes such as seawater erosion and crack development, facing frequent extreme and haphazard conditions. Hence, geological exploration drilling cannot easily reveal the formation characteristics, leading to difficulty in Beijing construction of the metro in the corresponding areas.

(b) The depth of the local rock strata is subject to environmental impact control.
As mentioned earlier, the thickness of the fourth series soil layer in Qingdao area is 0–20 m, and the thickness of the strata varies greatly. The depth of the strata in some areas is not much, which is significant for the stability of the tunnel itself. At the same time, the influence of blasting vibration on the surrounding buildings and pipelines is very important especially due to the shallow strata.

To sum up, the underground open-cut station in Qingdao is mostly of a hard rock type, and the upper layer consists of loose deposit of quaternary rock, with a geological type of typical "upper soft and hard" formation. In contrast to other cities, the fourth system in Qingdao is relatively thin, which is 4–9 m. The metro is generally built in the strong weathering or breeze strata of 2–10 m thickness, for which one needs to use drilling and blasting for excavation. This method is not utilized more in China, as station rock excavation impacts the stability of the surrounding rock, and such excavation methods have to be further studied.

4.2.2 *The evolution of excavating in the Qingdao area*

The presence of large-span hard rock in the underground area in a metro city is unusual. Excavation methods applied to hard strata are still those used for soft soil layers such as the CRD method, double-wall pilot tunnel method, the arch cover method, etc. In view of the characteristics of the hard rock strata, several methods, such as the step method, double-side wall guide pit method and arch cover method, are adopted in the underground open-cut station construction of the first metro line in Qingdao. Through multiple experiences during construction of Qingdao Metro Line 3, the Beijing arch cover method was chosen and is most widely used in the construction of Metro Line 2. The processes used for soft rock strata such as the PBA method, the mountain tunnel double-wall pilot method and other methods are combined with the characteristics of the subsurface-excavated station method, making it suitable for hard rock strata. The arch cover method is most suitable for hard rock strata, which makes full use of existing characteristics such as high strength and high bearing capacity for the surrounding rock, replacing the PBA method with great arch strength of the side pile or the side column and excavation blasting with weak blasting, thereby significantly reducing the station construction excavated volume and solving the degree of difficulty of hard rock strata excavation (see Fig. 4.5).

The first phase of the Qingdao Metro Line 3 was constructed at a shallow depth of 5.1–15.6 m, with an average depth of 11.07 m, strata thickness of 2.0–2.0 m and rock type of II–IV level. The Dunhua road station has a width of 20 m or so and a height range of 16–18.4 m. The section form of the underground station is mainly composed of a single-arch structure of straight wall, and the supporting form has compound lining. Qingdao Metro Line 3 was the first of the metro lines to be constructed. Therefore, the need to ensure construction safety was great, reflecting in the choice of the conservative Beijing method, such as the double-wall pilot tunnel method, and partially adopting the arch cover method, whereby pillar type for large-span station excavation are chosen.

Table 4.1 shows the main excavation method of the underground open-cut station of Qingdao Metro Line 3.

However, experience shows that in the hard rock stratum of Qingdao, there are five problems:

(1) The strength of the rock strata is high, for which blasting excavation is required, the construction process is considerable, there is no cross operation between the processes and the construction efficiency is low.
(2) The operating space is small and there is no space for the large machinery, and the efficiency of digging and discharging is low.
(3) There are many crews of temporary support personnel who are vulnerable during blasting construction.
(4) The closed loop time is long, and the effect of settlement control is poor.
(5) The construction seam is large and it is difficult to waterproof it.

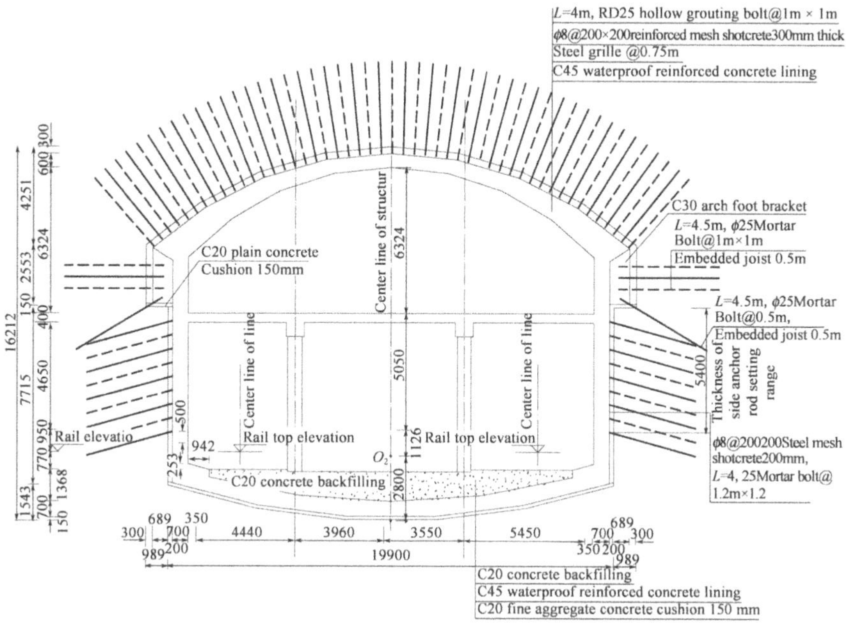

Fig. 4.5. Excavation support diagram of typical underground open-cut station (size unit: mm).

After the experience of Qingdao Metro Line 3, for Qingdao Metro Line 2, increasing the average depth to 2.63 m did not improve the condition of surrounding rock. The depth of the underground open-cut station is 9.28–17 m, with an average depth of 13.7 m. The span of the station is 0.42–0.78, and the thickness of the overburden is 0–18 m. The condition of the surrounding rock is improved, such that the main body side wall of the station is located in the rock layer, and the main body of the car station uses the arch cover method, which accounts for 70%, and the two-sided wall guide pit method accounts for 20% of the construction. Table 4.2 shows the main excavation method of the underground open-cut station of Qingdao Metro Line 2.

Through the construction of the underground open-cut station of Qingdao Metro Line 2, the arch cover method has been proved to be very suitable for the construction of a large metro station in the hard rock stratum.

Construction on the Qingdao Metro Line 1 started in 2016, with five rounds of digging, and the arch cover method was adopted in the selection of the construction method and the double-side wall guide method was officially eliminated as shown in Table 4.3.

Table 4.1. Qingdao Metro Line 3 underground excavation of the main excavation method.

S/n	Station name	Method	Overburden thickness (m)	Digging method	Notes
1	Zhongshan Park Station	Digging	10–12	Arch cover method	Straight wall, two single arches
2	Zhanshan station	Digging	Average approximate 13.4	Double-side wall guide pit method	The wall with the inverted arch, two single arches
3	Jiangxi Road Station	Digging	8.9–10.5	Double-side wall guide pit method	The wall with the inverted arch, two single arches
4	Dunhua Road Station	Digging	9–14	Digging column method	Straight wall, tower-style single arch
5	Qingjiang Road Station	Digging	6.3–8.3	Double-side wall guide pit method	Straight wall, two single arches
6	Wannan Quan Road Station	Digging	5.1–10.1	Double-side wall guide pit method	Straight wall, two single arches
7	Junfeng Road Station	Digging	7.4–15.6	Double-side wall guide pit method	Two single arches

It is worth mentioning that the arch cover method also enriched the construction of tunnels. The improvement of the excavation work method also promoted the development of tunnel excavation as shown in Table 4.3.

The arch cover method according to the secondary lining construction time can be divided into the early arch cover method (practice) and the second lining arch cover method (uniaxial), and a combination of the two methods is widely used in Beijing construction.

Table 4.2. Main excavation method of Qingdao Metro Line 2.

S/n	Station name	Method	Overburden thickness (m)	Digging method	Notes
1	Yan'an Road Station	Digging	7–12	Arch cover method	Arch double-side wall, lower step method
2	Zhiquan Road Station	Digging	0–9	Double-side wall guide pit method	Nine-step two-sided wall
3	Kaohsiung Road Station	Digging	5–13	Arch cover method	Arch double-side wall, lower step method
4	Xujiamai Island Station	Digging	6.75–14.2	Primary arch cover method	Arch step method (local CD), lower step method
5	Haichuan Road Station	Digging	6–15	Primary arch cover method	Arch step method (local CD), lower step method
6	Tongan Road Station	Digging	5–8	Arch cover method	Arch double-side wall, lower step method
7	Zaoshan Road Station	Digging	8–17	Traditional arch cover method	Arch CD, lower step method
8	Lee Tsuen Park Station	Digging	0–8.7	Double-side wall guide pit method	Nine-step two-sided wall
9	Automobile East Station	Partly bright, partly dark	6–10	Arch cover method	Upper double-side wall, lower step method
10	Huancheng South Road Station	Partly bright, partly dark	8–18	The steps method	Bench construction

- **Two-line arch cover method (reverse course):** At the beginning of the first round of construction after pilot drift through the supporting arch cover, the arch of the second line is modified, creating an opening above the medium plate and side walls, continuing the dig to the base, bottom and side walls under the plate.

Table 4.3. Qingdao Metro Line 1 underground open-cut station and part of the main excavation method.

S/n	Station 1 interval name	Method	Overburden thickness (m)	Digging method	Notes
1	Taihang road station	Digging	13.8–20.3	Primary arch cover method	Straight wall, two single arches
2	People value range	Digging	22.9–30	Arch cover method	Straight wall, two single arches
3	Hengshan road station	Digging	16.5–18.7	Arch cover method 法	Straight wall, two single arches
4	Xue Gudao station	Digging	16.5–18	Arch cover method	Curved wall with inverted arch, two single arches
5	Yongnian road station	Digging	17.8–38.9	Arch cover method	Two single arches
6	Ruijin Road Station	Digging	13.7–15.4	Arch cover method	Straight wall, two single arches
7	Gas interval	Digging	17.9–18.8	Arch cover method	Four-line section
8	Guizhou station	Digging	6.2–7.0	Arch cover method	Straight wall, two single arches
9	West town station	Digging	15.7–23.6	The arch is leading the small guide hole to expand the lower step method.	Straight wall, two single arches
10	Zhongshan Road station	Digging	13.4–18.7	Primary arch cover method	Single arch straight wall double-layer structure
11	Guangrao Road Station	Digging	11.5–21.5	Primary arch cover method	Single arch straight wall double-layer structure
12	Jiangsu Road Station	Digging	3.8–21.3	Primary arch cover method	Single arch straight wall double-layer structure

Table 4.3. (*Continued*)

S/n	Station 1 interval name	Method	Overburden thickness (m)	Digging method	Notes
13	Undercut section of haipoqiao station	Digging	15.2–15.6	Arch cover method	Straight wall (bottom corner arc), two single arches, local three-level single arch
14	Village station	Digging	9.5–17.5	Arch cover method	Single arch wall double structure
15	Beiling station	Digging	19–24	Arch cover method	Single arch wall double structure
16	Shuiqingou station	Digging	16.6–20.6	Arch cover method	Single arch straight wall double-layer structure
17	Kaifeng Road Station	Digging	7–8	Transformation method	Straight wall, two single arches
18	Anshun Road Station	Digging	15.5–17	Arch cover method	Single arch wall double structure

- **Arched two-bushing method (semi-inverse, semi-clockwise):** At the beginning of the first round, after pilot drift through the supporting arch cover, the arch of the second line stretches to the buckle arch construction, and then the arch structure of the subsurface-excavated station stretches directly to the basement, neatly arranging the main structure, medium plate and side wall.
- **Initial abutment method (cis):** After the opening of the guide hole, the initial support arch cover is deducted, the arch construction is not carried out, the foundation is directly excavated to the base, and the main structure of the body, side wall, middle plate and arch cover is made.

See Table 4.4 for a comparison of main works of underground open-cut station.

Table 4.4. Comparison table of main works of underground open-cut station.

Method names	Production processes	Waste engineering quantity	Cross construction	Construction period (d)	Notes
Double-side wall guide pit method	Many	Many	Many	26–3υ	The construction is relatively safe, the settlement control effect is poor and the ability to adapt to the complex formation is strong. High cost, a lot of temporary support needed, which is removed or shifted afterward.
Two-liner arch cover method	More	More	More	2υ–24	Low formation disturbance and low cost. The construction joint of arch foot easily leaks, and the proposed method needs to protect the middle plate structure.
Arch cover method	Less	More	Less	16–2υ	Low formation disturbance and low cost. It is necessary to pay attention to the stability of high wall surrounding rock at the foot of large arch; there is high dependence on formation.
Primary arch cover method	Few	Few	Few	12–16	The initial process has less conversion, low cost, good safety, good overall quality and high construction efficiency. It is necessary to pay attention to the stability of high wall surrounding rock at the foot of large arch; there is high dependence on formation.

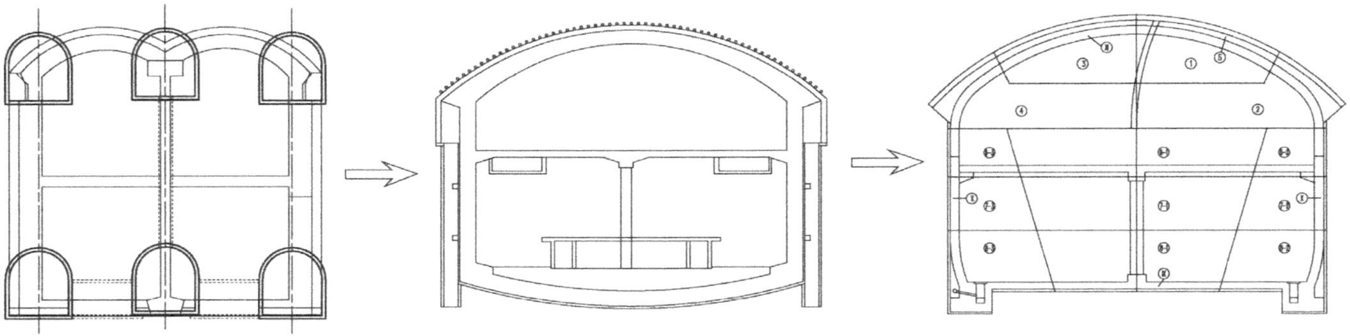

a) PBA method b) Arch cover method c) Initial support arch cover method

Fig. 4.6. The main evolution process of the initial arch cover method in the underground open-cut station.

The main evolution process of the initial arch cover method is shown in Fig. 4.6.

4.3 Subsurface-excavated station construction technology in hard rock stratum

4.3.1 *The main excavation method*

4.3.1.1 *The steps method*

4.3.1.1.1 Overview

In the Beijing steps method of excavation, each section is divided into several steps, for example, three steps to open the dig and inverted arch, and tighten and close it into a ring. To sum up, the applied method, under the stairs, which form a certain interval, with a new order of left to right work for surface shape leads to a phase fault, which can be seen in all the work surface per cycle footage of parallel flow of water. For industry applications, the Taiwan workers ordered the dig machine after checking whether the grilled slag is accessible in the next working procedure, such as in the line steel frame installation work, where the section can approximately be a loop. According to the surrounding rock

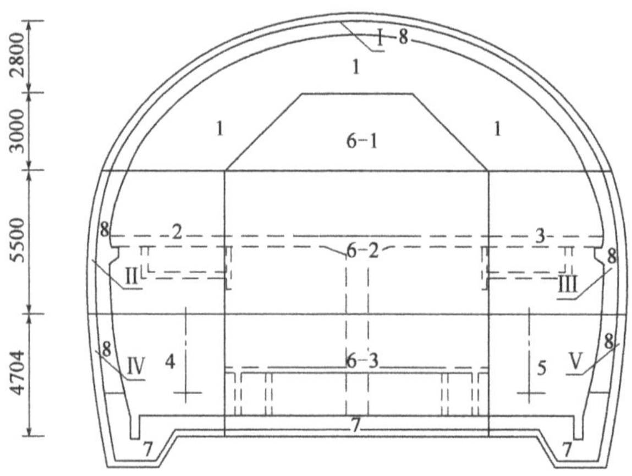

Fig. 4.7. Steps method of construction, step 1 (size unit: mm).

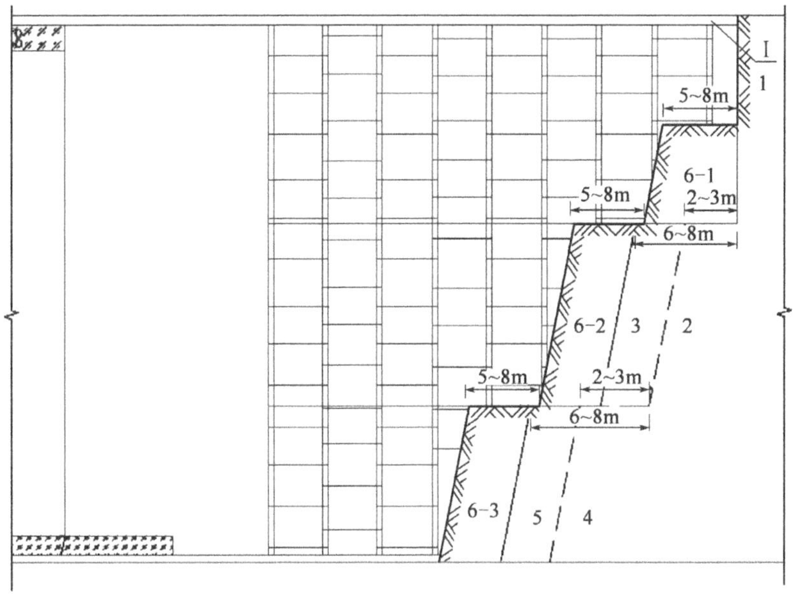

Fig. 4.8. Steps method of construction, step 2.

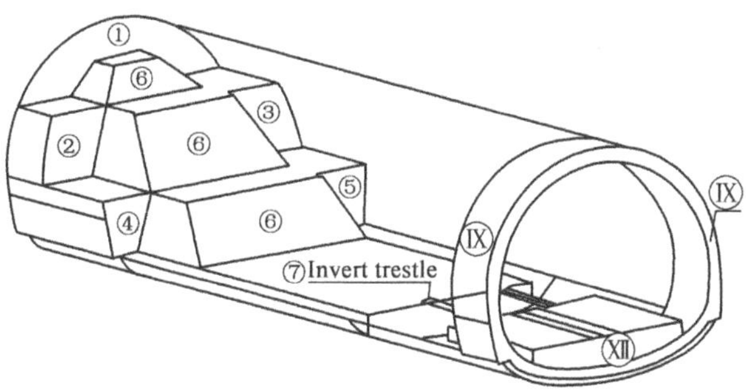

Fig. 4.9. Steps method of construction, step 3.

situation, the steps should be controlled from 5 to 8 m, and the working face of the same step should be staggered for no less than two steel frames.

The construction steps are shown in Figs. 4.7–4.9.

4.3.1.1.2 Upper guide pit construction

The micro vibration control blasting excavation is as shown in Fig. 1 (in circular excavation and reserved core soil, because of the large excavated section, core soil is retained; on the premise of guaranteed stable constraints, and as a work platform for grill installation, operating injection of coagulation soil is done). Immediately after the excavation at the beginning of 4-cm thick concrete spraying, in an enclosed operation, the steel rods are erected, reinforced steel mesh pieces are installed and connected, and a drill set radial system with anchoring bolt, lock foot and steel frame is welding firmly, after thickly spraying concrete as per the design, forming a stable bearing structure.

4.3.1.1.3 Construction of middle and lower guide pit

Under the upper guide bearing arch, the construction of the middle and lower guides is carried out. The middle and lower guides are dug in the following ways:

First, two parts of the ground are excavated, and then the three parts in the drawing is excavated, so that the working face of the same step is staggered to a certain distance (generally controlled in the space of a 2-phase steel frame). It is forbidden to excavate the side wall of the same step, so that the arch is suspended. The excavation of the lower guide is the same as that of the upper guide. Construction stages 6 and 7 complete the initial support of the full ring closure.

4.3.1.1.4 Construction control points

(1) Strengthen the training of technical and construction personnel, improve the safety and quality consciousness of all the staff members, and specially focus on the educational training work.
(2) The principle of "weak blasting, short penetration, strong support, early closure and frequent measurement" is adhered to. In the excavation of the station, the stability of the surrounding rock should be judged by observing and describing the structure of the rock fracture, the hardness of the rock body, the amount of water flowing out and the design of the surrounding rock. According to the geological

situation, the construction method should be adjusted in a timely manner, and not done blindly.

(3) When the excavation work is done, the length and width of the groove in the middle and lower guides (i.e., the middle part of the process) are reasonably determined according to the stability of the rock mass, so as to prevent displacement of the side wall and cause a cave-in accident. When the stability of the surrounding rock is better, the effect of optical explosion is improved and the construction process is accelerated.
(4) It is forbidden to excavate the side wall and the distance is not less than 2-phase steel frame.
(5) The length of each step is controlled, and each step is generally 3–5 m, which is convenient for mechanical operation, especially during the scraping of the slag of the upper step excavator, so as to speed up the construction progress.
(6) Each cycle is used to measure the lofting test, adjust the blasting parameters in time and strictly control the overshoot, requiring the average linear supersize to be no more than 10 cm, and the maximum overshoot is no more than 15 cm.
(7) In the process of moving the excavator, it is necessary to pay attention to the initial support of the construction and prevent a collapse.
(8) We should pay attention to the balanced production of each process so that the initial branch can be closed as soon as possible.

4.3.1.2 *Double-side wall guide method*

4.3.1.2.1 Overview of two-side wall guide pit

The double-side wall guide pit is a construction technique for excavation side support. The construction steps are shown in Figs. 4.10 and 4.11. The principle is as follows: The whole tunnel is divided into three small sections: left, middle and right. After the initial support of the back-arch ring, the two sides of the tunnel act as temporary support to form a full section. Both sides of the guide hole are inverted goose egg shaped, which is conducive to control the dome sinking. The method is widely used in leaching soil layers, sand layers, sand egg layers and shallow buried sections of

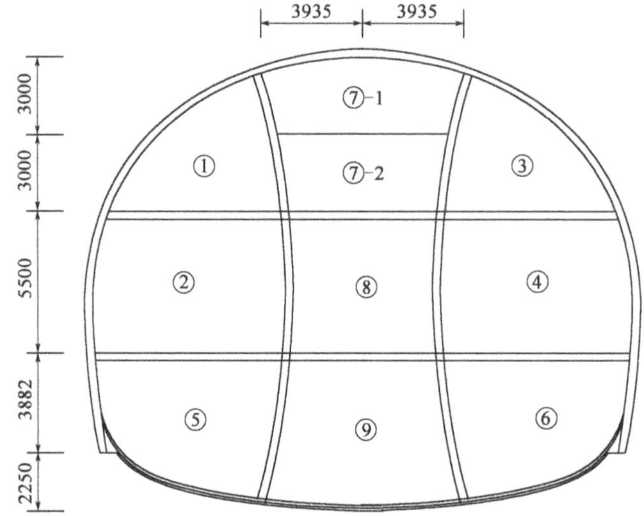

Fig. 4.10. Construction of double-side wall guide pit, step 1 (size unit: mm).

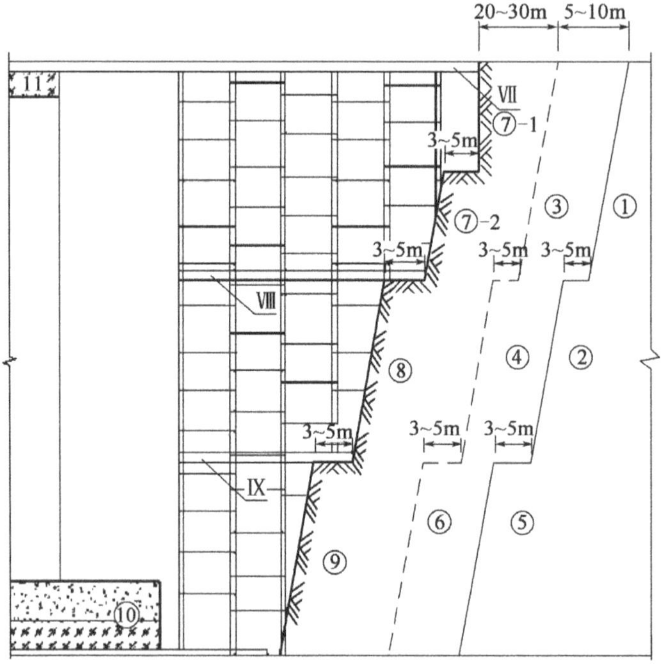

Fig. 4.11. Two-side wall guide pit construction, step two.

large tunnels. According to the different sections of the surrounding rock, the two-side wall excavation method can be divided into six steps and nine steps.

4.3.1.2.2 Construction process

- **Step 1:** Excavation of the first part is done, and a chute of 4-cm concrete is closed immediately after operation, involving erection of a steel frame and a temporary frame according to design requirements, drilling the system, lock foot bolts and consulting the cross arm, after adjusting the concrete to the designed thickness, thereby reinforcing the surrounding rock.
- **Step 2:** Excavation of the second part is done immediately after spraying 4-cm closed constraints of concrete, involving the pilot tunnel surrounding parts, long steel frame and the temporary steel frame, drilling the system, lock foot bolts, consulting the cross arm and adjusting the thickness of shotcrete as per the design.
- **Step 3:** Excavation of the third part involves the initial support, temporary support, steps and a construction procedure.
- **Step 4:** Excavation of the fourth part involves the initial support, temporary support, steps and construction procedure as per step 2.
- **Step 5:** Excavation of the fifth part involves the long steel frame, the temporary steel frame, drilling the system, locking the anchor bolt and then spraying the concrete to the designed thickness.
- **Step 6:** Excavation of the sixth part involves the initial support, temporary support in the surrounding areas, steps and construction procedure as per step 5.
- **Step 7:** The seventh part involves dividing the steps, installing the arch grill steel frame, drilling the system bolt and spraying the shotcrete to the designed thickness.
- **Steps 8 and 9:** According to the construction schedule, the excavation shall be carried out to shorten the distance from the upper step and provide suitable conditions for the construction of the inverted arch as soon as possible.

4.3.1.2.3 Construction control points

(1) Strengthen the training of technical and construction personnel, improve the safety and quality consciousness of all the staff members, and specially focus on the education training work.
(2) The principle of "weak blasting, short penetration, strong support, early closure and frequent measurement" is adhered to. In tunnel excavation, after each excavation and time observation, the structure of the surrounding rock fractures, the hardness of the rock body, the amount of water output, the design situation and the stability of the surrounding rock are judged. According to the geological situation, the construction method should be adjusted in a timely manner, and not done blindly.
(3) When the excavation work is done, the side wall guide pit and the middle slot shall be excavated with short steps. The distance of each part shall be reasonably determined according to the stable state of the rock mass. After the excavation of each department, the initial support and temporary support shall be carried out in time to prevent lateral wall displacement and a cave-in accident.
(4) The two sides of the tunnel lead a trough of 20–30 m, which can be independently synchronized to excavate and provide support. The steps method is used to excavate and maintain parallel operations.
(5) In the groove after excavation, the connection between the arch and side wall of gratings is difficult to maintain in the two lateral wall pilot tunnels, so the location of the grid steel frame must be accurately positioned to ensure that each setup after the grid connection is on the same vertical plane, avoiding steel distortion.
(6) To strengthen monitoring measurement, an information guidance system is constructed. In particular, it is necessary to guide the demolition time of the temporary support with data, in a step-by-step manner, after taking all precautionary measures.
(7) The first branch is closed as soon as possible.

4.3.1.2.4 Construction step optimization of double-side wall guide pit.

According to the general design standard step, for the excavation of the left and right guides, "(1)", "(2)", "(3)" and "(4)", temporary invert construction is followed.

In order to avoid blowing up the temporary inverted arch during blasting construction, a certain adjustment and optimization can be made to the excavation step in the construction process:

- **Step 1:** Excavation of the first part is done, and a chute of 4 cm concrete is closed immediately after operation, involving erection of a steel frame and a temporary steel frame, an anchor and drilling system, a lock foot, a complex chute and reinforcement of the surrounding rock, all according to the design requirements.
- **Step 2:** Excavation of the second part is done involving the concrete closed palm surface and the surrounding part of the guide pit, and then the steel frame and the adjacent part. The steel frame is equipped with the system and the anchor bolt, and the top transverse bracing is installed and the concrete is designed as per the required thickness.
- **Step 3:** Excavation of the third part is done involving the initial support and temporary support, the steps and the construction procedure.
- **Step 4:** Excavation of the fourth part involves the initial support and temporary support, steps and construction procedure as per step 2.
- **Step 5:** Excavation of the fifth part involves connecting the long steel frame and the temporary steel frame, installing the top transverse brace, drilling the system, locking the anchor bolt and setting the concrete as per the design thickness.
- **Step 6:** Excavation of the sixth part involves the initial support and temporary support in the surrounding area, steps and construction procedure as per step 5.
- **Step 7:** Excavation of the seventh part involves installing the grid steel frame of the arch, drilling the anchor rod and shooting the concrete according to the set thickness.
- **Steps 8 and 9:** According to the construction schedule, the excavation shall be carried out to shorten the distance from the upper step and provide suitable conditions for the construction of the inverted arch as soon as possible.

The above steps help to strengthen the measurement monitoring, enforce strict implementation of the guidance information and provide timely treatment for abnormal occurrences, with a strategy of everything.

4.3.1.3 *Two-line arch cover method*

The two-lining arch cover method is used, which is also known as the uniaxial scheme, wherein after the pilot drift occurs through the supporting arch cover, the arch of the second line is modified following the excavation of the contents below the medium plate, above the medium plate and along the side walls. Digging is continued under ideal base conditions and ideal dimensions of the base plate and based on the distance between the plate and the side walls. The second lining arch cover method is the original form of the arch cover method, which evolved for hard rock strata, with a full-fitting process under excavation and subject crossing, but as its construction efficiency was lower than that of other arch cover methods, so its application is less.

4.3.1.3.1 Excavation of the station arch

When the arch of the station is excavated, the distance between the upper and lower guide holes is 3–5 m, and the left and right guide holes are at a distance of 15 m. The loose surrounding rock should be cleaned before the construction of the arch foot, and the anchor rod is connected with the arch. The design is shown in Fig. 4.12, and the profile of the vertical construction step of the station is shown in Fig. 4.13.

(1) The arch of the CD is excavated
- **Step 1:** The rock mass on the left is excavated, and 40 mm of concrete is immediately sprayed to enclose the surrounding rock,

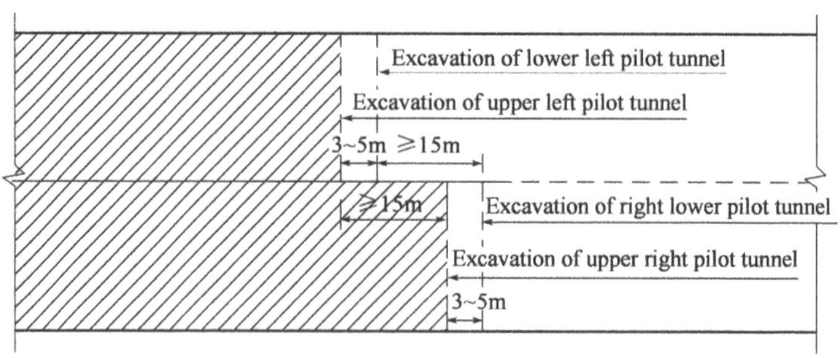

Fig. 4.12. Construction step plan of the arch part of the station.

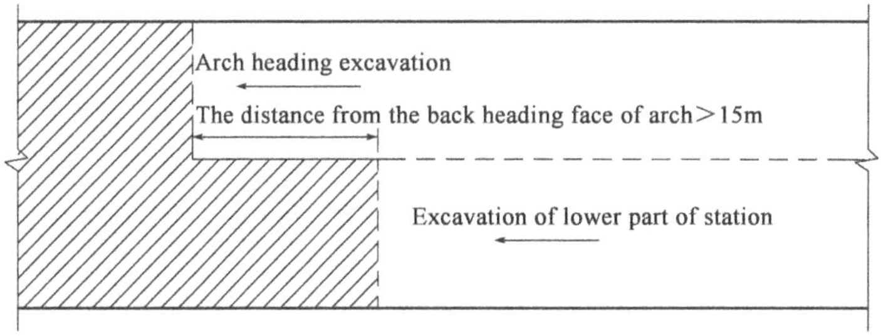

Fig. 4.13. Profile of station longitudinal construction step.

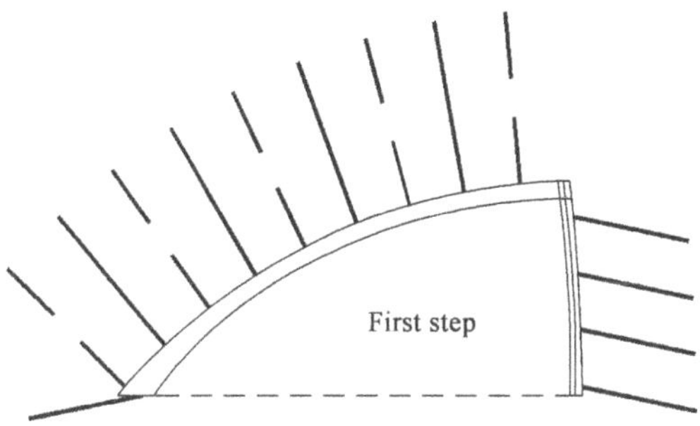

Fig. 4.14. The first step of the arch CD method.

fixing the bolt, grid steel frame and vertical temporary support, installing the steel mesh and spraying the concrete. The first step is shown in Fig. 4.14.

- **Step 2:** The excavation is elaborated at the bottom of the rock mass, involving the 40-mm closed surrounding rock, shotcrete and bolts, the grid steel frame and the vertical temporary support and installation of steel fabric staggered up and down the steps for 3–5 m, thereby expanding the settings of the anchor rod lock foot arch, which is a reliable connection with the grid steel frame. The second step is shown in Fig. 4.15.
- **Step 3:** Excavation at the upper rock drift involves the 40-mm closed surrounding rock, shotcrete and bolts, the grid steel frame and the vertical temporary support, installation of steel fabric, injection of coagulated soil and constraints on both sides of the

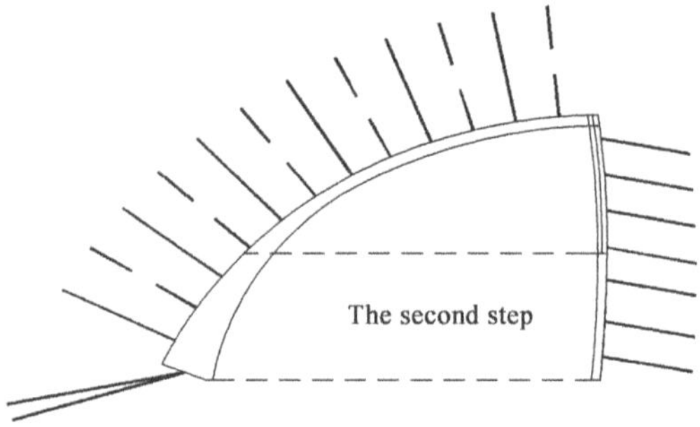

Fig. 4.15. The second step of the arch CD method.

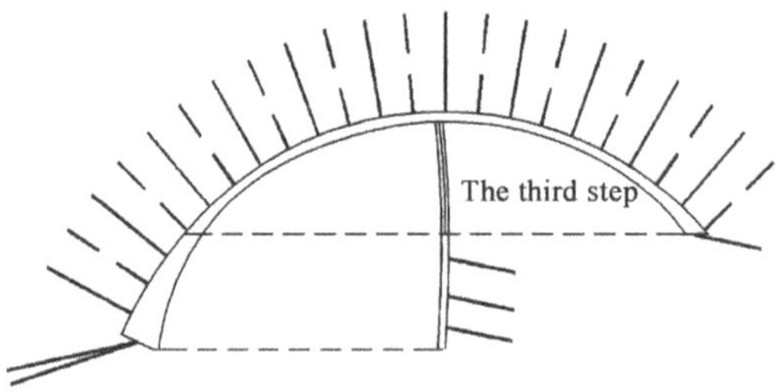

Fig. 4.16. The third step of the arch CD method.

staggered length that are not less than 15 m. The third step is shown in Fig. 4.16.

- **Step 4:** Excavation of the pilot tunnel rock mass at the lower right involves the 40-mm closed surrounding rock, shotcrete and bolts, the grid steel frame, and the vertical temporary support, installation of steel fabric, constraints on both sides of the staggered length that are not less than 15 m and the expanded the arch foot set bolt lock, which is a reliable connection with the grid steel frame, as shown in Fig. 4.17.

(2) The steps are dug

The arch part of the station is better than the step method. It is simple to excavate using the step method without removing the temporary structure, so the construction efficiency is high.

- **Step 1:** The central rock mass of the circular guide hole is excavated, the concrete is immediately sprayed at a thickness of

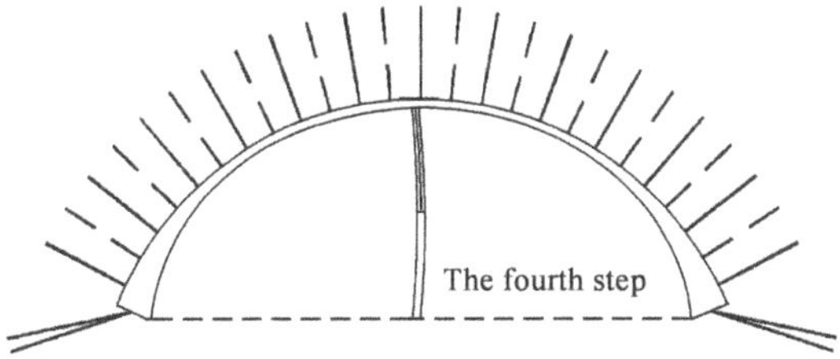

Fig. 4.17. The fourth step of the arch CD method.

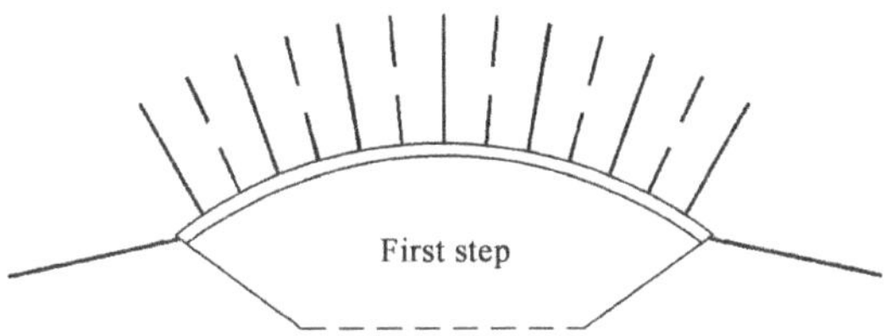

Fig. 4.18. The first step of the arch step method.

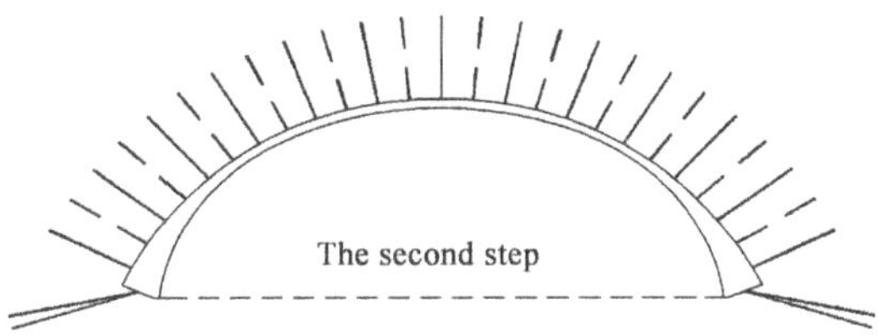

Fig. 4.19. The second step of arch step method.

40 mm to enclose the surrounding rock and the bolt, grid steel frame, steel mesh and shotcrete are fixed. Step 1 is shown in Fig. 4.18.

- **Step 2:** The circular bottom rock drift is excavated, and the surrounding rock is immediately sprayed at a thickness of 40 mm, and the process involves shotcrete and bolts, the grid steel frame, installation of steel fabric, expanding the arch foot set bolt lock and a reliable connection with the grid steel frame. Step 2 is shown in Fig. 4.19.

4.3.1.3.2 Construction of the arch part of the station

When using the CD method or double-side wall guide pit method to construct the arch part, the temporary support should be removed first, and the length of the first disassembly is not more than 6 m. The waterproof layer of the arch part and the secondary lining of the mold are reinforced, and it is necessary to strengthen the structural protection of the waterproofing plate (coil) and the stop water belt of the second lining of the arch. The process is shown in Fig. 4.20.

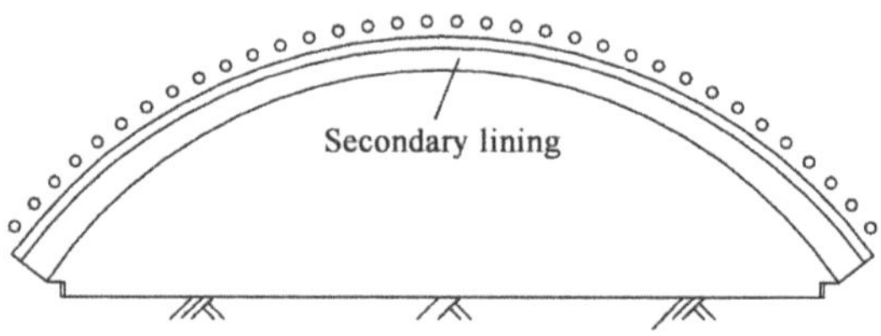

Fig. 4.20. Construction of secondary lining of arch.

4.3.1.3.3 Underground excavation and main structure construction

- **Step 1:** The step method is used to dig, with initial support for the excavation, to the middle plate and the above side wall structure and the middle plate of the car station. The process is shown in Fig. 4.21.

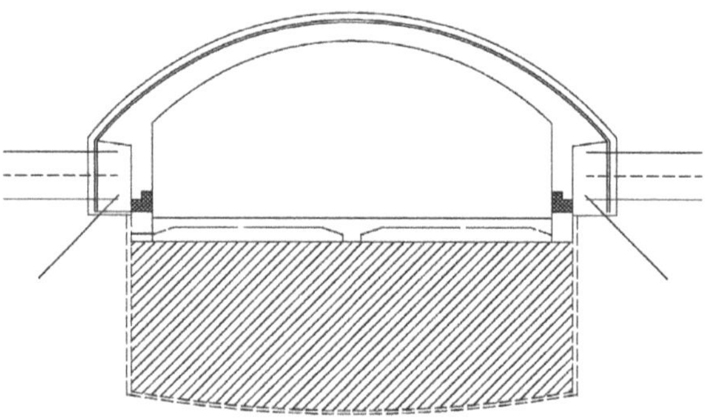

Fig. 4.21. Top down construction method for the side wall above the middle plate and the middle plate of the station.

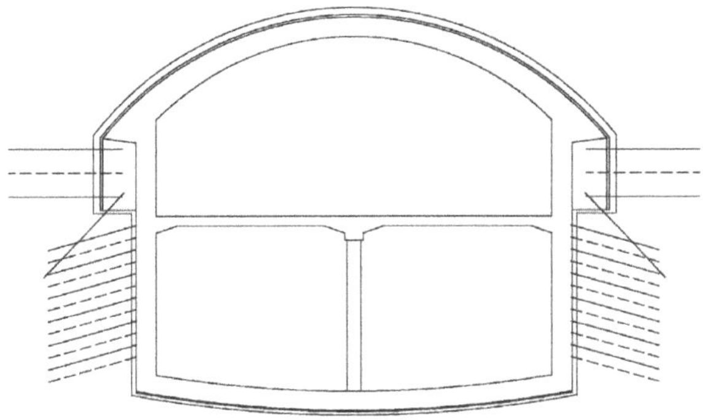

Fig. 4.22. Construction of side wall and bottom slab under the middle slab in the station by top-down method.

- **Step 2:** The step method is used to continue the excavation to the bottom layer and plate, and then the lateral wall waterproof layer and the side wall. The second lining arch cover method, namely, the whole uniaxial method, is less complicated relative to the PBA method, as it reduces the number of pilot tunnels, but the process is still relatively complex, especially for large-span stations or double-span structures, thereby needing the PBA method, which is applied first and then elaborated before the construction of the column.

4.3.1.4 *Initial arch cover method*

At the beginning, the arch cover method is elaborated in the primary support as the main breakthrough, without a second lining arch construction, involving arch department excavation directly to the basement, arranging the main structure, medium plate, bottom plate and side wall arch cover. There are several methods of excavation, such as two-side wall, CD and stair, which should be flexibly controlled according to the geological conditions.

4.3.1.4.1 Excavation of the station arch

At the beginning of the arch cover method, the station arch and the second lining arch cover are the same, but early support is provided to the long-span arch of the station structure under immense load, so much of the station requires strengthening of the primary support to ensure the safety during construction (Fig. 4.23).

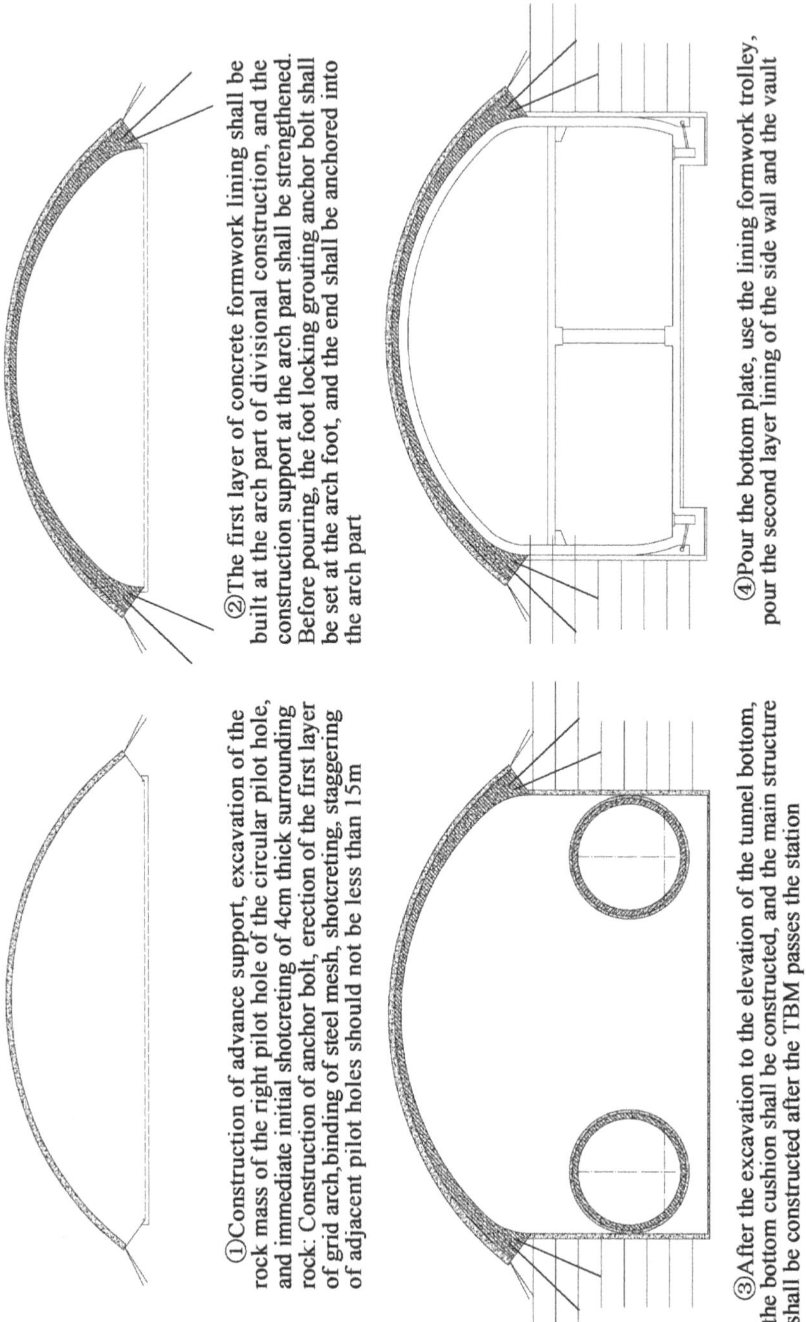

Fig. 4.23. The initial support of the arch is strengthened.

For Qingdao Metro Line 2 ocean-power road station, by using the arch cover at the beginning of construction, the station on both ends of the tunnel adopts the TBM method; the first part is tunneled using the TBM method and completed early and then the secondary lining construction using TBM steps is conducted. Support is provided to strengthen the arch after early excavation; after the completion of casting mold of the first layer concrete lining, the arch's temporary support is demolished by steps excavation down to the bottom, after the TBM, making it suitable for casting bottom, side wall and arch concrete structure.

4.3.1.4.2 Excavation at the bottom of the station

At the foot of the arch, the excavation should be done in layers, steps and sections. The Taixing road station of Qingdao Metro Line 1 can be taken as an example: Excavation is done at center after the first two sides, first in the central slot. After excavation, the process is continued on both sides, according to the anchor in the fractured place, to strengthen the stability of the arch cover foot under beam bedrock. Excavation on both sides of the wall, or loose blasting vibration, ensures that the integrity of the arch cover feet beam under the bedrock is maintained, thereby reducing the initial disturbance of the concrete arch department.

Construction control points:

(1) The excavation of the foot of the arch should be carried out in a layered and piecewise way.
(2) In the process of blasting, the stability and completion of the foundation rock are guaranteed by the loose movement or vibration reduction of the lower part of the lateral wall.
(3) It is forbidden to expose the rock to difficult conditions for a long time.
    - **Step 1:** Use the step method first in the central trough; the excavation depth can be set to two to four steps. The pull groove adopts the slope excavation. The length should not be too long. The process is shown in Fig. 4.24.
    - **Step 2:** Both sides of the straight wall structure should be picked at the beginning of the excavation process, according to the anchor rod in the fractured place. To strengthen the arch cover and the stability of the arch foot under beam bedrock after the

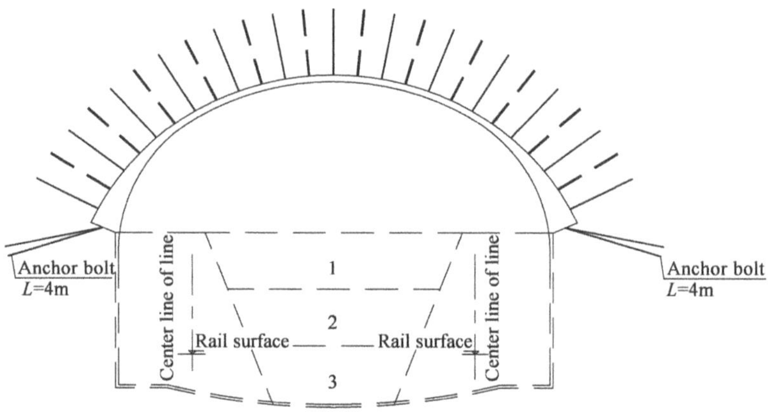

Fig. 4.24. Excavation step 1 of the lower part of the arch cover method.

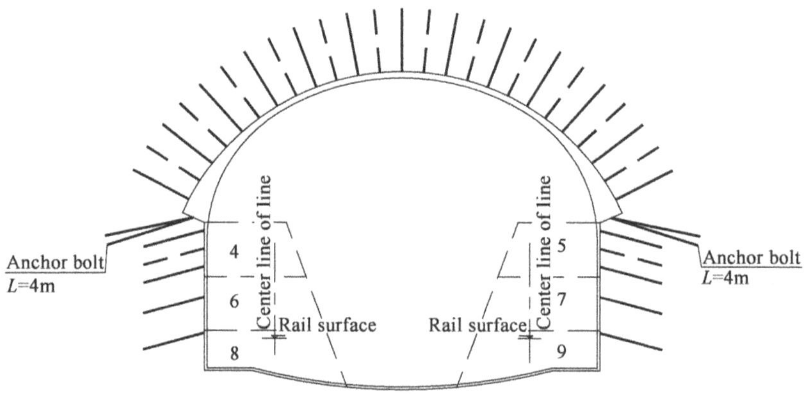

Fig. 4.25. Excavation of the lower part of the arch.

excavation, the process can be utilized in time. The process is shown in Fig. 4.25.

4.3.1.4.3 The main structure of the construction station

- **Step 1:** After excavating the elevation of the inverted arch, the construction pad, the waterproof panel (coil) and the protective layer shall be laid; and the secondary lining reinforcement shall be bound; and the concrete shall be fixed in the inverted arch and part of the side wall as shown in Fig. 4.26.

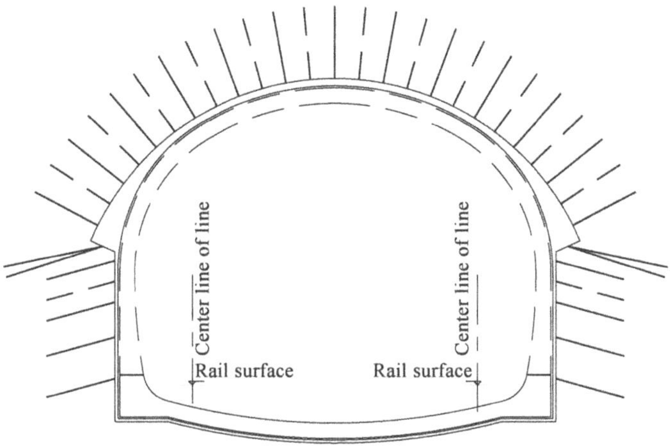

Fig. 4.26. Construction step of main body structure using the initial arch cover method.

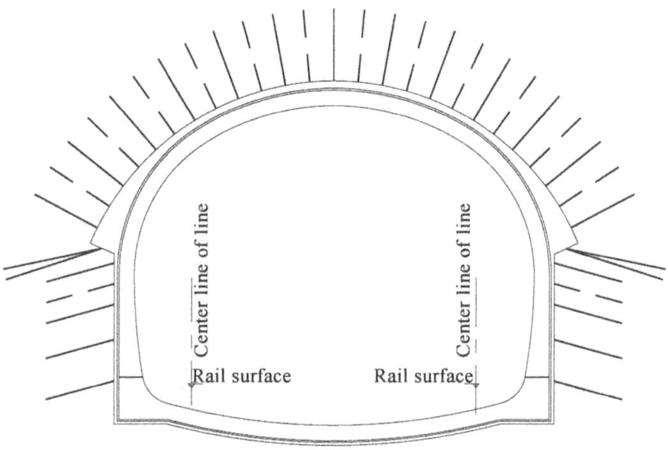

Fig. 4.27. Construction sequence of main body structure using the initial arch cover method ii.

- **Step 2:** The waterproof board (coil) is laid, secondary lining steel is tied, the mold is built and the arch concrete side wall, side wall and arch adopt a molding of second lining trolley, which can be combined using a small template during construction. The process is shown in Fig. 4.27.
- **Step 3:** The internal structure of the station is excavated and concrete construction is backfilled as shown in Fig. 4.28.

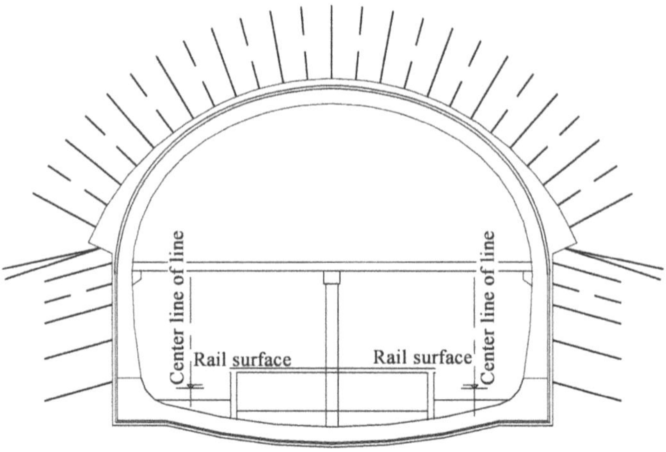

Fig. 4.28. The main body structure using the arch cover method iii.

4.3.1.5 *Two-line arch cover method*

4.3.1.5.1 Overview of the process of two-line arch cover

According to the formation of the Qingdao "soft–hard" characteristics, one has to make full use of the hard rock strata of high bearing capacity, with big arch foot walls, where a cross-section structure is given priority through the fault fracture zone and joint development that have taken place along a single arch of the straight wall composite lining. The surrounding rock under the arch cover is generally good, so the arch is widely used in the underground excavation of Qingdao.

The construction steps of the arch part are shown in Table 4.5.

4.3.1.5.2 Section excavation of the main body of the station

Generally, section excavation can be applied on both ends of the duct for supporting the station. This is done by opening the mouth to break the horse-head door. Arcing of the cross-section can be performed following the double-wall pilot tunnel method. Digging is done following the construction step sequence diagram. The two pilot tunnel section staggers should not be placed at a distance of less than 10 m. Promptly set up at the beginning of the arc line grating after excavation, step excavation is

Table 4.5. The construction of the arch part.

S/N	Construction procedure diagram	Say clearly
1	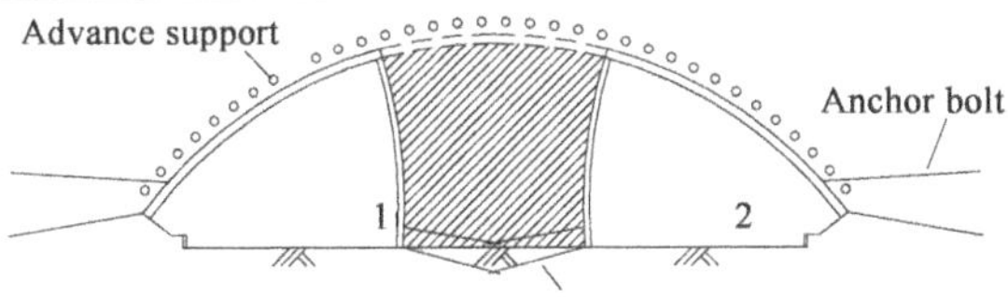	In advance, the rock mass on the left and right side of the fault surface is divided into two, and the arch and temporary .support of the grills are erected.
2	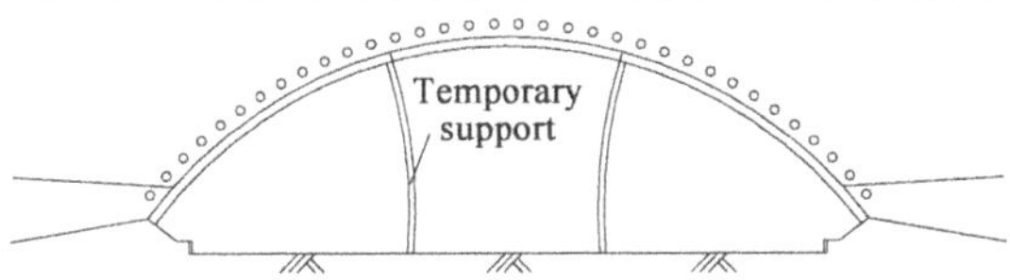	The upper part of the upper section of the rock body, the frame grid arch and temporary. construction support.
3	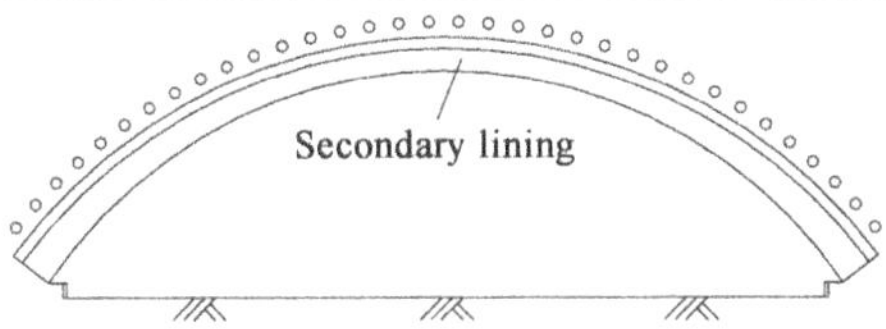	When the block is removed, the support (the length of one support is not greater than 6 m), the waterproof layer of the arch. part and the second lining of the mold are present.
4	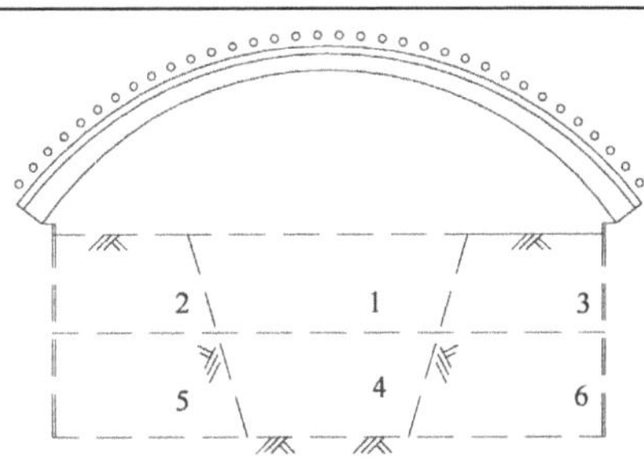	Control blasting technology is used to excavate the lower half section in order from section 1 to 6 and to support the initial construction in time.
5	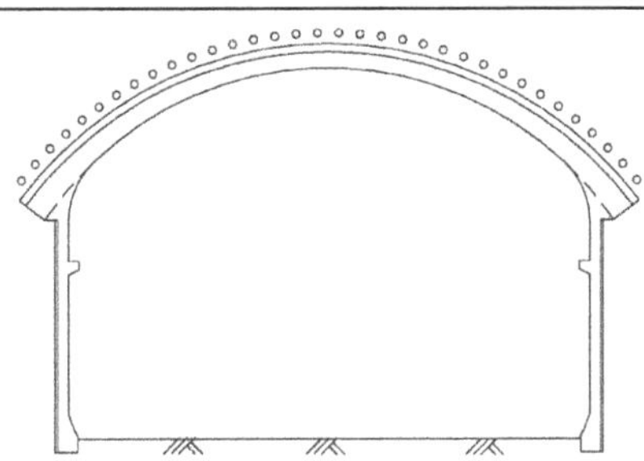	Waterproof layer of construction boundary wall and second lining.

(Continued)

Table 4.5. (*Continued*)

S/N	Construction procedure diagram	Say clearly
6	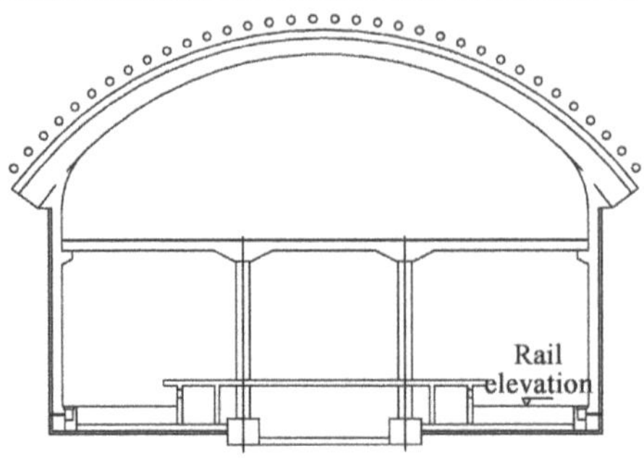	Construction station internal structure and decoration.

performed on the central arc and the temporary support is set up across at the bottom of the vertical, longitudinally reinforced grid that is connected to a whole ring, which is spaced 1 m away using the net-suspended spray concrete method. Through the underground pipeline and other sensitive areas, such measures as large pipe sheds, double-row small pipe grouting and tight grates are adopted.

Because the station is buried deep at about 10 m, the station arch is made up of strong weathered granite, the upper part is artificially filled with a coarse gravel sand layer, the formation is weak and the permeability is strong. It is very important to control the surface subsidence during the excavation of the railway station to prevent water inrush during construction. The measures adopted are as follows:

(1) Well-advanced measures are taken.
   (a) **TSP advanced detection:** TSP advanced detection of underground station structure is conducted, focusing on the fault fracture zone, fissure development zone and lithological interface.
   (b) **Advance probing:** In the tunneling process, each excavation ring, through the deep hole of the tunnel, is moved to the front of the tunnel for forward detection.
   (c) **Early geological drilling:** Survey found that forecast methods, advance geological drilling and advanced detection are used to

detect the abnormal areas, the fault fracture zone, fracture belt, shooting rock and extrusive rock contact zone.

(d) **Geological analysis:** The whole tunnel geological sketch, geological information records of the scene, and various detection methods of geological information are integrated, through the comprehensive analysis, forecasting the engineering, geological and hydrogeological conditions ahead.

(2) The arch part of the station is reinforced with a large pipe shed and a leading small catheter.
(3) Backfill grouting is performed behind the initial support.
(4) To control the dynamics of strata subsidence, the total subsidence control or target decomposition of each of the major construction process is done, so that each step in the construction has clear stratum settlement control standards; if a certain stage of the construction exceeds allowable values, strata subsidence is controlled to adjust the construction plan for follow-up.

4.3.1.6 *Construction of arch part ii*

When the initial convergence is less than 0.2 mmid, the arch construction is started. First, the temporary support is removed and the monitoring information is combined to determine whether anything needs to be changed. The following are used: buckle arch using stereotyped arch, steel template, bowl, every segment lining 9 m long and concrete by using pump pumping mode, symmetrical horizontal casting and vibrating plate. This ensures that the concrete is solid on the inside and beautiful on the outside. Attention must be paid to the protection of the waterproof plate (coil) when the reinforcement is tied up, and the wall connecting bar must be reserved.

(1) The lower part of the main body (blasting) and the side wall are provided with support during excavation. During bottom excavation, in the middle section first blasting is conducted, then the long vertical support is set up on both sides of the blasting excavation (using loose blasting, small vibration). The soil is excavated manually or by means of shallow eye loose blasting, and the residual soil is transported to the shaft by small machinery, which is hoisted to the storage yard for stacking

and external transportation. A shallow hole step is used to control the blasting in the central pulpit, and smooth blasting is used on the periphery to reduce the disturbance to the side wall and surrounding rock.

(2) Construction of the main body of the second lining adopts the arch reverse (first hat), and the lower structure is smooth. When one digs to the bottom of the hole, the waterproof plate (coil) shall be laid, and the reinforcement of the bottom plate shall also be laid and the concrete of the bottom plate shall be poured; then the side wall, the middle column and the middle plate shall be applied successively. The construction technology of the base plate, side wall and middle column of the main structure of the station is similar to that used for Ningxia road station.

(3) As for the construction of the main body of the station, the second lining of the arch is first done in the bottom structure as it is suitable for both the side wall and the top arch. The foot connection is a key point. The side wall construction joints should be waterproof, for which we should use good quality concrete casting. If the joint seam processing is not done properly, it will directly affect the overall waterproof effect of the whole station. Specific practices are as follows:

   (a) **Step 1:** The construction seams of the upper part of the wall are cut; the slurry and impurities are removed to expose the fresh concrete surface. The sealant is set on the clean concrete surface and the grouting pipe is buried. The grouting pipe and concrete surface must be fixed firmly.
   (b) **Step 2:** The side wall template is installed.
   (c) **Step 3:** Concrete should be poured on the steps of the side wall, and vibrated fully.
   (d) **Step 4:** After the concrete of the side wall reaches the curing strength, the mold is removed.
   (e) **Step 5:** The construction seam is excavated and the concrete is poured.

4.3.2 *Setting of air shaft (vertical inclined shaft) of underground excavation station*

The general Beijing construction method is similar in concealed excavation of tunnels, but usually it does not have a temporary vertical slope alone. Two side wind wells for construction of vertical shafts are available at the station and an air duct is used in the construction of channel. Due to the expansive excavation of the station, it is appropriate to construct the station from both sides, turning the construction direction, constructing the main structure junction of the wind tunnel, and finally constructing the wind tunnel and wind well.

The vertical shaft is constructed by the hanging wall method, and the initial support system adopts the supporting type of shotcrete and shotcrete steel frame.

After completion of the well and its accompanying parts, such as the pit, sump and floor, the shaft is made waterproof with secondary lining. The waterproof rating for the air duct is done. The wind well construction generally adopts a full-pack waterproof board (coil) which is laid without pinholes following the double-seam construction technology. The buffer layer adopts 400 g/m^2 of geotextile and horizontal construction joint is adopted for the expansion of the buried structure under the influence of water, and the vertical construction joints adopt a buried-type rubber water stop. Before pouring the secondary lining, one should check whether the waterproof panel (coil) is damaged or not and repair it in time. Then the waterproof board (coil) is laid according to the requirement of the construction drawings after binding the steel shaft hole along the lap length and location, paying attention to protect the waterproof board in the process of steel binding (coil). For perfusion of the secondary lining, the lining is divided into three parts from the bottom to the top. According to the construction of the cross-sectional rod, the ingate inch lap is set aside and ingate lining board (coil) is reinforced and then modified with the horse-head door lining the waterproof plate (volume) and reinforcement of the lap. The reserved hole or embedded casing in the shaft shall be buried before the concrete construction is completed.

4.3.3 *Air duct construction*

- **Engineering examples:** Case of wind tunnel construction of Jiangxi road station of Qingdao Metro Line 3
- **Engineering background:** Qingdao Metro Line 3 Jiangxi road station is located in the southeast and northwest two-wind tunnels, both of

which are single-span double-layer arched structures, with composite lining and CRD construction.

4.3.3.1 *Excavation earthwork process of wind tunnel*

(1) Multiple laser pointing devices are used to control the middle line and level of excavation, ensuring that the excavation is smooth and the grid position is upright.
(2) In the beginning of excavation, the principle of "pipe ahead, strict grouting, short penetration, strong support, quick closure and frequent measurement" is strictly followed in the excavation.
(3) The steps should be shortened as far as possible, and the ring should be closed as early as possible. The two sides of the hole must be excavated at the beginning, through the temporary support of the middle hole, to reduce the bias.
(4) The steps of each vertical adjacent small cavern are 3–m, and the distance of the palm of the adjacent small cavern is 8–10 m.
(5) Each cycle is drilled with a scale of 0.5–0.75 m, and the arch is used to excavate the core soil.
(6) In order to ensure the completion of the work in a safe manner, it is necessary to open the excavation face to brush the slope, which should not be steep at 2:1, and to seal the palm surface with the spray.
(7) Due to the multiple excavation steps, the soil is disturbed many times, and the anchor pipe is set in each small hole, and the grouting is strengthened to reduce the settlement.
(8) Due to the excavation of multiple parts, it is required that the grid connection plate should be properly arranged, the processing precision is maintained and temporary support is bolted, so as to ensure the quality of the connection.

4.3.3.2 *Construction technology of wind tunnel second lining concrete*

4.3.3.2.1 Floor concrete construction

The concrete shall be constructed only after the construction of the steel bar is finished, inspected and accepted by the supervisor. Each casting length is about 30 m. Pump concrete construction is adopted.

4.3.3.2.2 Side wall and arch (roof) lining

The inner lining of the cavity is constructed using self-made lining scaffold assembly. The length of the stand is 4–10 m. According to the actual section of the site, the template is specially designed with a thick steel mold with a smooth surface and smooth surface.

4.3.3.2.3 Platform assembly

After the concrete floor reaches a certain intensity, it can be assembled in stages, using work yu steel lining test bench, with a pull rod connection between arches spanning 1.0 m, and a head plate made of special steel, to prevent blocking of the leaked slurry from flowing away.

4.3.3.2.4 Precautions for concrete lining

The section size must be checked before lining construction to guarantee lining thickness. At the same time, it is necessary to check whether the waterproof layer meets the requirements, and whether they are damaged or not. In the case of any of the above problems, no lining construction shall be carried out, and construction shall be carried out after the specifications are met. The end block of the lining platform must be tightly packed with the formwork platform and the lateral arch wall, during which the pores must be blocked to prevent the slurry from leaking into the lining construction and affecting the quality of the concrete.

The concrete flows through the conveying pump. The side wall and the roof are vibrated with the trap vibrator, and the arch is vibrated with the attached vibrator. Flow of concrete should be continuous both symmetrically and horizontally along the inclined seam. In case interruption occurs when concrete is poured, before continuing with the casting construction, soft layers and the cement mortar thin films are eliminated by rinsing with high-pressure water.

When the concrete is disassembled, its strength must meet the specifications, and it shall be carried out immediately after removal of the template, and the curing time shall not be less than 14d. Before concrete construction, embedded parts must be prepared according to the design requirements, holes must be reserve and their positions must be reviewed.

Their positions and shape changes must be monitored during construction and measures taken to deal with it if necessary.

4.3.3.3 *Air duct*

At the intersection of the air duct and the main tunnel of the station, close to the side of the station, according to the initial support outline of the station, a reinforced ring frame is constructed, which is used for the stress conversion of the opening into the station. At the same time of the air duct construction, the reinforced ring construction is carried out at the side of the station, and the reinforcement ring adopts the field frame reinforcement and spraying C25 concrete. When it reaches the designed strength, it can be dug in the hole, and the opening of the hole strengthens the ring beam.

Concrete construction is done by first marking the air duct side on the wall in the cross-section of the station, within the duct before constructing the hole. Digging helps in supporting the cross-section of the station. The steel bar is erected following a side-by-side lap welding method, by welding the steel plate on the grid of the steel frame. The net-suspended spray C25 concrete is allowed to form a new structure as a whole.

After completion of the new support system, the wind grid steel frame is taken apart and the air duct grill is also removed. A prefabricated "L" shape steel should be placed at the top of the grid steel frame. According to the change of the arc, the vertical grid steel frame is constructed through the "L" shape steel bar and the vertical change of length in the form of a grid is used to vault the grill with the station section of the grid connected as a whole. In the front of the hole, the main part of the arch is used as a leading pipe shed. The distance between the pipe rings is 400 mm. According to the conditions of the hole and the performance of the construction machine, the pipe shed adopts 4-m pipe joints, which are connected by the interlocking wires.

PengShi excavation, advanced small pipe construction, drilling emplacement, check hole drilling, hole cleaning, installation of pipe roof steel, reinforcement of cage, sealing treatment of orifice and grouting process group are done. The process is shown in Fig. 4.29.

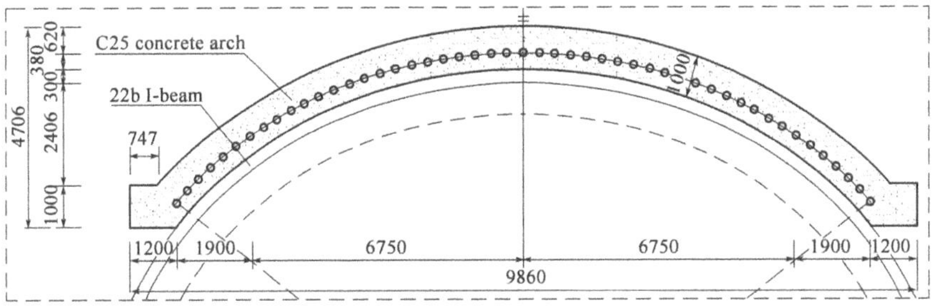

Fig. 4.29. Schematic diagram of the hole pipe shed (size unit: mm).

4.3.3.3.1 Excavation and support of the pipe shed

PengShi excavation is conducted and the construction on the waterproof and drainage facilities is done. The excavation sequence from top to bottom hierarchical divided, hard to dig attendants, timely closed slope, excavation to avoid rainy day.

4.3.3.3.2 Lead small catheters

In order to ensure the safety of the opening of the cave, the supporting measures of the leading small catheter are increased. Within the scope of the tunnel arch department at 110° along the steel flange, a small catheter forepole phi 42 is set for small catheters of 3.5 m length, connected to the steel frame for the whole body, with a ring spacing of 40 cm. In the slurry pipe ahead, the small pipe crossing angle is 15° with a vault profile line. Before excavation, the YT-28 pneumatic drill is used to drill through the outer edge of the arch, and then the leading small catheter is installed in the hole, and the end part is welded firmly with the steel arch. In order to reduce the disturbance of the surrounding rock structure, the small catheter adopts the double-liquid slurry for the grouting, and strengthens the surrounding rock system to enhance the stiffness of the small catheter. The construction process of leading small ducts is shown in Fig. 4.30.

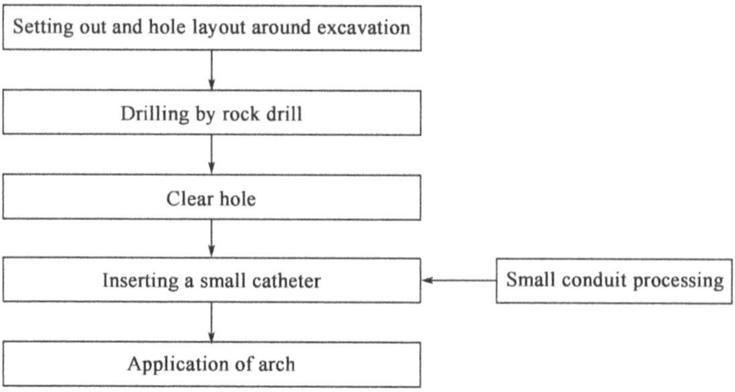

Fig. 4.30. Bar chart of lead small catheter construction process.

4.3.3.3.3 To make an arch

At the junction of the Ming and dark excavation, a C25 concrete arch with a length of 1.4 m and a thickness of 1 m is set. The specific construction method is as follows:

(1) In order to provide adequate lighting, when the concrete arch is lifted up to excavate the arch chamber line slope, the reserved core soil is used as a working platform. Using an artificial excavator, excavation is done on both sides of the wall, and the arch foot base is set at a reserved core soil depth of 30 cm using an artificial trim molding, based on the stability of the bedrock.
(2) Layout of the center line of the tunnel and the setting of the arch bottom elevation is done after proper checking is carried out according to the design requirements. Along the excavation contour line, I22b sections are erected 8 cm apart, with a phase spacing of 0.45 m. The back slope is set 25 cm apart, with the outermost edges of a set of arches being set 25 cm apart. Longitudinal welding is done to fix the C22 steel ring with a spacing of 1.0 m between the rings.
(3) According to the design requirements on the steel arch, the tubular steel pipe is made up of 150 holes, which can be fixed with the double-side welding (the length of the welding seam is greater than 5D), so that the steel arch and the orifice can form a whole.

(4) The external mold (with wood die) and the steel arch support are set. When the external mode support reinforcement is confirmed firmly and securely, the C25 concrete is poured, and the design strength is reached as early as possible in order to create the concrete arch, and a certain amount of early strength agent is added to the concrete.

(5) After the concrete pouring of the pipe shed is finished, it should be maintained in time.

(6) The installation of the arch formwork is done as follows:
   (a) The template adopts the wooden model. The board should be planted first and the edges are set at proper angles, so as to ensure smooth and tight joints in order to prevent leakage.
   (b) Along a set of arch contours, arch welding is done following a 3-ring template to support the C22 steel reinforcement ring. This is done to ensure that the concrete surface is bright and clean, the nails are set in the bottom die and edge die inside the floor, shunning paving.
   (c) The bottom die and the side die shall be supported by the square or steel pipe to prevent the expansion of the concrete.

(7) The following method is adopted to ensure that the tunnel limit is not invaded after the arch construction:
   (a) In order to ensure a solid foundation of the arch, it is necessary to ensure that there is no deviation and settlement during the construction of the pipe shed, and some temporary support should be added when necessary.
   (b) The concrete sleeve shall be used as the guide wall of the large pipe shed.
   (c) The orifice tube is used as the guide pipe of the large pipe, and the accuracy of the plane position, inclination angle and external angle of the pipe directly affects the quality of the pipe shed. Before welding, the whole station is used to set its plane position on the steel frame of the workpiece with the coordinate method. With the leveling ruler and the slope board, the angle of the orifice is set, and the outer corner of the orifice is set by the front and rear difference method. The orifice tube is firmly welded to the arch of the Gongyu steel to prevent the displacement when the concrete is poured.

4.3.3.3.4 The rig is in position

(1) The core soil is reserved to facilitate the construction of the arch and the drilling of the pipe shed as the principle.
(2) A mobile drilling rig is taken to the drilling site, the height of the drill is adjusted and the drill pipe is included in the set of arch orifice tube, with the total station, hanging wire and rod guiding the method of combining adjustment, for accurate verification of the drill position, ensuring the drill pipe axis is in conformity with the axis of the orifice tube.

4.3.3.3.5 Drilling

(1) The hole is drilled by two drilling rigs, which are symmetrical to the low-hole position, which can shorten the time of the mobile drilling and facilitate the positioning of the drill. For the smooth installation of tubular steel pipes, the diameter of the drill bit is should be 115 mm.
(2) The drill is positioned using the total station measurement, drill pipe direction and angle, which can be modified to meet the design requirements; while drilling a low speed is chosen at the beginning, and after drilling into a hole of 1.0 m at appropriate pressure, the drill must be withdrawn from the joint pipe and then continued.
(3) During the drilling process one has to pay attention to the change of drill pipe angle to ensure that the drill does not shift. One should check if the angle of the drill hole is correct for every 5 m, to ensure the correct drilling direction. If the deflection of the hole is too large, it can be corrected by a special drill. For example, for downward deflection, cement mortar is filling in the deflecting part, and then drilling is continued when the cement mortar is solidified. A special alloy bit is used for drilling again.
(4) The position of the drilling process is often determined with a slope ruler, and the hole quality is judged in a timely manner according to the drilling state of the drilling machine as are the holes made during the drilling process so as to avoid a card drilling accident.
(5) In drilling, the drilling machine ensures that the power device, centralizer and alloy bit are drilled in concentric circles.

(6) An original record of the drilling process is carefully made, and a geological judgment and description of the cuttings are made in time, which can be used as the advance geological forecast for the excavation of the hole.

(7) The following are the points to note during drilling:

   (a) If the drilling process has great resistance, it can be returned to about 1 m, repeated, and the resistance will be reduced after which one can continue drilling. The first drill is pushed in with low pressure to maintain the direction and prevent hole deflection.

   (b) The casing and the drill pipe shall be followed up to produce the function of the retaining hole, so as to avoid the collapse of the drill pipe after the hole is put forward and provide temporary protection hole to facilitate the injection of grouting in the hole.

   (c) After the hole is finished, the casing should be cleaned and then the drill pipe should be removed. The casing is still retained in the hole for protection.

   (d) The drill pipe is taken out with a hydraulic chuck with the drill stand, and the screw thread of the drill clamp is tightened and unscrewed, so that it can be used for operation.

   (e) After the steel pipe is inserted, one should take out the casing and drill into other holes. When the casing is taken out, the exposed rock will be pressed tightly in the hole. The steel pipe and the wall of the orifice are sealed with cement.

   (f) One must drill deep into the hole to meet the design requirements, and the drill pipe shall be connected to the length of the connecting sleeve until it is drilled to a depth of 0.5 m. One should drill down to the depth requirement and remove the drill pipe in turn.

4.3.3.3.6 Clearing of the hole

(1) After drilling the rig, the drill pipe should be used for drilling back and forth (115 mm) to remove the scum at the bottom of the hole to ensure that the hole diameter and hole depth meet the requirements and prevent blockage in the hole.The dregs from the bottom of the hole should be cleaned with high pressure wind.

(2) The whole station, slope ruler, etc., are used to detect the depth of the acceptance hole, the angle of the hole and the angle of the outside.

4.3.3.3.7 Installation of tubular steel pipe

(1) The piping and number of each drill pipe shall be arranged in advance according to the design of the steel pipe, so as to ensure that no more than 50% of the pipe connections are in the same section, with at least 1 m before and after the adjacent steel pipe joints. Steel pipe monoliths (that is, the length is 6 m or 9 m) are used under the socket length of 6 m and 9 m, and shall be of an odd and even number, respectively, and the even length is 2 m (if there are odd numbers of tubes of length 6 m in the first quarter, then there should be even number of tubes of length 3 m in the first quarter; the rest are 6 m and 9 m).
(2) The pipe shed with the material is 40 cm long and the outer diameter of the steel tube is 100 mm. When the steel tube is advanced, it is welded. The tube is drilled into 16 holes (no holes are drilled in the scope of the hole at 1.05 m). In order to facilitate grouting in the manhole and, the first section of the pipe head is welded into a cone.
(3) The pipe section of the large pipe shed is equipped with an orifice tube guide and reverse pressure top tube technology. When the resistance is increased and the manpower cannot be sent in, with 10 t and above, the winch is used to cooperate with the multi-direction of the pulley group to reverse the pressure, while always ensuring that the direction of the anti-pressure force is consistent with the hole axis.
(4) The steel tube shall be installed in a timely and fast manner so that it can be delivered to the bottom of the hole smoothly.
(5) The following matters needing attention when installing a shed steel pipe:
    (a) **Top pipe operation:** The first pipe is pushed into the manhole by connecting it with the drill pipe.
    (b) **Take over:** Holes are drilled on the outer side of the steel pipe at 30–40-cm intervals. The tongs are left on the pipe shed. Reverse

drilling of the rig is done. The steel pipe jacking is installed in the next section of the steel tube, with the last section of the steel pipe end suspended using artificial tongs. Even though the steel pipes are still connected together, the pressure on top of the two steel pipes is low.

(c) **Steel pipe end connections:** Ends are joined using a 0.4-m long phi 100 steel tube and a phi 108 steel pipe of 20 cm length, with an electric welding gap between the tube maintained at 20 cm. Both the front and the back edges of the pipe are welded. In order to prevent the backflow of slurry, the end of each pipe shed is welded with a slurry board, and the slurry board is made of a 2-mm thick steel plate, and the middle drill has 20 threads for grouting.

4.3.3.3.8 Steel cage

The reinforcing steel cage can meet the design requirements, and the joints are welded firmly.

4.3.3.3.9 Sealing of the orifice

Before grouting, the steel tube of each pipe shed shall be welded and sealed, and the steel plate shall be welded with a small steel tube with a set of grouting pipes, and the exhaust valve shall be set up.

4.3.3.3.10 Grouting

The grouting is performed by two kby-50/70 hydraulic grouting machine sections from the minimum hole position to the maximum hole position, and the grouting pressure increases from low to high. Initially, the pressure will be at 2 MPa, and then the grouting will be stopped after 15 min. Grouting hole quantity is 1.5 times the circular cylinder borehole; if grouting quantity is overrun, and it still does not meet the pressure requirements, one must continue to adjust the serum concentration of the grouting until the grouting quality standards are met to ensure that the borehole surrounding rock mass and the steel tube around the pore are filled with grout before termination of grouting. In order to reduce the adverse effect of wet depression of a large area caused by segregation and diffusion of the slurry, the slurry gel time is controlled by adding the accelerator at 3 PI min, and the maximum time is not greater than 1h.

(1) The application of the grouting process
   (a) When the pipe shed is installed, the small wood model is used to tighten the steel pipe and the surrounding rock wall, and then the waterproof cement (anchoring agent) is used to seal the gap, and the grouting shall be backed by the back grouting.
   (b) Grouting can be taken from the arch. The grouting speed is determined according to the volume of the grouting hole, generally from fast to slow. The gate valve is closed at the end of the grouting and the feed pipe is removed for the next cycle.
(2) Notes for grouting
   (a) One must always observe the change of the waterproof plaster around the grouting pipe and crack it when the slurry pressure increases.
   (b) Before the grouting, the foot tube connector must be prepared.
   (c) The pipe connector shall not be discharged after the slurry has solidified, and it shall be recycled after cleaning.
   (d) After the grouting is finished, the slurry is removed from some pipes and the pipes are filled with cement mortar to enhance the stiffness and strength of the pipe shed.
(3) Technical quality requirements
   (a) The diameter and quantity of the osmotic holes are ensured to guarantee the effect of osmosis.
   (b) The precision of drilling is high, and the final hole position is accurate. The holes are all falling on the same boundary surface to avoid large deviation and deformation.
   (c) Pipe roof elevation according to design requirements is 1°.
   (d) The co-axiality of the borehole is ensured to avoid anything being stuck in the pipe.
   (e) Steel pipe joints shall not be in the same section so as to avoid a weak surface force.
   (f) The connection between two steel pipes must be ensured.

(g) The steel tube hole position allows a deviation of 5 cm, and hole depth allows deviation of 5 cm.
(h) The axis deviation of steel pipe construction is less than 1%.
(i) The pipe length's allowable deviation is not less than the design length.

4.3.4 *The sound excavator is designed with blasting*

As the Qingdao area metro station is near residential and commercial areas, and according to the characteristics of the project and the factors that influence the degree of blasting vibration velocity, the general requirements are that blasting vibration velocity should be controlled below 1.5 cm/s. The engineering construction is determined by using the engineering analogy and calculations in the parameter selection process. The tunnel of the station is constructed by the drilling and blasting method. The blasting principle and the selection of blasting parameters are similar to that of the drilling and blasting method, which is not introduced here. The Beijing steps method and double-wall pilot tunnel method are used by excavating a section or by temporarily providing support. Certain requirements have to be fulfilled before blasting. Double-side heading method is divided into the six-step excavation method and nine-step excavation method. The construction process of a shed is shown in Fig. 4.31.

4.3.4.1 *The steps method*

The steps method is good for the surrounding rock of II or III class, which is given priority by the three-step method construction, with concrete blasting parameters as shown in Table 4.6

The blasting design is shown in Fig. 4.32.

4.3.4.2 *6-tunnel double-side wall guide hole method*

The sequence of the 6-tunnel excavation is shown in Fig. 4.33.

It can be seen from the figure that since the main structure of the station is symmetrical, 1, 3 and 2, 4 guide holes have the same size, and the hole layout is the same. Guide holes 5, 6 are excavated by

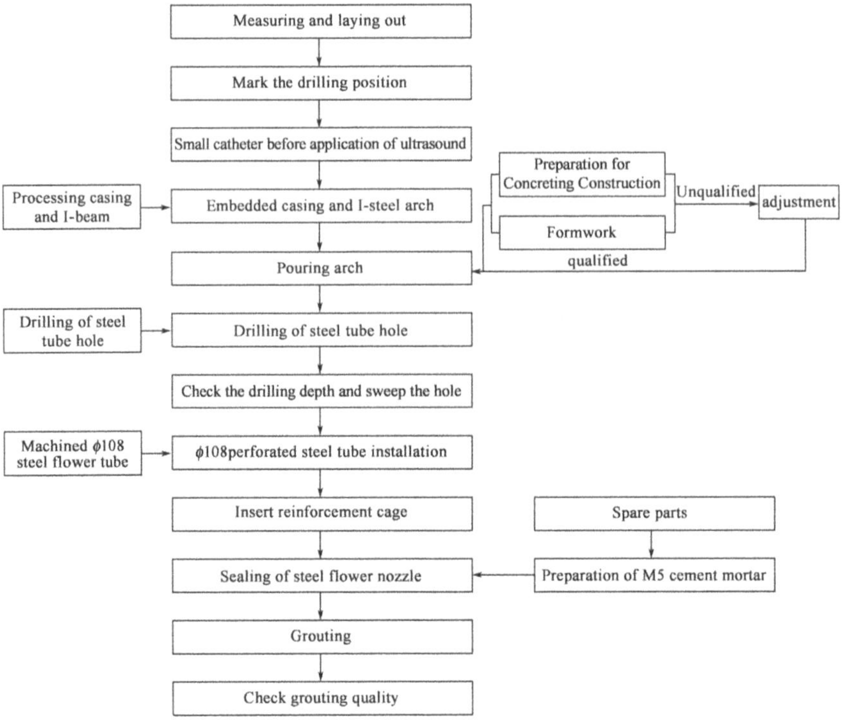

Fig. 4.31. Construction process block diagram of leading pipe shed.

Table 4.6. Steps method blasting parameters to excavate the station.

Open excavation method	Partial position	Area (m^2)	Hole number (No.)	Unit consumption (kg)	Total charge (kg)
Benching tunneling method	1	60.83	198	0.68	31.2
	2 (or 3)	28.05	42	0.81	17.1
	4 (or 5)	28.05	50	0.98	20.55
	6 — 1	22.47	22	0.61	10.35
	6 — 2	57.7	60	0.61	27.45
	6 — 3	50.92	50	0.6	22.95

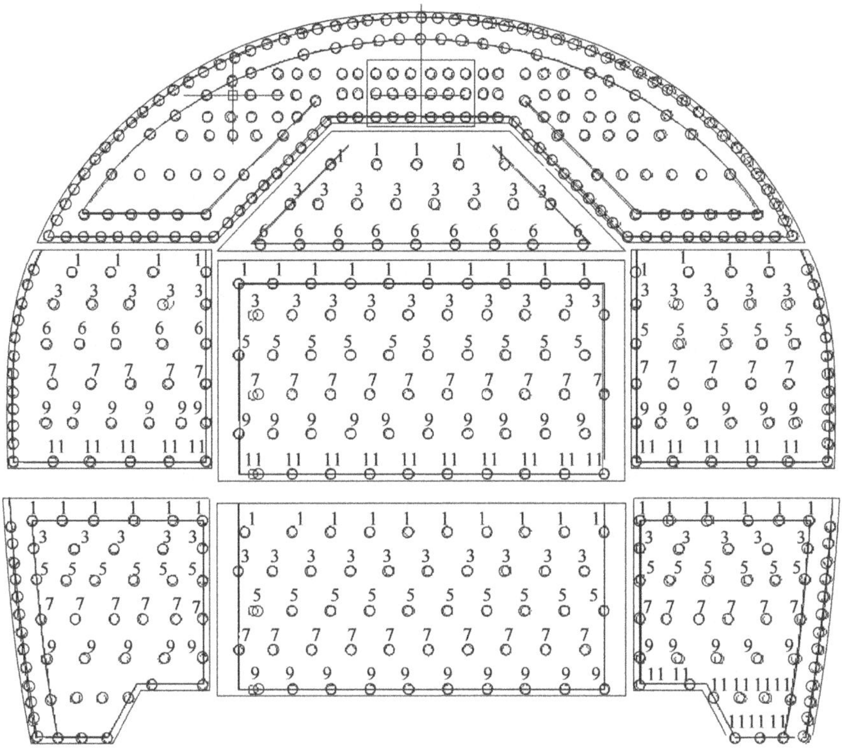

Fig. 4.32. Steps method.

the steps method due to the large section size. The specific perforation layout and blasting parameters are shown in Fig. 4.34 and Tables 4.7–4.9.

4.3.4.3 *Two-side wall guide hole method for main body of station*

When the station passes through the broken belt of the surrounding rock, the two-side wall guide pit is excavated for the main body of the station with the division 9, and the excavation sequence is shown in Fig. 4.35.

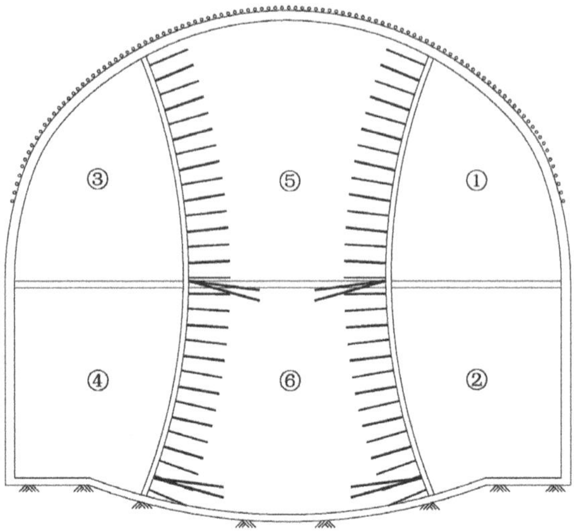

Fig. 4.33. Schematic diagram of the excavation sequence of the two-side wall of the guide hole.

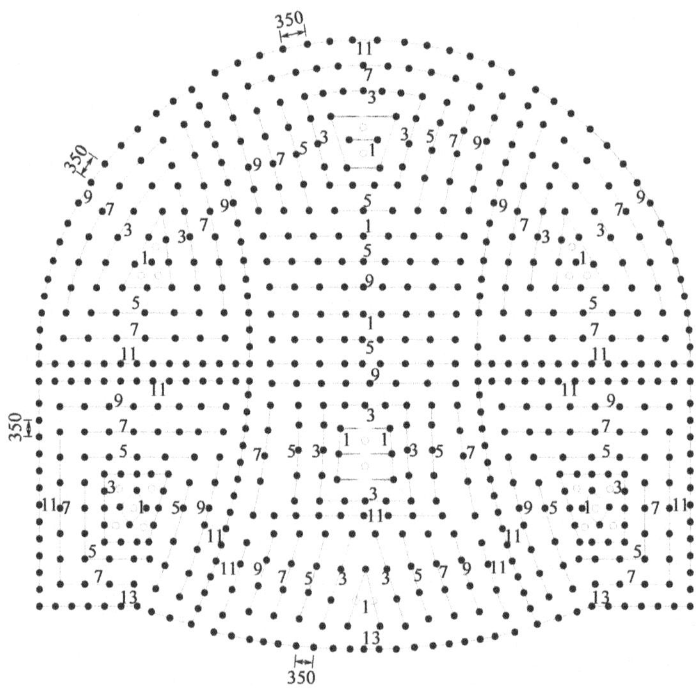

Fig. 4.34. The dimension of the hole layout in the guide hole (the size: mm).

Table 4.7. Blasting parameter table of 1/2 guide hole for the method of double-side wall guide pit in 6-pilot tunnel.

Gun hole name		Segment number	Aperture (mm)	Number of holes (no.)	Hole depth (m)	Pitch row (m)	Single-pore dose (kg)	Single-stage dose (kg)	Loaded length (m)	Loaded constitution
1	Cuthole	1	42	9	1.4	0.4	0.45	4.05	0.6	Continuation
Lead	Heading eye	3	42	11	1.2	0.55	0.35	3.85	0.4	Continuation
	Inner ring eye	5	42	5	1.2	0.7	0.35	1. 75	0.4	Continuation
	Inner ring eye	7	42	23	1.2	0.7	0.35	8.05	0.4	Continuation
	Periphery hole	9	42	23	1.2	0.35	0.3	6.9	0.4	Interval
	Bottom eye	11	42	21	1.2	0.5	0.3	6.3	0.4	Continuation
2	Cuthole	1	42	4	1.4	0.4	0.45	1.8	0.6	Continuation
Lead	Heading eye	3	42	15	1.2	0.5	0.35	5.25	0.4	Continuation
	Inner ring eye	5	42	17	1.2	0.7	0.35	5.95	0.4	Continuation
	Inner ring eye	7	42	17	1.2	0.7	0.35	5.95	0.4	Continuation
	Inner ring eye	9	42	15	1.2	0.7	0.35	5.25	0.4	Continuation
	Periphery hole	11	42	30	1.2	0.35	0.25	7.5	0.4	Interval
	Bottom eye	13	42	14	1.2	0.5	0.3	4.2	0.4	Continuation

(Continued)

Table 4.7. *(Continued)*

	Position	Excavated section (m^2)	Expected ruler (m)	Blasting stone square (m^3)	Total number of gun holes	Total explosive (kg)	Powder factor (kg/m^3)	Peripheral eye spacing *e* (mm)	Peripheral eye resistance line *w* (mm)	Position
Comprehensive parameters number	1lead	41.66	1.2	49.99	92	30.9	0.743	350	650	1lead
	2lead	44.29	1.2	53.15	112	35.9	0.811	350	650	2lead
Instructions	(1) The condition of the surrounding rocks can be adjusted as appropriate, with no less than 40 cm in the perforation length. (2) In 1 lead the total number of the guide holes is 92 and the total dose is 30.9 kg. (3) In 2 lead the total number of guide holes is 112, and the total dose is 35.9 kg. (4) The explosives are emulsion explosives, the detonator is a non-electric millisecond detonator and the specific blasting construction parameters are adjusted after the field test.									

Note: The guide holes are the same as 1 and 2.

Table 4.8. Blasting parameter table of guide tunnel for the method of double-side wall guide pit in 6-pilot tunnel 5.

Gun hole name		Segment number	Aperture (mm)	Number of holes (no.)	Hole depth (m)	Pitch row (m)	Single-pore dose (kg)	Single-stage dose (kg)	Loaded length m)	Loaded constitution
Upper stage	Cuthole	1	42	6	1.4	0.4	0.45	2.7	0.6	Continuation
	Heading eye	3	42	17	1.2	0.5	0.35	5.95	0.4	Continuation
	Inner ring eye	5	42	13	1.2	0.65	0.35	4.55	0.4	Continuation
	Inner ring eye	7	42	21	1.2	0.65	0.35	7.35	0.4	Continuation
	Periphery hole	9	42	12	1.2	0.6	0.3	3.6	0.4	Interval
	Periphery hole	11	42	14	1.2	0.35	0.3	4.2	0.4	Interval
Middle stage	Heading eye	1	42	9	1.2	0.65	0.40	3.6	0.4	Continuation
	Heading eye	5	42	9	1.2	0.6	0.40	3.6	0.4	Continuation
	Heading eye	9	42	9	1.2	0.55	0.40	3.6	0.4	Continuation
Lower stage	Heading eye	1	42	9	1.2	0.65	0.40	3.6	0.4	Continuation
	Heading eye	5	42	9	1.2	0.60	0.40	3.6	0.4	Continuation
	Heading eye	9	42	9	1.2	0.55	0.40	3.6	0.4	Continuation
	Position	**Excavated section (m^2)**	**Expected ruler (m)**	**Blasting stone square (m^3)**	**Total number of gun holes**	**Total explosive (kg)**	**Powder factor (kg/m^3)**		**Peripheral eye spacing E (mm)**	**Peripheral eye resistance line *W* (mm)**
Heteropsis parameter number	Up	32.79	1.2	39.35	83	28.35	0.86		350	650
	Middle	21.93	1.2	26.316	27	10.8	0.50		—	—
	Below	21.93	1.2	26.316	27	10.8	0.50		—	—
Instructions	(1) The state of the surrounding rock can be adjusted, and the length of the hole is not less than 40 cm. (2) The total number of shot eyes was 137, and the total dosage was 49.95 kg. (3) The explosives are emulsion explosives, the detonator is a non-electric millisecond detonator, and the specific blasting construction parameters are adjusted after the field test.									

Table 4.9. Blasting parameter table for six pieces using double-side wall guide pit method for 6-guide hole.

Gun hole name		Segment number	Aperture (mm)	Number of holes (no.)	Hole depth (m)	Pitch row (m)	Single-pore dose (kg)	Single-stage dose (kg)	Loaded length (m)	Loaded constitution
Upper oral step	Cuthole	1	42	6	1.4	0.4	0.45	2.7	0.6	Continuation
	Heading eye	3	42	16	1.2	0.5	0.35	5.6	0.4	Continuation
	Inner ring eye	5	42	10	1.2	0.65	0.35	3.5	0.4	Continuation
	Periphery hole	7	42	10	1.2	0.65	0.3	3.0	0.4	Interval
	Periphery hole	9	42	8	1.2	0.6	0.3	2.4	0.4	Interval
	Bottom eye	11	42	8	1.2	0.5	0.35	2.8	0.4	Continuation
Lower mouth step	Cuthole	1	42	3	1.4	0.4	0.5	1.5	0.6	Continuation
	Heading eye	3	42	8	1.2	0.55	0.35	2.8	0.4	Continuation
	Inner ring eye	5~9	42	24	1.2	0.60	0.35	8.4	0.4	Continuation
	Periphery hole	11	42	8	1.2	0.5	0.3	2.4	0.4	Interval
	Bottom eye	13	42	19	1.2	0.35	0.3	5.7	0.4	Continuation

	Position	Excavated section (m^2)	Expected ruler (m)	Blasting stone square (m^3)	Total number of gun holes	Total explosive (kg)	Powder factor (kg/m^3)	Peripheral eye spacing E (mm)	Peripheral eye resistance line *W* (mm)
Heteropsis parameter number	Up	36.11	1.2	43.33	58	20	0.554	500	650
	Below	31.88	1.2	38.26	62	20.8	0.65	500	650

Instructions

(1) The state of the surrounding rock at the junction of charge can be adjusted as appropriate, and the length of plugging is not less than 40 cm.

(2) The total number of holes was 120 and the total dosage was 40.8 kg.

(3) Emulsion explosives and non-electric millisecond detonators are used. The specific blasting construction parameters are adjusted after field test detonation.

It can be seen from the figure that since the main structure of the station is symmetrical, 1, 2 and 3 leads are the same size as that of 4, 5 and 6 guide holes, and they have the same gun holes.

The specific perforation layout and blasting parameters are shown in Fig. 4.36 and Tables 4.10–4.12.

4.3.5 *Summary*

(1) In general, the hard rock geological conditions suitable for construction for Qingdao Metro Line 3, Line 2 and Line 1 by means of the Beijing method require improvement and evolution. The double-wall pilot tunnel method of Line 3 is given priority, gradually shifting to the mainly subsurface-excavated station method of the design and construction scheme for Line 1. The buried depth also gradually

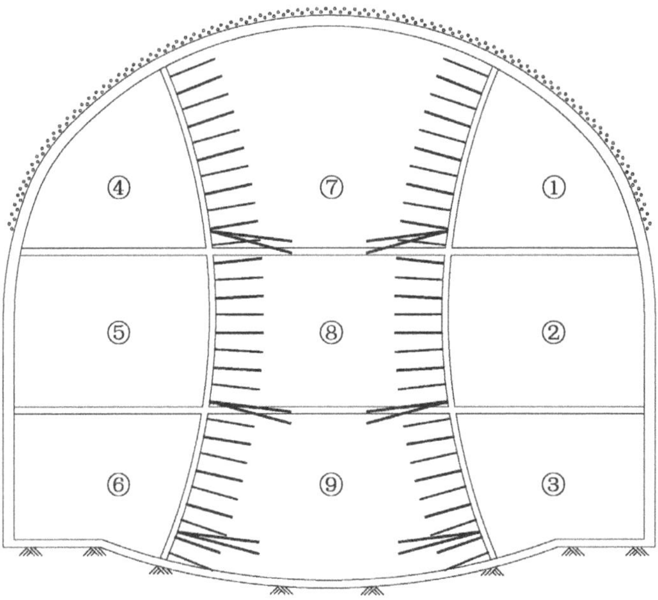

Fig. 4.35. The excavation sequence diagram of the guide hole of the two-side wall of the guide hole.

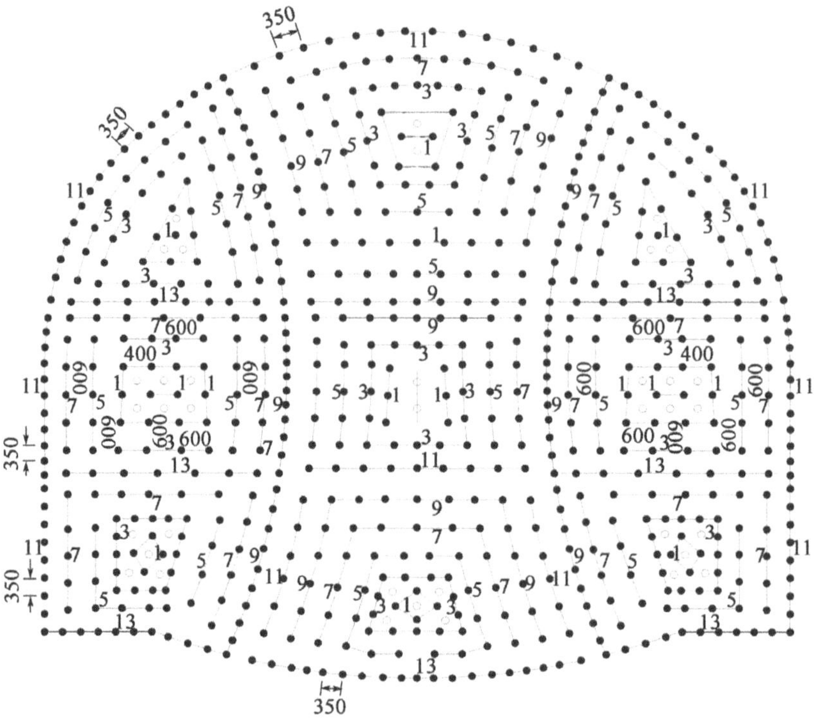

Fig. 4.36. The dimension unit of the hole layout of the guide hole (mm).

increases, and the hard rock geological Beijing program gradually matures.

(2) It is difficult to meet the requirements of land expropriation and relocation for the surrounding area, and it is difficult to meet the requirements of traffic flow. The advantage of this project is that the construction covers a smaller area, using a small amount of space (about 1,000 m^2), does not disturb the roadside green belt much and does not require traffic diversions or guides that will increase traffic pressure on the city.

(3) Hard rock geological conditions (grade II, III surrounding rock) are such that the station is buried deep. There are three appropriate methods. If encountering soft–hard formation, arch cover construction is

Table 4.10. Blasting parameter table for 1D 2 N 3-pilot cave using the method of double-side wall pit in 9-pilot tunnels.

Gun hole name		Segment number	Aperture (mm)	Number of holes (no.)	Hole depth (m)	Pitch row (m)	Single-pore dose (kg)	Single-stage dose (kg)	Loaded length (m)	Loaded constitution
1	Cuthole	1	42	9	1.4	0.4	0.45	4. Y5	0.6	Continuation
Lead	Boring eye	3	42	11	1.2	0.6	0.35	3.85	0.4	Continuation
	Inner ring eye	5	42	14	1.2	0.65	0.35	4.9	0.4	Continuation
	Inner ring eye	7	42	7	1.2	0.7	0.35	2.45	0.4	Continuation
	Periphery hole	9–11	42	30	1.2	0.35	0.3	9	0.4	Interval
	Floor eye	13	42	8	1.2	0.6	0.35	2.8	0.4	Continuation
2	Cuthole	1	42	12	1.4	0.4	0.45	5.4	0.6	Continuation
Lead	Boring eye	3	42	8	1.2	0.5	0.35	2.8	0.4	Continuation
	Inner ring eye	5	42	10	1.2	0.65	0.35	3.5	0.4	Continuation
	Inner ring eye	7	42	10	1.2	0.7	0.35	3.5	0.4	Continuation
	Periphery hole	9–11	42	22	1.2	0.35	0.3	6.6	0.4	Interval
	Floor eye	13	42	7	1.2	0.65	0.35	2.45	0.4	Continuation
3	Cuthole	1	42	4	1.4	0.4	0.45	1.8	0.6	Continuation
Lead	Boring eye	3	42	15	1.2	0.5	0.35	5.25	0.4	Continuation
	Inner ring eye	5	42	11	1.2	0.65	0.35	3.85	0.4	Continuation
Lead	Inner ring eye	7	42	11	12	0.7	0.35	3.85	0.4	Continuation
	Periphery hole	9–11	42	17	1.2	0.35	0.3	5.1	0.4	Interval
	Floor eye	13	42	11	1.2	0.6	0.35	3.85	0.4	Continuation

(Continued)

Table 4.10. *(Continued)*

	Position	Excavated section (m^2)	Expected ruler (m)	Blasting stone square (m^3)	Total number of gun holes	Total explosive (kg)	Powder factor (kg/m^3)	Peripheral eye spacing E (mm)	Peripheral eye resistance line W (mm)
Heteropsis parameter number	1 lead	25.27	1.2	30.32	79	27. Y5	0.892	350	650
	2 lead	30.25	1.2	36.3	69	24.25	0.729	350	650
	3 lead	27.40	1.2	32.88	69	23.7	0.865	350	650
Instructions	(1) The condition of the surrounding rock can be adjusted as appropriate, with no less than 40 cm in the perforation length. (2) The total number of holes is 217 and the total dose is 75 kg. (3) The explosives are emulsion explosives, the detonator is a non-electric millisecond detonator and the specific blasting construction parameters are adjusted after the field test.								

Note: The double-side wall method is divided into 9 holes. Guide holes 4, 5 and 6 are the same as 1, 2 and 3, respectively.

Table 4.11. Blasting parameter table for the method of double-side wall guide pit in 9-guide holes 7-guide tunnel.

Gun hole name		Segment number	Aperture (mm)	Number of holes (no.)	Hole depth (m)	Pitch row (m)	Single-pore dose (kg)	Single-stage dose (kg)	Loaded length (m)	Loaded constitution
Upper stage	Cuthole	1	42	6	1.4	0.4	0.45	2.7	0.6	Continuation
	Boring eye	3	42	17	1.2	0.6	0.35	5.95	0.4	Continuation
	Inner ring eye	5	42	10	1.2	0.65	0.35	3.5	0.4	Continuation
	Inner ring eye	7	42	21	1.2	0.7	0.35	7.35	0.4	Continuation
	Inner ring eye	9	42	12	1.2	0.5	0.35	4.2	0.4	Continuation
	Periphery hole	11	42	14	1.2	0.5	0.3	4.2	0.4	Interval
Lower stage	Boring eye	1	42	9	1.2	0.6	0.45	4.05	0.6	Continuation
	Boring eye	5	42	9	1.2	0.6	0.45	4.05	0.4	Continuation
	Boring eye	9	42	9	1.2	0.6	0.45	4.05	0.4	Continuation
	Position	**Excavated section (m^2)**	**Expected ruler (m)**	**Blasting stone square (m^3)**	**Total number of gun holes**	**Total explosive (kg)**	**Powder factor (kg/m^3)**	**Peripheral eye spacing E (mm)**	**Peripheral eye resistance line W (mm)**	
Comprehensive parameter number	Up	35.75	1.2	42.9	80	27.9	0.65	350	600	
	Below	18.43	1.2	21.76	27	12.15	0.55	—	—	
Instructions	(1) The condition of the surrounding rock can be adjusted as appropriate, with no less than 40 cm in the perforation length. (2) The total number of holes is 107 and the total dose is 40.05 kg. (3) The explosives are emulsion explosives, the detonator is a non-electric millisecond detonator and the specific blasting construction parameters are adjusted after the field test.									

Table 4.12. Blasting parameter table for the method of double-side wall guide pit in 9-guide tunnel of 8/9-guide tunnel.

Gun hole name		Segment number	Aperture (mm)	Number of holes (no.)	Hole depth (m)	Pitch row (m)	Single-pore dose (kg)	Single-stage dose (kg)	Loaded length (m)	Loaded constitution
8	Cuthole	1	42	6	1.4	0.4	0.45	2.7	0.6	Continuation
Lead	Boring eye	3	42	16	1.2	0.6	0.35	5.6	0.4	Continuation
	Inner ring eye	5	42	10	1.2	0.7	0.35	3.5	0.4	Continuation
	Inner ring eye	7	42	10	1.2	0.7	0.35	3.5	0.4	Continuation
	Periphery hole	9	42	9	1.2	0.5	0.3	2.7	0.4	Interval
	Floor eye	11	42	9	1.2	0.65	0.35	3.15	0.4	Continuation
9	Cuthole	1	42	4	1.4	0.4	0.45	1.8	0.6	Continuation
Lead	Boring eye	3	42	13	1.2	0.45	0.35	4.55	0.4	Continuation
	Inner ring eye	5	42	14	1.2	0.6	0.35	4.9	0.4	Continuation
	Inner ring eye	7	42	13	1.2	0.7	0.35	4.55	0.4	Continuation
	Inner ring eye	9	42	17	1.2	0.7	0.35	5.95	0.4	Continuation
	Periphery hole	11	42	12	1.2	0.6	0.3	3.6	0.4	Interval
	Floor eye	13	42	20	1.2	0.35	0.3	6	0.4	Continuation
	Position	**Excavated section (m^2)**	**Expected ruler (m)**	**Blasting stone square (m^3)**	**Total number of gun holes**	**Total explosive (kg)**	**Powder factor (kg/m^3)**	**Peripheral eye spacing E (mm)**	**Peripheral eye resistance line W (mm)**	
Comprehensive parameter number	8 lead	30.49	1.2	36.59	60	21.15	0.694	350	600	
	9 lead	38.29	1.2	45.95	93	31.35	0.682	350	600	
Instructions	(1) The loading capacity and the condition of the surrounding rock can be adjusted as appropriate, and the blockage length of the perforation is not less than 40 cm. (2) In 8 the total number of guide holes is 60, and the total dose is 21.15 kg; in 9, the total number of holes is 93, and the total loading is 31.35 kg. (3) The explosives are emulsion explosives, the detonator is a non-electric millisecond detonator and the specific blasting construction parameters are adjusted after the field test.									

used; for unfavorable conditions, the double-side heading method is used; and for hard rock (in general with weathering above), one should ensure that the arch foot is used, all of which ensure the integrity of the structure. Therefore, the arch cover method on the soft–hard formation should be further investigated.

(4) When decorated, the vertical slope makes full use of the station wind well, air duct and adjacent structures. The wind well has set priorities for the station main body excavation of vertical shafts, and the ducts can be used as channel construction, as long as they are conditionally adjusted to the temporary population slope gradient that is set for construction, greatly reducing the construction degree of transformation, improving the overall station construction efficiency and reducing the need for a temporary junction structure.

4.4 Waterproofing drainage construction of subsurface-excavated station

4.4.1 *A general description of the waterproofing of the sound*

Tunnel waterproof design should be formed in advance taking the following into consideration: geological forecast system analysis for the geological fracture zone ahead + advance grouting behind and early slurry injection + waterproof membrane + secondary lining concrete waterproofing system + multi-channel lines + a variety of methods. All these methods were designed in order to resolve the leakage point processing fault or fracture zone construction joint and deformation joint waterproof weak positions. Since the establishment of a reinforced concrete structure waterproof system, in order to structure the waterproofing, one had to strengthen the reinforced concrete structure, the crack seepage prevention ability and the crack resistance of concrete, and take effective measures to ensure the concrete density, good integrity and decreased cracking of the structure, thereby improving working conditions for reinforced concrete structures and further improving its long-term resistance.

The chosen flexible waterproof material should possess excellent durability, high property index, adapt to the expansion of concrete structures' changed shape, be convenient in construction and have a certain microbial resistance and corrosion resistance. One should avoid the use of poor construction and waterproofing materials.

It is preferred to choose waterproof materials and systems which are not conducive to channeling water.

In view of any congestion that the tunnel drainage system may produce, one must consider the maintainability of the drainage system in order to avoid water pressure rise due to drainage clogging the behind the lining, which is a safety hazard for the structure.

4.4.2 *Waterproof standard*

The waterproof level of the underground station and the centralized section of the electromechanical equipment shall be of the first grade; water seepage shall not be allowed, and the structure's surface shall not be wet. The waterproofing should be of a level 2 for the tunnels, connection channels and subsidiary tunnel structures, and the top is not allowed to drip or leak. However, the junction structure surface can have a small wet stained area that is not more than 2/1,000 of the total waterproof area, or a 100 m^2 arbitrary wet stain in no more than 3 places, or a single wet stained area not greater than 0.2 m^2. Leaking in the tunnel can be no greater than 0.05 $L/m^2{\cdot}d$ as the average amount, or 0.15 $L/m^2/d$ in any area of 100 m^2. For the fault fracture zone, the pre-grouting method is used to block the water or seepage of the surrounding area from the tunnel. The main tunnel should not be larger than 0.2/d·m.

Besides the waterproof structure, the concrete waterproof layer between the lining and modulus of the tunnel should have strong anti-aging ability, high tensile strength, high tensile rate for the waterproof board (coil), good corrosion resistance and bacterial resistance. The thickness of the waterproof panel shall not be less than 1.5 mm, width shall not be less than 2 m and welding shall be minimized.

The thickness of the waterproof concrete structure is not less than 250 mm. The crack width of the concrete is not greater than 0.2 mm and no penetrating cracks are allowed.

4.4.3 *The running structure is self-waterproof*

Using compound lining and secondary lining for the concrete structure of the tunnel keeps it waterproof, strengthening crack seepage prevention ability and thereby ensuring that the level of permeability is not less than P10. The early support design adopts anticorrosive high-performance anti-seepage shotcrete and adopts wet shotcrete technology.

The construction mix ratio for waterproof concrete should be determined through tests of the anti-seepage level, which should be improved by one level (0.2 MPa) more than the design requirement. The waterproof concrete should meet the requirements of anti-seepage resistance, and it should also meet the durability requirements for compression, crack resistance, antifreeze and corrosion resistance.

4.4.4 *Station waterproofing system*

The waterproof measures include the analysis of the geological fracture zone ahead of the geological prediction system. Advance grouting improves the permeability coefficient of surrounding rock and control permeability. Between the first and the second lining, the structure of the tunnel prevents water infiltration. The specific waterproof measures are as follows:

(1) First, the geological fracture zone is analyzed by a geological prediction system. Based on the advanced geological prediction system of TSP and advanced geological drilling, this chapter comprehensively analyzes the geological and water inflow conditions of the front excavated surface.
(2) By means of pre-grouting, the water seepage in the surrounding area of the tunnel is blocked out of the structure. The grouting of the fracture zone is carried out. The injection material is mainly composed of ordinary cement single-liquid slurry, superfine cement single-liquid slurry, special sulfur aluminate cement slurry, etc.

    In the local crushing section, advanced grouting is formed around the tunnel chamber (or the whole section curtain grouting), and water space in the bedrock is closed.

According to the advance grouting osmotic quantity of the groundwater, by adjusting the lining of the primary support ring to the bolt grouting system, it further closes the diameter of the underground water channel and decreases groundwater seepage. Supplementary grouting treatment should be carried out before the application of the waterproof board (roll material) for the initial leakage support, especially during the construction period.

(3) Design of the waterproof layer of the inverted arch tunnel is as follows: 400 g/m^2 of the arch wall and 1.5-mm thick + 0.5-m wide ECB waterproof plate (coil) and 1.2-mm thick concave and convex drainage belt.

Double-seam welding technology is adopted for the lap joint of the waterproof board (coil), and the waterproof plate (coil) is welded with the same material gasket to ensure good construction quality of the waterproof plate (coil).

(4) The waterproof function of the structure is strengthened, a small amount of water seepage in the flow of primary support and secondary lining is closed and the secondary lining is of waterproof concrete, with a permeability grade level no less than P10 and structure thickness not less than 300 mm. Strict control of the mold and mold temperature of concrete is required to reduce the effect of temperature difference shrinkage and drying shrinkage. The crack width of concrete production is not greater than 0.2 mm, and penetrating cracks are not allowed.

4.4.5 *Station drainage*

4.4.5.1 *Drainage design*

During the operation, the cranny in the surrounding rock of the tunnel was the main water source, with a requirement for a drainage system, vertical drains from the ring to the drainage plate, horizontal drain pipes and tunnel drainage. In order to maintain the convenience of the construction, the drainage ditch is located at the side wall of the station. Between the tunnel primary support and secondary lining arch wall waterproof board (coil), yuan spinning cloth and a permeable layer are fixed, such as a concave and convex drain structure through the washable DN100 double-wall corrugated longitudinal drainage tube; then the water drains through the phi 100 horizontal PVC drain pipe on both sides of the station, through the drainage ditches, transfer stations and waste water pools via the pump discharge hole.

4.4.5.2 *Maintenance of drainage system*

In view of the possible jamming of the tunnel drainage system the maintainability of the drainage system must be considered so as to avoid drain clogging leading to a rise in water pressure. In order to ensure the maintainability of the drainage system, it is necessary to meet the mechanical performance index of high-pressure flushing and to have good permeability. An inspection well is installed every 50 m along the tunnel to maintain the vertical drainage blind pipe. The overhaul well is maintained by high-pressure water jet cleaning equipment and a special chute head, with high-velocity flow broken by the structure, the high-pressure water discharged into the overhaul well and the special mechanical joints being cleared of obstacles when necessary.

4.4.6 *The station has a drainage structure*

Figures 4.37 and 4.38 show the details of the drainage structure.

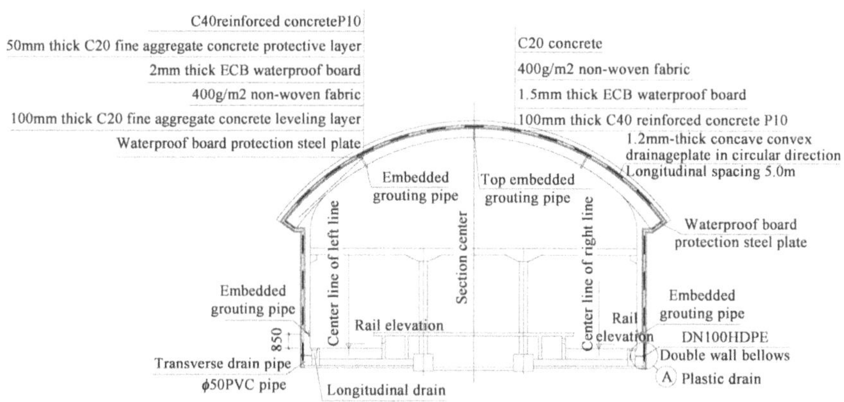

Fig. 4.37. Anti-drainage structure of large arch foot section (size unit: mm).

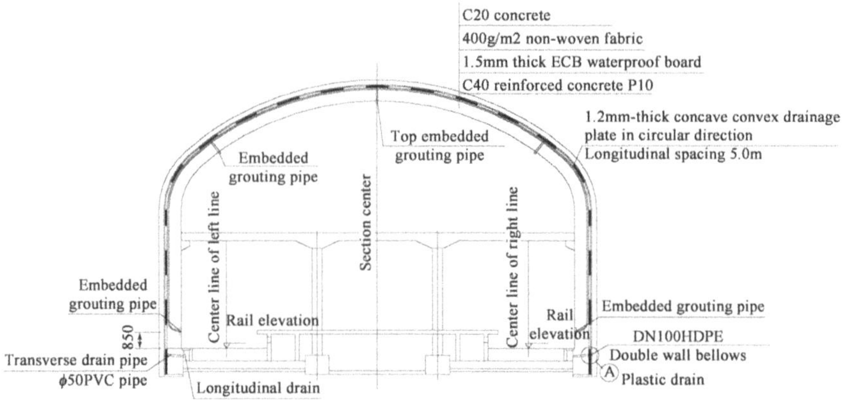

Fig. 4.38. Anti-drainage structure of compound lining section (size unit: mm).

4.4.7 *Flexible waterproof layer construction*

(1) The cushioning material is made of earthwork non-woven fabric. The lining of the cushion is aligned with the middle line of the tunnel. A special plastic washer made of the same material as the waterproof panel (coil) is fixed on the PE cushion, and the nail or expansion pipe screw is used to anchor it.

(2) The anchor points should be vertical and must not exceed the flat surface of the round cushion. The anchorage points are arranged in a plum shape. The spacing is as follows: the arch part is 0.5–0.7 m, the side wall is 1.0–1.2 m and the anchorage point is appropriately increased in the bump.

(3) The waterproof board (coil) is laid with a nail-laying method. First, a fixed buffer layer is laid on the concrete surface, and then the waterproof layer is welded or bonded to the buffer layer to prevent the waterproof layer from penetrating the nail hole.

(4) It is equipped with a 1.5-mm thick ECB waterproof board (coil) to provide the function of half a layer of waterproofing, and the non-woven protective layer of 400 g/m^2 is set on the surface of the waterproof layer in the plane part of the bottom plate, and the protective layer of C20 fine concrete with a thickness of 7 cm is poured.

(5) The waterproof board (coil) has a thermal double seam, and should not be less than 15 mm, and the width of a single weld

should be compact, without virtual welding. The weld surface should be smooth without ripples to withstand waterproof sealing detection.

4.4.8 *Waterproofing special parts of station*

(1) The waterproof treatment of the construction joints is done to maintain the grouting pipe system by using the backside stop water belt, and the slurry is filled into the compact space of the construction seam through the grouting of the embedded groove.
(2) The longitudinal construction joints of the wall shall not be left at the top or the bottom of the bending moment or the bottom plate and the side wall, and shall remain on the wall not less than 300 mm above the surface of the floor. The spacing between the construction joints of concrete structures has a great impact on the structure cracking, and the spacing between the ring and construction joints is not greater than 18 m.
(3) In order to control the longitudinal uneven settlement of the tunnel, the deformation joints are set up at the tunnel entrance, where there is maximum formation of mutation and the structural change. Three waterproof lines are designed for waterproofing: The first line is for waterproofing the outer layer of the structure, the second line is for embedding the central structural concrete in order to prevent the water from seeping into the surface and the third line is to install the check hose and sink (structure template construction according to design the reserved grooves).
(4) Through wall pipe fittings (such as earth electrode or wall tube), parts of the waterproof layer adopt the water-stop flange, and the water expansion article putty rubber (water) is used for waterproofing. Different waterproof materials must be chosen as appropriate to prevent leaking.

4.4.9 *Summary*

The waterproofing construction of the station structure is based on the principle of "avoiding the mains, a combination of rigid and flexible, multi-line defense, alteration for local conditions and comprehensive management". The following aspects should be focused on:

(1) The treatment of the initial base surface should refer to the initial acceptance standard and acceptance system in the tunnel, and ensure that the waterproof board (coil) is applied in water-free environment.
(2) The waterproof board (coil) is made of a real-name system, and the quality can be traced back.
(3) Concrete self-waterproofing is the foundation, which should be quality controlled during transportation, pouring and construction.

Bibliography

[1] Wang Minggao. Research on underground excavation of metro stations. *Theoretical Research on Urban Construction* (Electronic Version), 2016 (34): 86–89.

[2] Guijin Ben, Liu Wenbing, Zhou Jinqiang *et al*. Research on the structural anti-deviation technology of the construction of the column hole of the underground railway station. *Railway Standard Design*, 2016 (12): 114–118.

[3] Wang Mengshu. *China Tunnel and Underground Engineering Construction Technology*. Beijing: People's Transportation Press, 2010.

[4] Ma Linlin. The construction technology of the PBA in the four-six guide hole combination. *Railway Construction Technology*, 2016 (07): 10–13.

[5] Li Xiansong. Construction technology of shallow buried open-cut station adjacent to buildings. *Journal of Hunan University of Arts and Sciences (Natural Science Edition)*, 2016 (02): 54–59.

[6] Wang Jun. Key technology research on the construction design of underground open-cut station of rock texture. *Journal of Underground Space and Engineering*, 2016 (02): 442–449.

[7] Lu Yaobang, Gao Junwei, Liu Dongliang. The application of BIM technology in underground excavation of metro stations. *Construction Technology*, 2015 (S2): 679–682.

[8] Liao Qiuilin, Ma Jingdong, Zhang Pengfei *et al*. Key technology for the construction of the tunnel pile method in the underground excavation of

water-rich sandy pebbly stratum. *Construction Technology*, 2015 (23): 49–53.

[9] Yang Huijun, Kong Heng. Analysis on the surface subsidence of metro station construction by shallow buried deep excavation. *Journal of Railway Engineering*, 2015 (05): 81–85.

[10] Liu Na. Research on the prevention and drainage technology of underground open-cut station in Zhongshan Park of Qingdao Metro. *Shanxi Architecture*, 2015 (10): 153–154.

[11] Feng Yheng. The technique of the two-side wall guided pit in the pre-core soil of the ground iron and dark excavation site. *China National Defense Communication Engineering and Technology*, 2014 (05): 52–55 +51.

[12] Lu Bo. Key construction techniques for underground excavation of metro station. *Modern Tunnel Technology*, 2014 (03): 181–187.

[13] Huang Zuo Qiong. Research on the construction technology of underground tunnel crossing existing lines in metro station. *Modern Tunnel Technology*, 2014 (02): 133–139.

[14] Sun Xudong. The key technology for the construction of the cable station under the secret tunnel of the tunnel. *Tunnel Construction*, 2013 (05): 412–418.

[15] Wang Lixin. Method and safety analysis of a certain dark open-cut station and an analysis of the safety of an underground railway. *Underground Space and Engineering*, 2012 (04): 857–862 +868.

[16] Wu Xuefeng, Lv Wenjie, Zhang Liming *et al.* Stability analysis and support optimization of surrounding rock in underground railway station. *Journal of Underground Space and Engineering*, 2012 (05): 1059–1064.

Chapter 5

Grouting Reinforcement Technology of Poor Geological Sections

5.1 Overview of grouting reinforcement of poor geological sections

Grouting technology occupies a very important position in the field of poor area reinforcement. Grouting technology has been applied in the areas of civil engineering, especially in mine engineering and hydropower engineering where it has been widely used, and is an integral part of the construction method and its application has the following nine aspects: Building foundation reinforcement, improvement of the stability of the soil slope, reinforcement of the soil-retaining wall, construction repair of concrete cracks and flaws, dam foundation and anti-seepage reinforcement, reinforcement of underground structures and the water stop, reinforcement engineering of mines, broken belt body reinforcement and fractured rock mass-based seismic strengthening of water and power lines.

5.1.1 *Development of grouting technology*

Grouting technology can be roughly divided into four stages: the historic, original leaching soil slurry phase (1802–1857), the primary cement injection slurry phase (1858–1858), the intermediate chemical grouting stage (1920–1969) and the modern grouting phase (after 1969).

According to the relevant literature, the historical development of grouting technology is summarized as Table 5.1.

Table 5.1. Historical development of grouting technology.

Time interval		Country	Matter
Primitive clay slurry stage	1802	France	Charles Berigny used a slurry pump to repair the gravel soil foundation of a tidal gate that had been eroded by the current.
	1824	Britain	Joseph Aspdin invents cement.
	1838	Britain	The cement fills the Thomson tunnel.
	1845	America	W. E. Watson infuses cement mortar with a steep groove on a spillway.
	1854	America	W. E. Watson reinforced block masonry, and grouting is not considered a construction method, and is only used as a remedy for defective foundations.
Primary cement grouting stage	1886	Britain	Hosagood built bridges using chemical grout for the first time, and as a result it appeared in India. Kinniple used clay cement blocks for the underground seepage of the Rosetta and Damietta dam foundations. The British developed compressed air grouting pumps, which sped up the development of the cement grouting method.
	1887	Germany	The original silicification method was created which used two techniques: one drilled in water glass and the other bordered on calcium nitride.
	1909	Belgium	By increasing the concentration of dilute acid in the water glass, le Malmonte discovered the principle of gelation, and proposed that two kinds of slurry were used in a grouting system to obtain the patent.
	1914	Belgium	Albert Frankois used water glass and aluminum sulfate grouting.

Intermediate chemical grouting stage	1920	Holland	E. J. Joosten first demonstrated that the chemical slurry was feasible, using water glass and $CaCl_2$ to grout the same medium in two systems, and it patented in 1926.
	1934	France	Langge improved the Eusidon method, and made the slurry with superior diffusion radius and penetration capability.
	1940s	Japan	A method was used to create the conditions for using the slurry with a short gel time.
	1950s	America	The acrylic resin slurry (AM-9) was developed, and its leaching degree was close to water, and gel time could be adjusted arbitrarily.
Modern grouting stage	1974	Japan	In Fukuoka, the chemical grouting of acrylic acid was contaminated, leading the Japanese general office to stipulate that the slurry could only be used in water glass.
	1978	America	The production of AM-9 was stopped.
	1980s	Japan	The Mc-500 superfine cement grouting material was created.

Although the grouting technique was recognized and applied in China earlier, systematic research on it was conducted later than in other countries. Until the 1950s, we had little research on it. Since then, it was applied to all kinds of civil engineering projects. The main development process is shown in Table 5.2.

In recent years, with the development of the grouting material to green pollution-free environment-friendly pulp material, many scholars have used industrial waste residue to completely replace cement or have utilized the cement-added industrial waste residue method, developing a large number of new green grouting materials. Zhang Gaozhan from the Wuhan University of science and technology developed water glass industrial waste residue (fly ash, slag, steel slag, etc.) double-liquid grouting material and successfully used in the Wuhan Yangtze river tunnel. The method not only improves the performance of grouting materials but also saves the dosage of cement and reduces the environmental pollution caused by industrial waste slag, as it is a high-performance grouting material and a method of energy conservation and environmental protection.

Our country, besides creating many varieties of grouting material, has developed the construction technology process rapidly, forming a complete set of grouting equipment and detection methods in a short period of time, especially in terms of the grouting application field.

5.1.2 *Grouting reinforcement process*

The grouting reinforcement method is used by curing or injecting rock and soil foundation cracks or pores, set it in, filling it up, which is an extrusion process used to improve the physical and mechanical properties. According to the principle of grouting, grouting technology is divided into four categories: osmotic grouting, splitting grouting, compaction grouting and injection grouting. Among them, there are several common techniques of grouting reinforcement in hard rock formation.

5.1.2.1 *High-pressure rotary jet pile advance guard*

The advance guard of the high-pressure jet grouting pile reinforcement beam has principle effects on shell, such as improving the physical and

Table 5.2. Development of grouting technology in China.

Time interval	Advance path
1953	Water glass was used as the grouting material, with the first use of water glass slurry in Jiamusi, Heilongjiang province.
1960	The three gorges project team pioneered the formation of coagulation.
1964	The study of chemical slurry in laboratories and classrooms began. The Guangzhou institute of chemistry at the Chinese academy of sciences successfully developed the propylene cool glue (known as propyl coagulation), and the industrial test for the same was completed, which has added a new chapter to the history of chemical grouting in China.
1967	The cement grouting method was developed successfully, which was a big step forward.
1968	The chemical research institute of Guangzhou developed an epoxy resin chemical slurry with rice as the diluent.
1973	Tianjin university and other units developed the chemical slurry (oxygen coagulation).
1974	Water-soluble polyammonia vinegar slurry (HW) was developed by the science institute of east China at the survey and design department.
1979	The Chang Jiang academy of sciences and the Guangzhou institute of chemistry developed an elastic polyurethane slurry for unique chemical pulp materials in China, and made a significant contribution to the water-retaining capabilities of the Gezhouba dam. conservancy project.
1980s	The China institute of water resources and hydropower research developed sk-e, and JX was developed by the scientific research institute of China water conservancy and hydropower engineering bureau. The chemical research institute of Guangzhou developed a series of modified epoxy slurries, such as "sinochem-798". The Nanjing institute of water science and technology developed a solvent-free grouting material MU-acrylic vinegar.
1990s	The China water conservancy and hydropower research institute developed the underwater grouting material PU/EPIPN polyurethane epoxy resin. The Yangtze academy of sciences developed a highly permeable, highly pro-run CW epoxy, which has been successfully used in construction.

mechanical properties of soil, its high strength, large bearing capacity, good permeability, strong controllability, economical quality and practicality, and the cost is low, but the pile length is limited, with general horizontal jet grouting pile in the 8–15 m. Due to weathering of the broken surrounding rock, the hole efficiency is low. The size of the slurry diffusion is small, and it is impossible to attain the required reinforcement when there are stones or hard objects.

The water ash is large and the consolidation volume is larger, which may cause problems in the sandy uneven soil layer. It is mainly suitable for processing of silt, silt soil, plastic flow, Ruansu or plastic soil, powder soil, sandy soil, loess, and grain-filling soil and the gravel soil stratum; for big stones and a large amount of plant organic matter, the groundwater flow rate is too high and there is poor water field applicability, which was determined according to the results of the field study.

5.1.2.2 *Surface grouting reinforcement*

Tunnel construction is often prone to instability due to landslides, especially in the rainy season when surface soil is vulnerable to the rains, thereby washing away the outer and inner portions of the constructed tunnel, reducing the tunnel's roof covering and increasing the bias load. Ground grouting can improve the overall strength and rigidity of the surface rock mass, reduce the permeability of the surrounding rock, improve the tunnel arch and thus achieve the purpose of protecting water resources and guaranteeing the safety of the project. The effects of grouting treatment on shallow, buried, surface subsidence areas are remarkable, so the actual geological condition is dependent on the construction. The surface reinforcement technology is adopted for the external environment of the construction. The surface grouting reinforcement process mainly includes sleeve valve tube grouting, WSS grouting and steel tube grouting.

5.1.2.3 *Inner curtain grouting*

Curtain grouting in holes is utilized in the tunnel vault, with two constraints, a reasonable sized drill hole and the injection of grout, making each grouting body overlap one another to form a structure similar to the concrete diaphragm wall of the curtain, to truncate the flow of water, so as to achieve the aim of plugging. Curtain grouting in the hole can be divided into full-section curtain grouting and half-section curtain grouting. As the name implies, full-face heavy curtain grouting refers to the whole section that is closed deep, with shallow hole grouting consolidation, and water stop at the half section of the curtain grouting, referring to the weak part of the ground section of semi-closed deep water, shallow hole grouting consolidation. The purpose is to form a certain thickness of water wall curtain around the tunnel excavation, so as to ensure the smooth excavation of the tunnel and reduce the chance of accidents.

5.1.2.4 *Small pipe grouting reinforcement*

Small duct grouting reinforcement strata technology is used through the longitudinal lean forward along the tunnel excavation contour line, achieving the purpose of the reinforcement of the surrounding rock and water in advance; at the same time, a small catheter can also advance the protection effect.

Small pipe grouting construction technology is simple, easy to operate, safe in construction, and the soil layer reinforcement attains fast results, with slurry loss and low cost, and is one of the most commonly used reinforcement soil layers in tunnel construction. Small catheter grouting is only an auxiliary means to prevent the collapse and subsidence of underground engineering construction. Different design parameters should be selected according to the geological conditions, the size of the tunnel section and the type of supporting structure. It is mainly applicable to sand layers and gravel (egg) stone layers which are unreachable and self-stabilizing. The construction of small ducts is only to reinforce the local soil layer of the excavated palms, which should not be exposed for a long time, and the principle of excavation should be adhered to first. At the same time, small catheter grouting can also be used for various temporary formation reinforcements.

5.2 Conventional grouting material

With the development of grouting technology and the complex types of grouting engineering, the types and quantities of grouting materials are increasing. Among them, leaching soil and lime materials, the earliest type of grouting materials, are still widely used in some projects. In Britain in 1826 after the invention of Portland cement, based on the leaching degree of cement slurry, a sodium silicate cement slurry was used in grouting engineering on a large scale; at present, it is still the main material used in grouting engineering. Along with the development of the preparation technology of the chemical industry, cement-based materials made out of modified inorganic and organic chemical components are used in grouting engineering. In 1970, after the invention of sulphoaluminate cement, the rapid hydration and expansion of materials also occurred. In order to overcome the shortcomings of the cement materials, the relevant researchers at home and abroad developed solution-type inorganic or organic chemical grouting materials, and obtained good application results for the same.

At present, although there are many kinds of grouting materials, according to their composition and performance characteristics, they can be divided into inorganic grouting materials and organic grouting materials. Among them, the inorganic series grouting materials are classified into pure cement materials, water glass materials and cement matrix composites. Organic series grouting materials can be divided into simple organic polymer and organic compound materials. No machine is used for cement grouting using single liquid, and a sodium silicate cement grout is given priority; the commonly used organic series includes mainly polyurethane grouting material, epoxy resin, acetate resin, etc.

5.2.1 *Inorganic grouting material*

5.2.1.1 *Single-liquid cement slurry*

Single-liquid slurry is mainly cement, adding a certain amount of additive, and the slurry with water transfer that is injected into the rock layer is called single-liquid cement slurry. Additive refers to the accelerator, early strength agent, plasticizer and suspending agent which can improve the performance of cement slurry.

5.2.1.1.1 Basic properties and characteristics of single-liquid cement slurry

(1) **The basic properties of pure cement slurry**

The basic properties of pure cement slurry without any additives are shown in Table 5.3.

Table 5.3. Basic properties of cement slurry.

Water cement ratio (weight ratio)	Leaching degree ($\times 10^{-3}$ Pas)	Density (g/cm^3)	Gelatin time interval		Stone rate (%)	Compression strength (MPa)			
			Initial set	Final set		3d	7d	14d	28d
0.5:1	139	1.86	7 h 41 min	12 h 36 min	99	4.14	6.46	15.3	22.00
0.75:1	33	1.62	10 h 47 min	20 h 33 min	97	2.43	2.60	5.54	11.27
1:1	18	1.49	14 h 56 min	24 h 27 min	85	2.00	2.40	2.42	8.90
1.5:1	17	1.37	16 h 52 min	34 h 47 min	67	2.04	2.33	1.78	2.22
2:1	16	1.30	17 h 7 min	48 h 15 min	56	1.66	2.56	2.10	2.80

It can be seen from Table 5.3 that, with the increase of the water cement ratio in pure water slurry, the leaching degree, density, stone rate and compressive strength of cement slurry are significantly reduced, and the initial coagulation and final coagulation times increase.

(2) **The basic properties of single-liquid cement slurry**

According to the requirements of actual engineering, some additives are added in cement slurry to adjust its performance to meet the requirements of the grouting effect in engineering. The commonly used additives and their improvement on the performance of cement slurry are as follows:

a. **Cement accelerator:** The accelerator can shorten the chemical agents such as "711", "red star", "Yang spring", "calcium chloride", "water glass", "soda", "sodium chloride", etc. When the above-mentioned accelerators are added to the cement slurry, it will soon be able to make a cement slurry that is not suitable for the single-liquid cement grouting. In general, use the old single-liquid slurry grouting method, for the slurry with less than 3% of sodium silicate or below 5% calcium chloride, adding calcium chloride and water glass slurry and significantly increasing the speed setting, which will have little influence on the strength of cement stone body.
b. **Quick hardening of cement:** It not only shortens the cementing time but also improves the early strength of the cement stone body, and

has a good effect on the grouting. The commonly used rapid coagulant and early strength agent has three ethanol gluconates, triisopropanol gel, chlorinated sodium chloride, dihydrate gypsum, sodium chloride, etc. In general, the dosage of early strength accelerator is as follows: The cement content of triethylene glycol (or triisopropyl alcohol) is 0.05–0.1%, and the content of chlorination is 0.5–1.0%.

c. **Other additives:** In order to meet the special needs of engineering, in addition to the above cement admixture, sometimes we need to add other additives in the slurry, for example, retarders, early strength agents, expansive agents, water flow reduction agents, air entrained agent, SAP, etc.

(3) **The characteristics of single-liquid cement slurry**

Single-liquid slurry of cement has a rich material source, low cost, high intensity and good permeability resistance; it can be used in the single-fluid injection mode. The craft is simple and convenient, and it is so far the most used and its effect lasts quite a long time after the application of a kind of grouting material. However, due to the cement being a granular material, it can be difficult to inject in fine sand, silt layers and tiny crack or gap formations, and the cement solidification time is long, and it easily causes loss of serous material, thereby precipitating drainage, slow growth, low strength and stability, which are some of its shortcomings; so the slurry applications have some limitations. Therefore, in recent years, a lot of work was done both at home and abroad to improve cement slurry performance, including research in superfine cement to expand cement application scope and improve cement slurry, for which breakthrough progress is needed.

5.2.1.1.2 Superfine cement slurry

The maximum particle size of ordinary cement is within the range of 44–100 m, and these particles make it difficult to inject the cement paste with a permeability coefficient less than 5×10^{-2} cm/s. The width of coarse sand or cement is less than 200 m. In order to solve the problem of poor injection of water slurry, Japan first developed an ultrafine cement grouting material (MC) in the early 1980s. Then the nations of the world adopted superfine cement grouting material for treatment of building foundations and construction of underground engineering in a widespread manner., as it is a good chemical grout. At the same time, due to the mechanical properties of the cement slurry concretion body, the groundwater and environmental pollution, and the low price, it is considered a kind of superior performance grouting material. The superfine cement grouting material is composed of extremely fine cement particles with an average particle size of 4 m, and the maximum particle size is about 10 m, and the surface area is above 8,000c m^2/g. This property makes the superfine cement slurry a good injection material. The injection permeability coefficient is 10^{-3}–10^{-4} cm/s. It can be used to meet the requirements of chemical slurry, and the strength of the stone is much higher than that of the chemical slurry, and it is free of pollution to the groundwater and the environment.

(1) **Chemical composition**

The chemical composition of superfine cement is shown in Table 5.4.

(2) **The characteristics of superfine cement slurry**

a. Under the same water cement ratio conditions, the superfine cement slurry station is lower than that of ordinary cement and colloidal cement.
b. The test results on the stability of superfine cement slurry and other cement slurries show that the superfine cement has good stability.
c. The strength of the slurry is calculated. The ultrafine cement particles have high chemical activity, can coagulate and harden well, and obtain high early and late strength. The compressive strength of the ultrafine cement stone at the age of 3d is not less than 20 MPa, and at 28d can reach up to 30 MPa (WIC = 0.6).
d. The determination of solidification time or setting time of superfine cement slurry is in the range of 45–150s.

Table 5.4. Chemical composition of ultrafine cement (wt.%).

Loss on ignition	SiO_2	Al_2O_3	Fe_2O_3	CaO	MgO	SO_3	Total
0.3	29.0	13.2	1.2	19.2	5.6	1.2	99.7

Because of the adjustable expansion rate of ultrafine cement, it is very easy to fill the entire fissure of the stone body, so that its interface is closely combined, which greatly improves its anti-seepage ability. The performance of superfine cement slurry can be significantly improved by adding some additives. The appearance of this new type of cement has opened a new technique for grouting. Under certain conditions, it can replace the slurry. However, due to its larger surface area, it is necessary to have a large amount of liquid to make it into liquid. Therefore, when the ultrafine cement grouting is adopted, the water cement ratio of the slurry should be controlled above 0.8. If the specific surface area is very large, the water retention of the slurry is strong, the water in the slurry is not easily eliminated and the strength of the stone is reduced. In order to solve this contradiction, an effective water-reducing agent can be used to improve the fluidity of the slurry.

5.2.1.2 *Cement water glass slurry*

Cement water glass slurry is also called c-s slurry, where c (cement) represents cement, and s (silicate) represents water glass. It is the main agent of water mud and water glass, where the two are injected in a certain proportion, and the grouting material with the additive is added when necessary. This kind of slurry overcomes the longer coagulation time of single-liquid cement slurry and cannot be controlled; thus, the stone rate is low, the effect of cement grouting is improved and the application scope of cement grouting is expanded. It can be used to prevent seepage and reinforce grouting, and it can be used to quickly plug the leakage in the stratum with high groundwater velocity. This is a very versatile and effective grouting material.

5.2.1.2.1 Water glass

Water glass is not a single compound, but a chemical that combines uranium oxide (Na_2O) with anhydrous silica (SiO^2) at various ratios, and its molecular formula is $Na_2O{\cdot}nSiO_2$. According to the pH of sodium silicate itself and in the gel, it can be divided into alkaline sodium silicate and non-alkaline sodium silicate. Alkaline sodium silicate is ordinary sodium silicate, itself a strong alkaline substance; when mixed with a gelling agent, it behaves like an alkaline gel. Due to strong alkalinity, a strong destructive effect will occur in the grouting process, which will gradually dissolve the formed silica gel and greatly reduce the durability of the treatment. Non-alkaline water glass is generally acidic; it is in the near-neutral range of gelation, avoiding dissolution of alkali, and thus increasing the durability.

Non-alkaline sodium silicate acid could be added directly to acidify ordinary sodium silicate as acid sodium silicate (pH = 1.5–2.0), and then to alkaline compounds near its neutral gelling scope to take off the reimbursement of ordinary sodium silicate, and it will also be made from silica sol, reoccupying its position as gelling agent.

There are two functions of water glass in cement paste: The first function is as a rapid coagulant, with less content, generally about 3–5% of cement weight. The second function is to use it as the main material, in a large quantity, which can be determined according to the purpose and requirements of grouting, and the water glass used for grouting has certain requirements for the modulus and concentration.

Modulus M is an important parameter that describes the performance of water glass, and is defined as the ratio of the number of moles of silica (SiO_2) in water glass to the number of moles of sodium (Na_2O).

The size of water glass modulus has great influence on grouting. When the modulus is small, in the mold hours, the silica content is low, the coagulation time is long and the stone body strength is low. When the modulus is large, the silicon dioxide content is high, the coagulation time is short and the stone body strength is high. The mold is too large or too small for the grouting. When grouting, the modulus of water glass is generally required to be 2.4–3.4.

5.2.1.2.2 The performance of cement in water glass slurry

For cement water glass slurry, according to the need of grouting engineering, it can generally be divided into two aspects: reinforcement and water shutoff. For water shutoff, water pressure is especially greater, and the water flow rate is faster in order to fill the big gap in rock soil. It is required that the coagulation time of slurry is short and has certain compressive strength. For the reinforcement foundation, the slurry should have sufficient compressive strength, and the following points are discussed.

(1) **Time of gelation**

Water glass can significantly accelerate the gel time of cement slurry. The gel time changed with the concentration of water glass, cement slurry concentration (water cement ratio), volume ratio of water glass and cement paste, and temperature.

In general, in a certain range, the concentration of water glass decreases, the gel time is shortened, and the two are linear. The smaller the water cement ratio, the faster the reaction between the cement and the water glass, and the shorter the gel time. In general, the thicker the slurry, the faster the reaction; the thinner the glass, the faster the reaction.

(2) **Compressive strength**

For cement slurry, the main factor determining the compressive strength of the slurry is its concentration (water cement ratio). When other conditions are certain, the thicker the cement slurry, the higher the compressive strength. Generally, the compressive strength of the slurry stone can reach 5–10 MPa.

When the cement slurry concentration is higher, the compressive strength increases with the increase of water glass concentration. When the cement slurry concentration is lower, the compression strength decreases with the increase of water glass concentration. However, when the water slurry concentration is in the intermediate state, its compressive strength is not changed much.

The volume ratio of cement slurry and water glass has a great influence on the compressive strength. When the volume ratio of cement slurry and water glass is 1:0.4–1:0.6, the compressive strength is the highest, and in this ratio range, the reaction is the most complete and the intensity is the highest. In fact, thick water slurry needs strong water glass; dilute cement slurry needs thin water glass. The excessive amount of water glass will have an unsatisfactory effect on its compressive strength.

Considering gel time, compressive strength, construction and cost factors, a suitable sodium silicate cement grout formula is given as follows: cement for 425 or 525 ordinary Portland cement, cement slurry water cement ratio of 0.8:1–1:1, cement and sodium silicate of volume ratio of 1:0.5–1:0.8, modulus sodium silicate by 2.4–2.4, the concentration of 22–40°Be′.

5.2.1.2.3 The holding point of cement water glass slurry

(1) The gel time of the slurry can be controlled accurately in a range of tens of minutes.
(2) The compressive strength of the calculi is 5–10 MPa.
(3) The calculi rate can reach 100% after coagulation.
(4) The permeability coefficient of the calculus body is 10^{-3} cm/s.
(5) It can be used for a sand layer with a fracture width of more than 0.2 mm or more than 1 mm.
(6) The source of the materials is abundant and the price is low.
(7) No pollution to the environment or groundwater.

5.2.2 The machine is grouting material

5.2.2.1 *Acrylic salt slurry*

Acrylate is a kind of grouting material with similar properties to propylene, but its toxicity is far lower than that of propylene. Aqueous acrylate grout has low leaching degree, low toxicity, low freezing point and adjustable gelling time; the gel has a certain strength, can be used in rock and soil and concrete microfissure grouting to stop water seepage.

Acrylate solution is dark green and transparent, with a proportion of 1.185 (22°C), viscosity of 0.01244 Pa·s (22°C), pH 6 and acrylic salt content not less than 35%.

Slurry is pale blue, with specific gravity of 1.081 (18°C), leaching degree of 0.0028 Pa·s (18°C), pH 7–8 and gel time from instantaneous adjustable to several hours between scope; for iron chloride, with an increase in the amount added and gel time is extended. The main factors affecting gel time are temperature, initiator and promoter concentration, slurry concentration, pH value, etc. The permeability coefficient of the gel was 1.7×10^{-7} cm/s, and the compressive strength of the solid sand body was 0.3 MPa.

5.2.2.2 *Polyurethane slurry*

Polyurethane slurry is prepared by using polyisotropic vinegar and poly-enzyme resin as the main raw materials and adding various additives. Due to the presence of non-reactive polyamide groups in the slurry, chemical reaction occurs in the water, and the crosslinking generates an insoluble polymer, which can be used to strengthen the foundation and prevent the water from seeping. During the reaction process, carbon dioxide is generated, which expands the volume and increases the consolidated volume. Moreover, it produces great expansion pressure, which causes the slurry to spread twice as quickly, thus increasing the diffusion range.

Polyurethane slurry can be divided into non-water-soluble polyurethane (PM) and water-soluble polyurethane (SPM).

5.2.2.2.1 Non-water-soluble polyurethane slurry

(1) **The composition of raw materials**

a. **Polyisonine vinegar:** The commonly used ones are polyisoaminoacetic acid (TDI), diphenyl methylaminic acid vinegar (MDI), polyphenyl methionine (PAPI), etc. Among them, toluene diisonine vinegar has the least degree of leaching, which is combined with the prepolymer, with low viscosity, high activity and fast water reaction. However, the prepolymer synthesized by MDI or PAPI was large, and the solid strength was high.

b. **Multiple base compounds:** To make a mixture, vinegar and several compounds have an enzyme of two kinds, because vinegar is easy to hydrolyzes, and the key enzymes are relatively stable, with the viscosity of the same molecular weight of poly (enzyme resin is smaller than poly resin vinegar, so generally polymerase resin as is used as grouting material).

(2) **Admixture**

In addition to the prepolymerization of the main agent, there are additives such as catalysts, diluents, surfactants, emulsifiers and retarding agents.

a. **Catalyst:** The purpose of a catalyst in slurry is to accelerate the reaction rate of slurry and control the foaming time. Catalyst has two kinds of glues and tin salts, because the electronegative nitrogen atom in the tertiary glue is very strong, and thus it improves the

NCO catalytic effect on the reaction with water. The glue catalyst has three b glue, three ethylene glue, two alcohol glue, etc. Tin salt is the main catalyst if it promotes NCO and water action chain growth and furthers gelation hardening, commonly used in dilaurate dii, tin and so on. In general, two kinds of catalysts are used together to play a synergistic role and get better results.

b. **Thinner:** The effect of the diluent is to reduce the viscosity of the prepolymer or slurry, and improve the injection ability of the slurry. Generally, the diluent used is propylene, xylene, dichloroethane, etc. The effect is the best, but the contraction of the polymer is greater. When the xylene is diluted, the shrinkage of the polymer is smaller, but when the xylene is too large, the hydrophobicity of the slurry will be increased, and the gel speed will be significantly reduced. The dosage of diluent should not be too much, otherwise it will reduce other physical, mechanical properties.
c. **Surfactant:** The addition of surfactant can enhance the stability of foam and improve the structure of foam; generally a foaming spirit is used, which is a non-ionic surfactant, and is a derivative of organic silicon wan and polycool. The dosage of surfactant is generally below 1%.
d. **Emulsifier:** Emulsifier can improve the dispersibility of the catalyst in the slurry and the dispersibility of slurry in water. A commonly used emulsifier is twin-80, the scientific name of which is polyoxyethylene sorbitol wt oleic vinegar, and the dosage is about 0.5–1%.
e. **Retarder:** The addition of a retarder in the slurry can reduce the support of the prepolymerization reaction, prevent the gel from forming, slow the reaction rate during the grouting process and increase the diffusion distance of the slurry.

5.2.2.2.2 Water-soluble polyurethane slurry

The water-soluble polyurethane prepolymer is also synthesized by the reaction of poly (cool) resin and polyisonine vinegar. However, this kind of clustering is usually obtained by the polymerization of cyclo-oxide-Anhui open-loop, and it can also be obtained by copolymerization of epoxide and epoxy-propylene. Water-soluble polyurethane can be mixed with water in various proportions to form emulsion, and it reacts with water to form an aqueous gel.

(1) **The slurry composition**

Water-soluble polyurethane slurry is composed of prepolymer and other additives. The additive is the same as that of non-water-soluble polyurethane, and there are two main types of prepolymer.

a. **Prepolymer of high-strength slurry:** The prepolymer was prepared by the reaction of epoxypropylene glycol and propylene glycol. Top polymers are toluene 2 different amino acid vinegar (80/20, the main agent), epoxy propyl anhui poly cool (604, the main agent), epoxy ethyl Anhui poly cool (molecular volume 1,500, the main agent), phthalate I vinegar (solvent), xylene (solvent) and sulfuric acid (polymerization), composed using the clustering method, and the water-soluble polyurethane prepolymers of vinegar are composed using 604 cool basic, but the reaction temperature should be controlled in 40–50°C. A temperature of 90°C will ensure that in the end the prepolymers mixture match with the other admixture slurry.

b. **The prepolymer of low-strength slurry:** The prepolymerization of epoxy propylene Anhui, epoxy-Anhui, and the prepolymerization of toluene diisonine vinegar (80/20, main agent) were conducted. When performed, the mixture will get cool after first heating and melting, and toluene and glycine vinegar are added and shaken; the mixture is maintained for 4 h under 80°C for obtaining waxy solid prepolymers. For proportioning, the mixture should be heated to melt, followed by the addition of additives.

(2) **The main performance of the slurry**

a. **Deterioration degree:** High-strength water-soluble polyurethane slurry is between 25–70cp, and the strength of consolidation is reduced by adding more solvents. As for the low-strength water-soluble polyurethane slurry, it depends on the amount of water added or the amount of solvent, and of course the strength of the solid.

It should be noted that the water-soluble polyurethane slurry has a good injection quality. If it is combined with a size of 10cp, it is better to injected than acrylic.

b. **Gelation time:** The gelation time of water-soluble polygels is adjusted according to the dosage of catalyst or retarder from a few seconds to tens of minutes. But the gelation time of low-intensity water-soluble polyurethane slurry is very short (usually within a few minutes).
c. **Compressive strength of consolidated body:** The low-strength water-soluble polyurethane slurry gel is an elastic gel containing water. The compressive strength of the solid sand body is 0.1–5 MPa (with the increase of water content).

   The compressive strength of high-strength water-soluble polyurethane slurry is similar to that of non-water-soluble polyurethane, and the compressive strength of the consolidated body under different pressures changes little.
d. **Anti-permeability:** The permeability of water-soluble polyurethane slurry was higher than that of non-water-soluble polyurethane slurry. Generally from 10^{-6} to 10^{-8} cm/s, the anti-seepage strength of high-strength water-soluble polyurethane slurry can reach 1.5 MPa.

5.3 Surface jet-rotating reinforcement technology

In the late 1960s, Japan applied high-pressure water jet technology to the grouting project, and created a new method of applying high-pressure injection. It is also called the "Chemical Churning Pile". In 1972, the ministry of railways was the first to develop high-pressure jet grouting. In 1975, China's metallurgy, hydropower, coal, construction and other departments, and some colleges and universities, also carried out experiments and construction. At present, the technology has been successfully applied to the foundation treatment of existing buildings and new construction projects, the support and protection of deep-foundation underground pit works, and the construction of underground waterproof curtains.

5.3.1 *Process principle*

High-pressure jet grouting is a kind of application process which combines high-pressure hydraulic jet-cutting technology and chemical grouting technology for foundation reinforcement. Its principle is to use the rig equipment that is installed in the grouting pipe at the bottom of the special nozzle, placed in the soil due at a deep degree, with high-pressure water (or water mud) or compressed air flow in underground chopped soft soil, and with high pressure to note water slurry mixed with chopped soil, reinforcing the cylinder form. According to the different ways of nozzle rotation and lifting in the construction process, swirl, spray and directional injection can be achieved, and the underground structures such as spinning jet pile, water-stop curtain and retaining wall can be formed accordingly.

5.3.2 *The construction method*

5.3.2.1 *Construction preparation*

The equipment required at the construction site should be checked according to the conditions of the layout and installation; the gas is checked, two big plasma systems are normal, the line is clear (ground line spraying) and the test monitoring instrument is ready and in good condition, making sure the correct equipment is used in the next process.

5.3.2.2 *Measuring lofting*

First, it depends on the total station position, the high-pressure jet grouting pile and then the steel tape "First, use total station to set out the position of high-pressure jet grouting pile, then use steel tape and hemp line to set out the position of jet grouting pile, use small bamboo stick to mark and sprinkle lime to mark, so as to ensure the accurate positioning of pile driver."

5.3.2.3 *Drilling in place*

Drilling emplacement is the first step of jet grouting; the drill should be placed on the number of the design and aimed at the center of the hole. At the same time to ensure that the design requirements are met after drilling of the vertical degree, after drilling emplacement, a correction must be leveled and the pipe's vertical alignment to the hole center position should be made, adjusting until the vertical degree requirements of the drill pipe are met. To guarantee the accuracy of the pile location, a positioning card is used, and the pile location error should not be more than 5 cm.

5.3.2.4 *Drilling*

In the process of drilling, the drilling parameters should be carefully controlled, and the drilling speed should also be controlled to prevent accidents in various holes, such as drilling and sticking. To avoid borehole tilt, in the process of drilling and the drilling emplacement, always pay attention to the vertical degree checking tool, and if it is found to lean or incline, it should be corrected in a timely manner, to ensure borehole stability within the scope of the design: the drilling rate should be fixed at a low gear, so as to prevent hole inclination.

5.3.2.5 *Grouting pipe*

The nozzle should be protected against the grouting pipe, so as to prevent the nozzle from being blocked, and when the grouting pipe is not in place or is subjected to different conditions, it should be treated to ensure that the hole does not collapse, and it must be ensured that the spray rod is under the expected depth, and the square can be injected with injection grouting.

5.3.2.6 *High-pressure swirl grouting*

After placing the grouting tube in the predetermined position, the slurry and air supply will be sent in order to adjust pump pressure and wind pressure as per the design value which will start to rotate as the hole is in a normal position, according to the design to determine the high-injection parameters for construction. In the process of high spray, the proportion of cement slurry is often tested. When it fails to meet the design requirements, the spray rod should be suspended immediately and the water cement ratio and proportion should be adjusted. After that, the spraying operation will be restored quickly. During the construction process, it is required to check and record the lifting speed, the pressure and flow rate, the pressure and air volume, the slurry and the proportion of the slurry, etc. Each hole shall be used for pulping and consumption (cement) as per statistics and records.

5.3.2.7 *Jet finish, rig shift*

After the construction of single-pile injection, the next pile of construction or the cleaning of the grouting pipe is carried out. When cleaning, the grouting tube and other machine tools should be washed and there should be no residual cement in the inner tube.

5.3.3 *Verification of reinforcement effect*

Engineering examples: Verification of the effect of high-pressure spin-jet reinforcement in the 1# inclined shaft of Jialing Section of Qingdao Metro Line 13

5.3.3.1 *Project overview*

The Qingdao Metro Line 13 Jialing Section 1 # is in the inclined well open section with a double row of ϕ800@600 high-pressure rotary jet pile, all of which adopt the triple-pipe process, and the swirl pile enters the impermeable layer at 1 m. There are 498 spin-jet piles, according to the geological data, the elevation of the pile bottom is –7.4 to –8.0 m (Fig. 5.1).

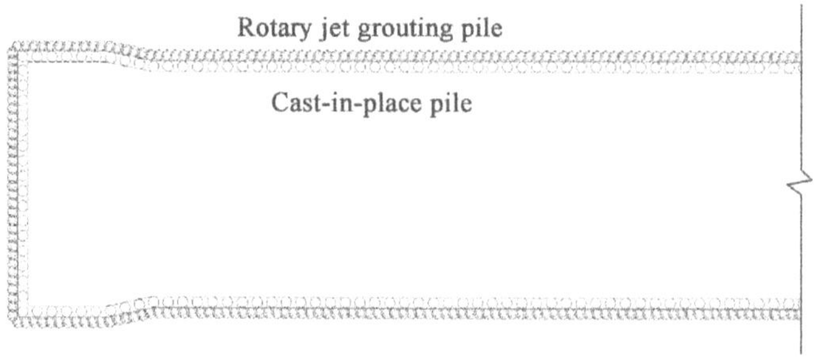

Fig. 5.1. 1# plot of high-pressure rotary jet pile in inclined well.

5.3.3.2 *Scheme design*

According to the drawings using triple-tube technology and the high-pressure jet grouting pile with a magnitude of 42.5, ordinary silicon acid salt water mud is used and a grouting pipe drilling rig with special nozzle must be used in the soil in a predetermined position. With high-pressure pump pulse, the drill pipe is used under the spray device, to level the cement slurry which will be sprayed at a high speed into the soil. With the help of the cutting soil layer, the impact of the liquid jet range soil body cutting is assessed. At the same time the rotating drill pipe on one side will be at a low speed, mixing the slurry. After cementation hardening of the base diameter is at an even strength of the cylinder (called jet grouting pile), interlocking between the pile occurs, in the deviated well open section to form a circle around the closed portion and stop the water effect, ensuring the safety of the construction process of the inclined well.

5.3.3.2.1 Construction parameters

On the afternoon of April 6, 2015, a high-pressure rotary jet pile test was conducted in the 1# inclined well, and three piles were tested, and the test pile was determined by summing up certain parameters. The construction parameters are shown in Table 5.5.

Table 5.5. Construction technology parameter table.

Heading		**Technical parameter number**
Compressed air	Pressure (M Pa)	0.75
	Tolerance (m^3/min)	5
Water	Pressure (m pa)	28
Cement paste	Cement dosage	320 kg/m
	Pressure (M Pa)	1.2
Water cement ratio		1.1
Cement slurry-specific gravity		1.5
Hoisting speed (cm/min)		15
Spin velocity (r/min)		20

5.3.3.2.2 Construction method and process flow

(1) **The principle of triple tube**

Three-pipe spinning is a method of water, gas and slurry injection. Using a three-layer injection pipe, high-pressure water and air are sprayed simultaneously, and the ground soil is cut, and the broken soil is excluded from the surface by the rising force of the air. At the same time, the other nozzle will inject at low pressure a water slurry into the base of the soil, cutting and stirring it, so that the cement slurry will be mixed with the soil to reinforce the structure.

Using the three-pipe spinning spray, one should first spray high-pressure water, resend cement slurry and compressed air. When spraying, it should first reach the predetermined injection pressure and spray amount, then gradually improve the grouting pipe. The length of the lap joint of the grouting pipe shall not be less than 100 mm. When the top height of the juice pile is reached or the slurry phenomenon occurs on the ground, one should immediately stop the spinning of the current pile and pull the nozzle out and clean the pipe path. The triple-pipe method is to spray the cement slurry and compressed air simultaneously. Besides extending the spraying distance and increasing the cutting ability, it can also promote the removal of waste soil and reduce the weight of the solid unit.

(2) **Construction technology of triple-pipe spinning**

See Fig. 5.2 for the process flow of triple-pipe spinning and spraying.

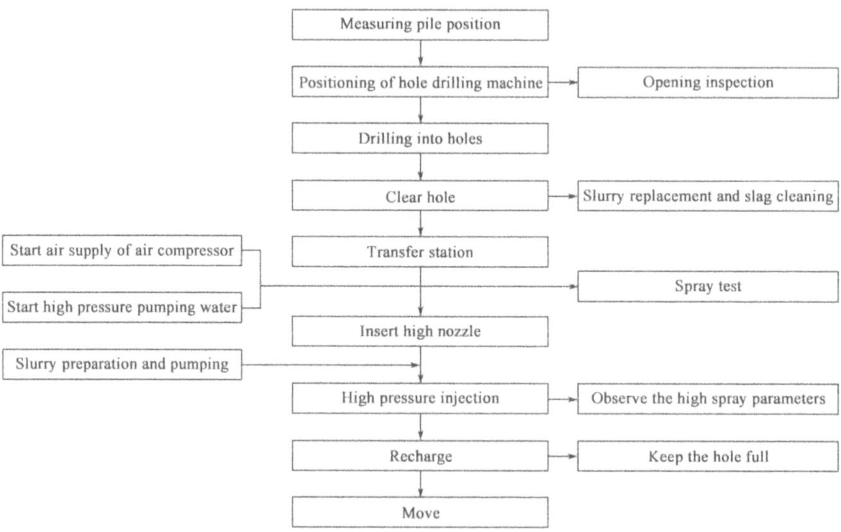

Fig. 5.2. Process flow chart of triple-pipe spinning and spraying

(3) **Specific construction procedures**

A. **Construction preparation:**

a. *Site formation*: Before the formal start of construction, after conducting pipeline investigation, one should remove the obstacles in the construction site below 2 m, and make sure that the protection measures cannot be removed, and then the layout is clearly visible. At the same time, reasonable arrangements of construction machinery, transportation pipeline and power line location should be made, ensuring the construction site is at the "three general and one level".

b. *Pile lofting*: Before the construction of the jet grouting pile is measured by total station construction control points and buried stone markers, after qualified test lines, use the steel ruler and line the real cloth pile, and using a nailed-down bamboo sign, one can confirm the pile hole in the heart to remove a partial difference of 50 mm.

c. Building the drainage and mortar mixing system, the jet grouting pile construction process will produce 10–10% Fan Jiang quantity, and the spent slurry will remain in the tank; after precipitation of water, pollution-free emissions can be checked based on conditions in the field. The deposited soil was carried away in the excavation. The mortar mixing system is mainly set in the vicinity of cement, which is easy to operate. It is mainly composed of mortar mixing equipment, mortar storage equipment and mortar conveying equipment.

B. **The rig is in position:** After the drilling rig is in place, adjust the vertical degree of the pile machine to ensure that the drill pipe is consistent with the pile position. The deviation should be within 10 mm, and the vertical degree error of drilling should be less than 0.3%. Before drilling, the air compressor and mud pump should be adjusted to make the equipment operate normally. Check the length of the drill pipe and mark the depth line using red paint on the side of the drill tower to ensure that the depth of the hole meets the design depth.

C. **Drilling:** According to the geological data and field survey, the coastal road is the backfill area of the beach, and there are gravel backfill layers in the stratum, which need to be drilled by a geological drilling rig. The drill pipe length of 18 m can be drilled into the design elevation at one time. Make a depth mark on the drill stand for easy control. Drilling into the strong weathering layer for 1 m can help determine the hole bottom elevation according to the geological data. According to the geological data, the top elevation of the strong weathering layer is –6.2 to –6.9 m, and the elevation of the hole bottom is –7.2 to –7.9 m.

D. **Spiraling:** When the drill pipe is drilled into the design depth, the high-pressure water pump is started, then the soil is cut by high-pressure water jet, and the mud is cleaned out by high-pressure air. In the process of slurry discharge, the slurry pump starts grouting. In order to ensure the quality of the bottom end of the pile, when the nozzle is sunk to the design depth, the nozzle is rotated in the original position for about 60 s, and then the nozzle will be injected into the created nozzle hole for improvement. The rotation of the drill pipe and ascension shall be continuous and shall not be interrupted. With the drill down, one should stop the ascending pipe and rotation, in case of a broken pile and immediately perform repairs or troubleshooting. To improve the quality of the pile bottom, one should appropriately increase the scope by 1.0 m at the bottom of the pile drill pipe spray jet grouting.
E. **Rig shift:** The rotary spray shall be lifted to the height of the design pile to stop spinning, lifting the drill hole, cleaning the grouting pump and conveying the pipe, and then shifting the rig.
F. **Grouting process control:** There are many quality factors that influence the consolidation of high-pressure tube pile grouting, and when certain forms of high-pressure tubes are used, the grouting process is one of the important factors that affect the consolidation.
G. **Jet grouting:** The high-pressure spin grouting is a continuous process from the bottom up. If there is an outage in the construction, after being repaired, the equipment needs to be put down to the length of 500 mm to ensure the integrity of the consolidated body.
H. **After spraying:** Without changing the spin chute technical parameter conditions, the same soil layer grouting spray is repeated (spray to drill down again the parts), which can increase the damage of the soil's effective length, thus increasing the

diameter or length of the concretion body and improving the consolidation strength. The more number of times the spray is repeated, the better the consolidation effect.

I. **Control of cement dosage:** Controlling the amount of cement is the key in the process of spraying. The amount of cement is directly related to the spraying pressure, nozzle diameter, lifting speed and water cement ratio, and the specific control method is as follows:
   a. For the cement quantity residual measure: appropriately increase the pressure of the spraying; increase nozzle diameter and reduce the speed.
   b. If the amount of cement is not sufficient, the pressure should be reduced appropriately in the case of pile diameter; the diameter of the tube is reduced appropriately.

      In the case of pile strength, the speed of lifting should be increased appropriately. Increase the water cement ratio.

      According to the project-specific conditions, each pile is mixed several times, with construction water measured to precision, to ensure the quantitative mixing of water and cement with mud. Through the above parameters one can ensure the cement quantity to meet the design requirements. In response to the sinking phenomenon of cement slurry (the density of water slurry is relatively high); it can be used to recover from the top of the pile, and the soil strength of the upper part of the pile can be guaranteed.

J. **Pulp treatment:** In the rotary spray process, often there is a certain quantity of soil particle; as part of the slurry along the grouting pipe wall comes out of the ground through the observation of the pulp in a timely manner, one can understand the status of the soil layer, the assessment of the effect of the general spin chute rationality and conclude the following parameters according to the test: slurry (soil particles, water and slurry) is less than the grouting amount, 20% is a normal phenomenon, if more than 20% with or without plasma, one should take corresponding measures in a timely manner to find out the reason.
   a. When the flow is constant and the pressure drops suddenly, the leakage of the site should be checked, and the grouting tube should be set aside when necessary to check the sealing performance.

b. It is considered a normal occurrence when there is no slurry or intermittent pulping, or when the soil is soft. Grouting is not improved if it is near a hole or a conduit pipe; grouting is performed until slurry appears or the serous solidification happens in the grouting pipe. This is done to avoid the risk of pulp grouting. When necessary, the quick-setting slurry is poured near the grouting pipe to achieve solidification.
c. Measures should be taken to reduce the slurry. The main reason for the excessive amount of slurry is that the effective ejection range is not suitable for grouting, and the grouting amount is much higher than the amount of pulp needed for the consolidation.
d. Effective measures should be taken. To increase the pressure (the amount of tubes), properly reduce the diameter of tube; increase and rotate speed. For the slurry that comes out of the ground, it can be used after selecting and adjusting the concentration to return to the back of the pile to prevent cavitation.

K. **Consolidation control:** The shape of the consolidated body can be controlled by adjusting the swirl pressure and grouting amount, changing the direction and speed of the nozzle. Due to the engineering design of consolidation shaping the cylindrical body, improvements while grouting are used in construction, such as the forming of a deep hole at the bottom of the injection, increasing jet pressure, repeating rotary spray or reducing the rotation speed of the nozzle, Moreover, it is necessary to increase the pressure and reduce the rotational speed of the tube, so as to achieve the symmetry of the consolidated body and ensure that the pile diameter is not very different.

L. **Ensure the height of the pile:** When using cement slurry for the chute, after mixing the slurry and soil grain during the solidification process, due to the slurry drainage, generally all parts have different degrees of shrinkage; due to the appearance of consolidation at the top, recess depth is commonly designed to be 1–1.5 times the pile diameter. This cavitation phenomenon is extremely unfavorable at the end of water curtain and must be eliminated by effective measures.

In order to prevent cavitation caused by solidification and contraction of slurry, it is convenient for the foundation to achieve the desired effect.

5.3.3.3 *Implementation effect*

The construction technical standard of rotary jet pile is shown in Table 5.6.

Table 5.6. Technical standard table of rotary jet grouting pile.

Order	Item name	Technical standard	Checking square method
1	Allowable deviation of perpendicularity of borehole	1/150	Measured or theodolite drill pipe
2	Hole position allowable deviation	50 mm	Dipstick metering
3	Allowable deviation of borehole depth	±200 mm	Dipstick metering
4	Allowable deviation of pile diameter	≤50 mm	Excavation scale
5	Pile center allowable deviation	≤0.2D	Design of pile diameter with ruler D at 500 mm under excavation pile top
6	Initial setting time of cement slurry	Not exceeding 20 h	
7	Cement soil strength	Qu (28) ≥1.1.0 MPa	Test inspection
8	Water cement ratio	1.1–1.3	Test inspection

5.4 Surface grouting reinforcement technology

Surface grouting, as a kind of reinforcement method with a lower cost, is widely used in underground engineering construction. At present, the application of the more extensive surface grouting reinforcement technology is as follows: sleeve valve tube grouting process, WSS grouting process and steel tube grouting process. Sleeve valve tube grouting is a kind of more advanced grouting construction technology, which first started in France; it has better ability to control the grouting pressure and grouting range, can be repeatedly performed, and has the possibility of taking pulp and string of pulp from a small point. WSS dual-liquid grouting technology originated in Japan and is widely used in Taiwan.

5.4.1 *Sleeve valve grouting*

5.4.1.1 *Process principle*

The sleeve valve tube grouting is done using a high-pressure injection in the rock and soil layers using the core grouting tube. Blocking the upper and lower layers can prevent layer grouting, but one needs to choose whether to use contact or jumping grouting. This method is indeed very useful in grouting construction. Segmented grouting is followed to loosen the strata and results in compacted formation, which leads to a good grouting reinforcement effect, thus avoiding further grouting work. Unconsolidated formation and relatively dense formation exist at the same time, with a large amount of loose strata grouting. More compact formation does not lead to pulp formation.

By using the principle of sleeve valve tube grouting mainly through the larger pressure grout, the sleeve valve pipe pressurizes the rock and soil layers through the infiltration and fracturing effect, which happens between particles of soil moisture and air, to reinforce the outside circle, thus filling the blank space between the soil particles. The scope of grouting surrounding the soft soil particles, through slurry condensation as a whole, fully improves the physical and mechanical properties of soil, thus improving the stability of the strata and the strength of the soil. It also reduces the permeability of the strata. The groundwater does not easily permeate under the reinforced layer, thus protecting the tunnel after excavation and providing stability of the strata and also ensuring the safe and reliable construction, thereby protecting the environment.

In the sleeve valve tube grouting method, serum flowing into the grouting pump after the application of pressure is gathered in the sleeve valve tube grouting pipe section, and then discharges through the plasma PVC pipes (i.e., sleeve valve pipe), under the action of external pressure that is applied by the expansion of the wrapped PVC rubber rings. Thus the set of shell material is crushed. When the pressure gradually increases to a certain degree, pressurized slurry can produce filling along the stratigraphic structure, with permeability, pressure and split flow; at this time due to the slurry amount being less than the required quantity, pressure will automatically return to equilibrium state and slurry will remain under pressure for use, afterward causing the splitting crack to extend outward, with slurry formed in soil consolidation, so as to increase formation strength and lower the formation permeability mesh. By lowering the grouting inner tube step by step one can realize the segmented grouting. The principle of sleeve valve pipe grouting is shown in Fig. 5.3.

A rubber ring is used when bilging occurs in the hole pressure grouting rubber ring. Slurry that flows from the discharge plasma hole into the stratum stops when the rubber ring on the outside of the sleeve valve tube grouting the slurry closes, thus preventing soil and groundwater from flowing backward into the sleeve valve pipe.

The set of shell material that is formed around the sleeve valve pipe has a certain strength from the protective layer, so that slurry leaks out of a hole in the sleeve valve pipe grouting areas or crowded broken casing material; the upper and lower casing materials still have a certain strength, preventing slurry from flowing up and down. In this way, the slurry flows horizontally

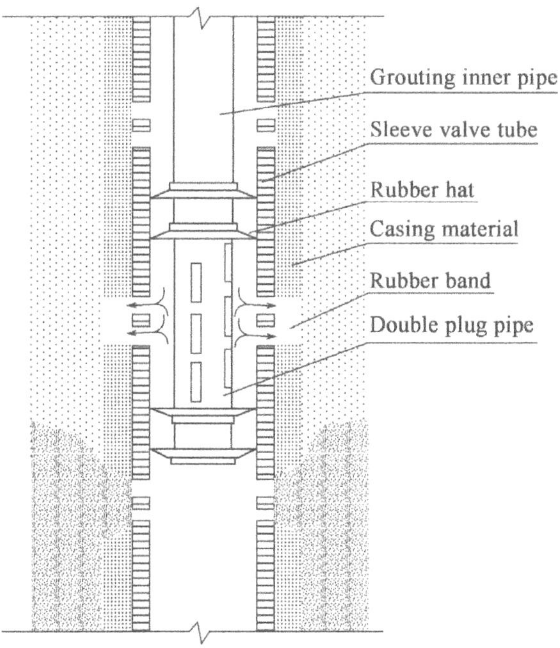

Fig. 5.3. Sleeve valve grouting principle diagram.

across a small area to increase the radius of the formation reinforcement. The function of the double-plug tube is to pressurize, and when the slurry enters the double tube through the injection of the inner tube, the slurry flows out from the outlet hole of the inner tube. When slurry is injected into the sleeve valve pipe and double plug in the middle, under the effect of pressure force, the rubber cap is jacked up; with the accumulation of the grout pressure after reaching a certain degree, the outside of the sleeve valve pipe rubber ring is expanding, the set of shell material is crush, and thus serous fluid is squeezed into the formation. The sleeve valve pipe grouting needs to be segmented, and each round of grouting should be completed once.

5.4.1.2 *Construction method*

The main construction process of the sleeve valve pipe grouting method includes setting and drilling, replacing shell material, inserting sleeve valve, grouting and so on.

5.4.1.2.1 Location and drilling

According to the design requirements of the whole station by using the site data, the holes can be measured and checked by the relevant personnel, and the drilling can be carried out after confirmation. Drilling holes and drilling depth are strictly designed. The hole position should be determined before drilling, and the horizontal degree of the fuselage of the drilling machine and the perpendicularity of the drill pipe should be checked, along with the basic required level of the drilling rig; and the vertical deviation of drilling is less than 1%. In the process of drilling, the drilling speed is recorded, the geological condition is analyzed carefully, and the pressure adjustment is provided for reference during grouting. The pore-forming standards are as follows: for a deep hole or drilling, it is more difficult to determine the drilling depth. Also, we need accurate records when determining the drilling depth and need to pay attention to the orifice and hole wall protection. The drill points are as follows: (1) Make sure the drill is horizontal or vertical before opening the hole. (2) When drilling, maintain medium speed. When encountering a hard layer, slow down to prevent sticking. (3) The mud wall should be done well to prevent the collapse of the hole.

5.4.1.2.2 Replace the shell material

The shell is usually made of fly ash, bentonite and cement by making a hole through the drill pipe with a set of shell material that is replaced immediately after making a hole in the mud. The method is to cycle through the mud pipe by receiving the extrusion-type grouting machine under the action of grouting pressure. Through the drill pipe hole, mud is displaced as are the complete sets of shell material. Under the action of pressure, the casing material flows through the bottom of the pipe into the hole. The shell material flows into the mud from the surface orifice and replacement of mud occurs through the orifice as mud is ditched into the mud pools. When a shell is found in the discharged mud, the replacement is stopped.

5.4.1.2.3 Insert the sleeve valve tube

The sleeve valve should be buried immediately after the shell filling is finished. The connection in the tube should be strong when the sleeve valve is lower, and the rubber sleeve should be closed to cover the grouting hole. The bottom end of the sleeve valve tube should be fitted with a conical plug. The top should be protected with a protective cap to prevent the mud or shell material from entering the sleeve valve tube. When the tube is filled with water to reduce the influence of buoyancy, the sleeve valve will sink smoothly and reduce the curvature of the tube. According to the requirements of the mixture configuration set of shell material of the sleeve valve pipe, the sleeve valve pipe is as far as possible in between the borehole and is upright; the set of shell material can be uniformly consolidated around the sleeve valve pipe, and the grouting effect is more obvious.

5.4.1.2.4 The grouting

Segmented grouting is adopted, and the length of each grouting is called the grouting step. The length of the open steel pipe is the length of the grouting step. The grouting step is generally selected to be 0.6–1 m, which can effectively reduce the influence of uneven formation on the grouting effect. For the sand layer, a low value should be used for the grouting step. For pebbles or broken rocks, a high value should be used for the grouting step. In the process of grouting, the length of the core tube is moved up or down one step after each round of grouting is completed. It is advisable to use lifting equipment to move it, or manually adopt a 2-pipe clamp symmetrically clamped in the core tube, and both sides uniformly exerting force to move the core tube. To complete 3–4-m length of grouting, a section of grouting core should be removed. The following aspects should be noted in the grouting process: (1) Before each formal round of grouting, it is necessary to use the water test pump to see whether the performance of the grouting machine is good. (2) In the process of grouting, continuous operation is made to ensure that the grouting attains the design effect and prevents the slurry from blocking the pipe. (3) The grouting pressure should be controlled at 0.15–0.5 MPa. (4) The principle of grouting should be followed from the bottom up. (5) The grouting flow is strictly controlled at 7–10 L/S; it should not be greater than 20 L/S for filling perfusion. (6) When the pressure reaches or exceeds 0.5 MPa after the official grouting, or when the pulp is smaller than 1 L/min, the grouting should be stopped, and the slurry of the flameproof tube should be broken. After the grouting is finished, cover the grouting tube to facilitate the construction. The grouting process is shown in Fig. 5.4.

5.4.1.3 *Verification of reinforcement effect*

In recent years, the sleeve valve grouting method has been widely used in grouting reinforcement engineering, and in the construction of the Qingdao Metro, it has been used to reinforce the stratum, which has achieved good results.

Engineering examples: Verification of grouting reinforcement effect of sleeve valve pipe in 1# inclined well in No. 13 Line of Qingdao Metro

5.4.1.3.1 Project overview

The sleeve valve pipe of the Jialing interval 1# inclined well passes separately through the fourth system is a new series of artificial soil layers, the Holocene series, and the fourth series of the new series of alluvial deposits, the keratinite, the strong weathering zone, the weathering belt and the breeze belt. This Jialing range 1 # slope dig of sleeve valve pipe surface reinforcement used dark tunnels and surrounding soft rock of inclined well of the consolidation grouting section, and is located at Binhai avenue south,

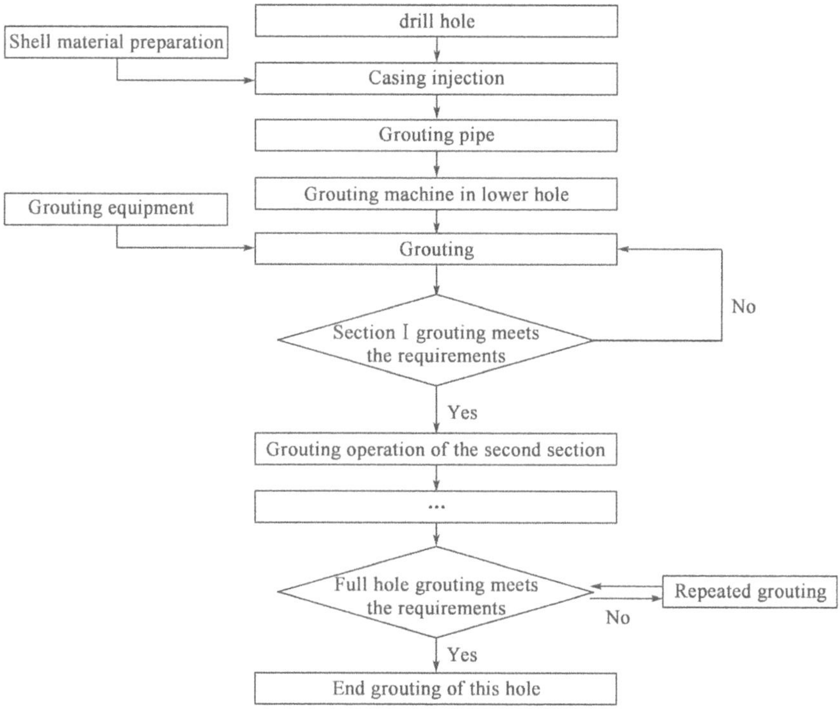

Fig. 5.4. Sleeve valve grouting process flow chart.

at the boundary between the area in light and the area under the shade of the inclined shaft, with a longitudinal length of about 90 m. The standard width of the sleeve valve pipe construction scope of slope is 5 m, with a width of 6 m round the corner, a wrong lane width of 7 m, and the in-plane grouting width of the sleeve valve hole diameter about 2 times (outside the excavation contour line of 3 m), for an excavation process of gradual transition.

5.4.1.3.2 Scheme design

The grouting section is divided into two groutings of full-section grouting and half-section grouting. The grouting length of the whole section is about 40 m, and the grouting section layout is shown in Fig. 5.5.

Sleeve valve flat surface spacing is 1.0 m × 1.0 m, with plum flower arrangement. When encountering a large tree, avoid it and drill down from both sides (Fig. 5.6).

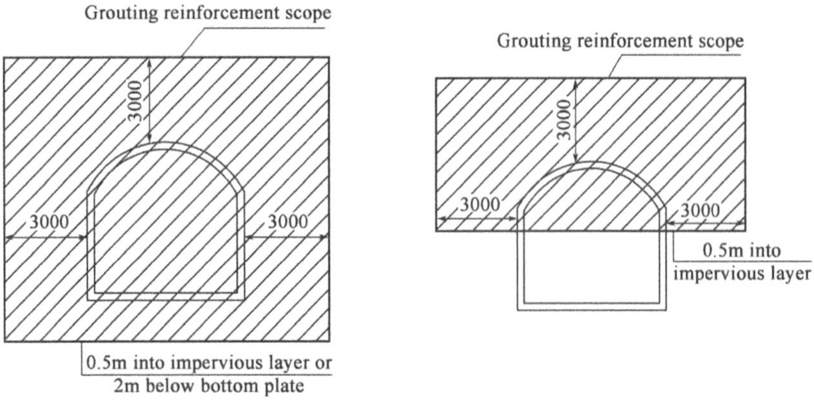

(a) Schematic diagram of Full Face Grouting (b) Schematic diagram of half section grouting

Fig. 5.5. Jialing interval 1# inclined shaft sleeve valve pipe construction full- and half-section grouting drawing.

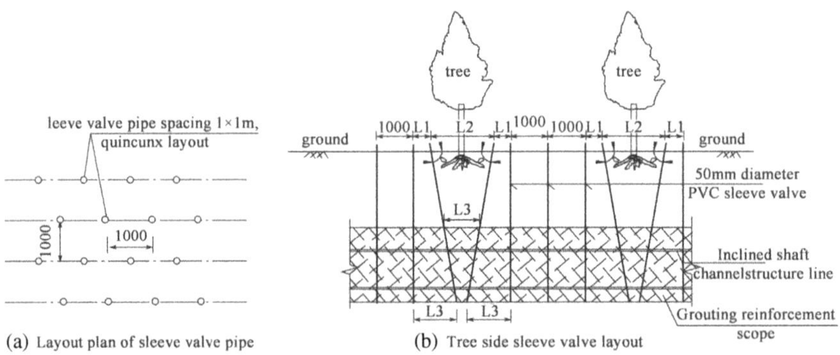

(a) Layout plan of sleeve valve pipe (b) Tree side sleeve valve layout

Fig. 5.6. Jialing interval 1# inclined shaft sleeve valve pipe construction plan and tree side layout.

In accordance with the requirements of design drawings, the sleeve valve pipe is combined with long tube grouting using a DTH drill and drilling hole grouting construction technology, by using 42.5 ordinary Portland cement and sodium silicate for high-pressure double-fluid grouting, so as to realize the purpose of consolidation due to formation, whereby the safety of the construction process of inclined well is ensured when a certain water effect is reached. Some geological conditions are good, hard and complete, and drilling technology is adopted. The diameter of the hole is about 91 mm, and the hole is about 0.5 m in the impervious layer. The drilling depth should reach the height of grouting. After the hole is formed, the corresponding grouting tube shall be connected according to the design, and the bottom of the hole shall be lower than 50 cm above the ground.

Part of the stratigraphic geology is poorer, with the existence of a gravel and sand layer, which causes mechanical sticking in the gravel layer, and is not easy to form into a hole; an anchor drilling machine (model YGL-C150 m) can be used with a pipe drilling hole formation process, drilling about 0.5 m into the impermeable layer, creating a hole of 150 mm diameter. The soil layer adopts a full hydraulic pressure drive. The impacts of the following are assessed: power head type, double-pipe drilling, with the drill pipe to outer pipe set inside the double-tube pipe, tube pressure water injection using impact drilling and bit rotary drilling, circulating water flow from outside between tubes and out of the mud into the rocks beyond the tube to continue drilling, and the depth to grouting consolidation section height. To prevent the grouting pipe from being damaged by extrusion, remove it after drilling a hole in the pipe. According to the corresponding grouting pipe connections and the design requirements, the upper part of the drill should be more than 50 cm above the ground.

The grouting process is used. The length of each grouting round is called the grouting step. The length of the tube is the length of the grouting step, and it is generally selected as 0.33–1.0 m, which can effectively reduce the influence of the heterogeneity on the grouting effect. For sand layers, a low value should be used for the grouting step. For pebbles or broken rocks, a high value should be selected for the grouting step.

According to the sequence of hole forming, after the casing material has a certain strength, the ϕ12.7 mm grouting steel pipe with double plugs is moved from the sleeve valve pipe to the grouting section. From bottom to top section grouting, segmented spacing shall not be more than 2.0 m. The grouting pressure is 0.8–1.2 MPa and the open-loop pressure is 0.5 MPa. If the grouting quantity is increased, the grouting can be made by the "intermittent quantitative grouting method". The ground does not produce crack and uplift.

5.4.1.3.3 Implementation effect

In order to check the quality of grouting, check the core of the sleeve valve. By checking the core sample and checking the strength and

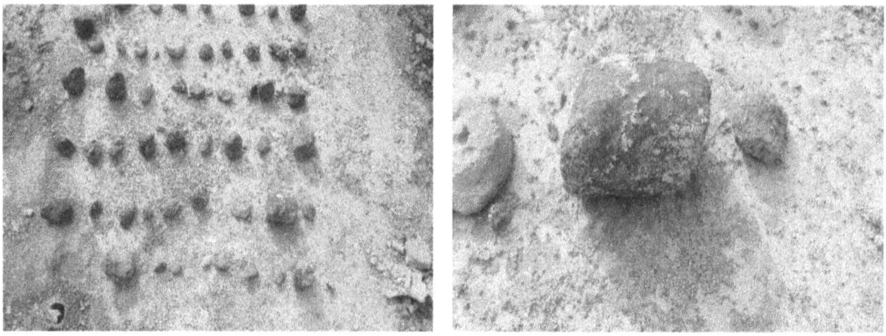

(a) (b)

Fig. 5.7. Core sample photos.

anti-seepage coefficient of the core sample, the design requirements are met (Fig. 5.7).

5.4.2 *WSS grouting*

5.4.2.1 *Process principle*

WSS is the Chinese abbreviation for free, double-liquid grouting. Double-pipe non-shrinkage double-fluid grouting is conducted in the double-tube mixing chamber with two kinds of mixed slurry, namely, A and B liquid or C, and is needed to reinforce the ground water-stop position, without changing the layer group, under the conditions of the formation of water between particles, with serous consolidation, for the purpose of improving soil properties. The characteristics of the shotcrete cause the formation of C, and the internal friction angle (φ) value increases, which makes the formation bond strength and density increase, and the reinforcement effect is achieved. When the formation particles are filled with non-flowing and consolidated slurry, the permeability of the formation is reduced, thus forming a relatively waterproof layer.

The characteristics of WSS grouting technology include the following:

(1) The double pipe used by the drilling rig can be used directly as the drill pipe to reach the predetermined depth or location, while the double pipe can be used for direct grouting, and the pipe head is equipped with a 30-cm mixer to fully mix the double liquid.

(2) During the grouting process, the grouting pipe can rotate (both positive and negative), and there will be no stuck drill pipe and slurry overflow, which saves the cost of other grouting pipes in one time, and also helps to protect the environment from pollution.
(3) There is a solution type (A, B liquid) and a suspension type (A, C liquid). The slurry has a strong permeability to the soil layer, and the grouting range can be controlled by adjusting the slurry ratio and grouting pressure. The setting time can be adjusted, and can be combined into construction to meet different requirements.
(4) The slurry mixer of the double pipe end can completely mix the two slurries in the outlet pipe, which can make the slurry even, and there is no conventional method to easily exit the tube phenomenon.
(5) The normal reinforcement can be grouted from the ground, and the surrounding area of the tunnel can also be inclined to the grouting, adjusting the grouting pressure, and horizontal advance grouting can be carried out.
(6) From drilling to grouting, it is a continuous operation.
(7) The grouting materials are cement, water glass, glacial acetic acid, silica gel negative body, etc.
(8) The materials are of small size to facilitate convenient movement in difficult construction environment.
(9) The technology can be used in various soil layers, and rock formations can also be applied, provided that additional drilling RIGS are used for early searching.

5.4.2.2 *Construction method*

The construction method of WSS grouting technology mainly includes the following points:

(1) The hole spacing for drilling is determined according to the actual project, generally 1–1.5 m.
(2) The drill pipe (grouting pipe) of the grouting system is set at the predetermined depth, and the grouting system is connected, and water is injected and discharged from the end of the slurry mixer.
(3) The back plate is used to prepare the slurry and ensure continuous supply.

(4) During grouting, when applying pressure, it must be carefully controlled. Once the pressure at some point reaches the predetermined value, slowly elevate the drill pipe (grouting tube) to reduce or increase the pressure by 30–50 cm.
(5) According to the construction requirements, each hole can be made from top to bottom, and can also be carried out from bottom to top. When porous, it is necessary to divide the pore order into the principle of interval grouting.
(6) After grouting is finished, the grouting pipe should be flushed clean, and the grouting hole shall be sealed to restore the original condition.
(7) The slurry strength, hardening time and permeability can be adjusted according to the actual needs of the project.
(8) The slurry does not loosen and does not shrink after consolidation; the hardener is non-toxic and does not cause pollution to the surrounding environment and groundwater resources. The grouting process is shown in Fig. 5.8.

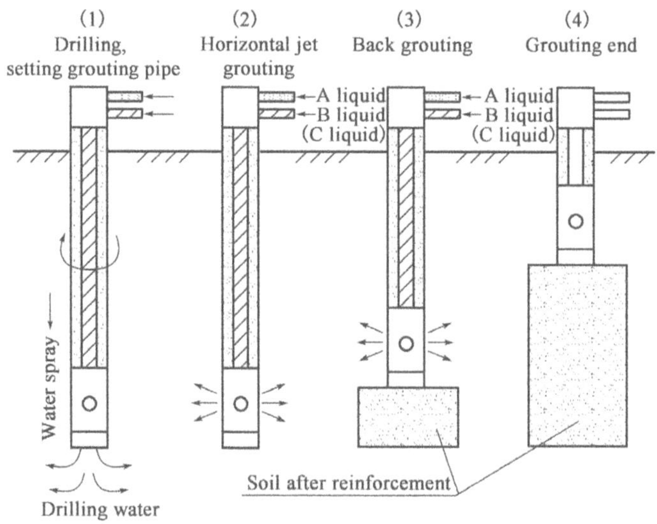

Fig. 5.8. WSS grouting process flow chart.

5.4.2.3 *Verifying the reinforcement effect*

WSS grouting technology adds a new method to the filling and cementing technology, and it has a good effect in practice.

Engineering examples: Validation of WSS grouting reinforcement effect for 4# horizontal passage of shaft in the Jingjia section of Qingdao Metro Line 13

5.4.2.3.1 Project overview

The carnival station at Jinggangshan road is near the Jinggangshan road station. The proposed shipping center will then be transferred to the west of the Lijiang River, which will reach the carnival station on Tangdao bay, sea carnival, and the total length of the area is about 7333.72 m. In the middle of the interval there are three construction shafts, a wind well, a temporary shaft and an inclined shaft. The center distance of the horizontal channel of 4 # is YSK11 + 419.309, and the distance interval is about 809 m. The shaft is located in the section about 147 m away from the carnival subway station. The clear space in the shaft is 5 m × 9 m, and the clear width of the cross passage is 6 m. The cross passage passes through the green belt of the marine carnival community on Lijiang West Road, with a total length of 57.8 m and a buried depth of about 9 m. The bolt and shotcrete support structure is The shaft is located in the section about 147 m away from the carnival subway station. The clear space in the shaft is 5 m ×9 m, and the clear width of the cross passage is 6 m. The cross passage passes through the green belt of the marine carnival community on Lijiang West Road, with a total length of 57.8 m and a buried depth of about 9 m. The bolt and shotcrete support structure is adopted.

The transition section of the transverse channel is excavated to 28 m, and an underground water bursa is encountered, and water can gush out from the gap between the water capsule and the closed grill. The field has been filled with concrete through the filling holes, but the body and the surrounding area should be further filled with grouting, and the loose soil between the body and the raised section should be strengthened. In order to ensure the safety, the WSS grouting process is adopted for reinforcement.

5.4.2.3.2 Scheme design

According to the geological data, this WSS grouting pile is passed through the asphalt surface layer, backfilled concrete layer, mixed filled soil, soil filling, silty clay, strong weathering diorite rock and so on. Because the formation is more complicated after a collapse, it is necessary to use a drill hole in combination with the guide hole, and the WSS double-liquid grouting machine is used for grouting construction. For WSS grouting construction across the muddy clay powdery layer, the strata is not less than 0.5 m in length, with a diameter of 1.2 m, a reinforcement range of the horizontal channel contour outside enlarged to 3.5 m, with 7-m south constraints (without casing) and 5-m north constraints (protective). The four sides are arranged with 0.8-m spacing and a double row of plum flower, with the middle range having a 1-m spacing. The two liquid grouting piles have 230 roots, and the outer row spacing is 0.8 m, the pile diameter is 1.2 m, the pile length is 12 m, the inner row spacing is 1 m and the pile diameter is 1.2 m.

The project uses two zlj-350 coal mine tunnel drilling machines to drill the holes, and two syb-60/5 dual-fluid variable grouting pumps for low-pressure grouting. The grouting flow rate is 2.4 m/h, the grouting pressure is about 0.5–0.8 MPa, the density of cement slurry is 1.52 kg/cm^3 and the density of water glass is 1.15 kg/cm^3. The cement slurry and water glass diluent are mixed by a 1:1 volume ratio in the grouting tube head. According to the proportion of cement slurry, it is estimated that 380 kg of cement should be used for each cubic meter.

In construction, grouting is arranged from the outer row to the inner row, as is perforating grouting. After the grouting drill pipe reaches the bottom of the hole, the grouting is started: the hole bottom is grouted to the ground, the drill pipe is raised slowly and the average speed is controlled at 2.0 m/h. In the near concrete backfill layer of 1 m, the lifting of the drill pipe is accelerated, the grouting amount is reduced, and the ground uplift and cracking are minimized. The density of cement slurry, water glass slurry and the double-liquid mixture is measured regularly. Combined with the field grouting effect, the mix ratio is adjusted in a timely manner. The coagulation time of the double-liquid mixture is measured regularly, and the mixture ratio is adjusted so that it does not cause blockage, and it can guarantee the effect of water plugging and reinforcing.

5.4.2.3.3 Implementation effect

In order to check the quality of grouting reinforcement, the core construction is carried out in the unreinforced area. The core is drilled with a diameter of 90 mm, and the core depth is 13 m, and the hole is located in the 0.6-m range of the grouting hole. Through the coring, it can be seen that the filling soil, silty clay and sand layer of the concrete backfill layer are well reinforced, and the core sample is intact. At the same time, the soil in the tunnel has been well reinforced, mainly in two aspects: The original loose silty clay in the tunnel is better integrated with WSS grouting and has good strength. There are almost no leakage points in the tunnel, and the soil moisture content is very low. No drainage is needed in the excavation.

5.4.3 *Steel tube grouting*

5.4.3.1 *Process principle*

The spraying mechanism of the molten steel pipe grouting is through pressure; for the cement grout injection into the slope rock mass, the cement slurry is injected under the actions of pressure, compaction, filling, sealing rock pore and fracture, and the process of condensation around the particles causes the physical and chemical properties of the rock and soil mass to change. After grouting, the steel tube is permanently implanted into the slope, and it acts as a reinforcing anchorage for the rock mass. Spraying of the molten steel pipe grouting technology and grouting reinforcement are two kinds of slope reinforcement technologies that increase the compactness of the rock and soil mass, causing it to form a force as a whole, thereby lifting the waterproof properties of the slope, and the slip of shear resistance, effectively improving the crushing, joint fracture of the slope's weak mechanical properties. In the slope reinforcement engineering, the steel tube grouting technology is mainly applied to three aspects:

(1) It is used for the pre-excavation and pre-reinforcement of the slope.
(2) The grouting anchor pipe is substituted for the anchor rod on the side slope.
(3) It is used to improve the anti-slip resistance of the stress concentration area.

5.4.3.2 *Construction method*

(1) The grouting pump is generally used at a rated pressure above 6 MPa. The pressure gauge on the pump should be installed correctly to ensure its normal operation.
(2) The rated pressure of the high-pressure grouting hose used for grouting should be above 20 MPa, and it should be ensured that there is no porosity in the pipe. A control valve shall be used for the closed slurry, connecting the steel tube and the pressure grouting head of the high-pressure grouting hose.
(3) Before grouting using the steel pipe, it is necessary to seal the orifice in advance, and the C30 concrete that is mixed with the accelerator is used to lift the setting speed and strength, and an exhaust pipe shall be embedded before sealing.
(4) The first round of grouting adopts a 1:1 ratio of cement mortar, with a PVC pipe being used to wash the wall to the bottom of the hole grouting. The spray between the outer wall of the tube and the rock mass is rinsed with water and the gap is filled, in order to clear the slurry from the orifice, and if the cement slurry injection is colored at the same time, then one must stop to wash the wall. The purpose of washing the wall is to fill the cement slurry between the outer wall of the steel tube and the wall of the rock mass, and increase the number of knots. If a steel tube with a diameter of 50 mm × 3.2 mm is used, then the wall should not be washed (Fig. 5.9).
(5) After washing the wall for 45–60 min, the outer space of the steel tube is sealed, and a water pipe with a nozzle is used to catch the

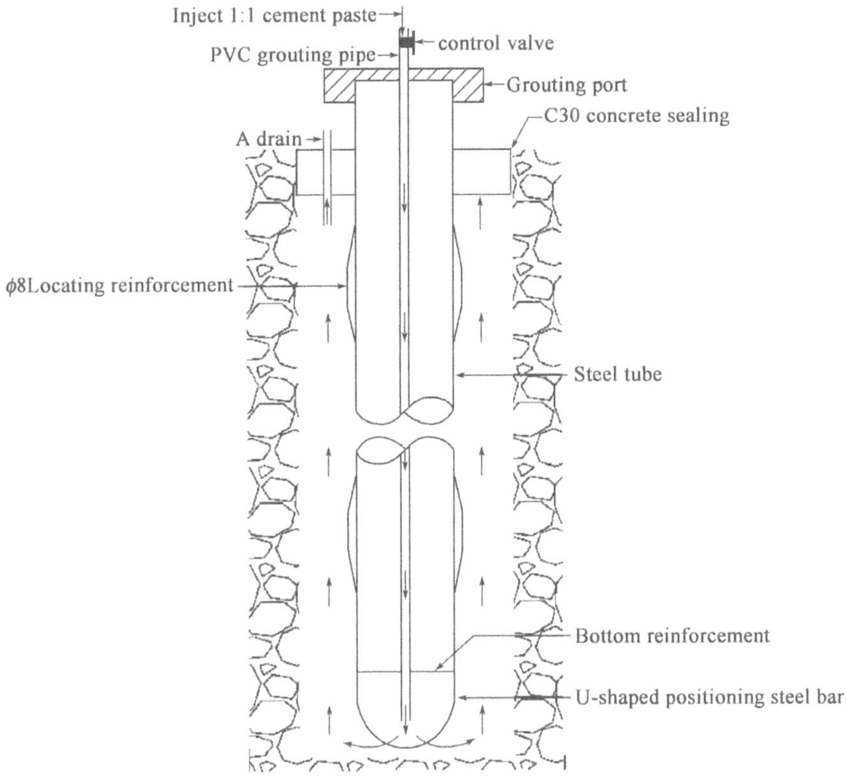

Fig. 5.9. The first grouting diagram.

bottom of the steel tube, and the cement paste is washed in the pipe with high-pressure water to clear the hole.

(6) The second grouting material is the pure cement slurry of M30, with intermittent grouting. After the first grouting is stabilized, the vent hole is sealed and pressure grouting in the second steel tube hole is carried out. The grouting pressure should be stable at 1.5–2.0 MPa. According to the design requirements for general grouting, the quantity used is 200–300 L/m; if the dosage of grouting volume has exceeded design requirements, the grouting pressure gauge will not be stable, or the local water seepage might create mud, thereby suspending grouting, or

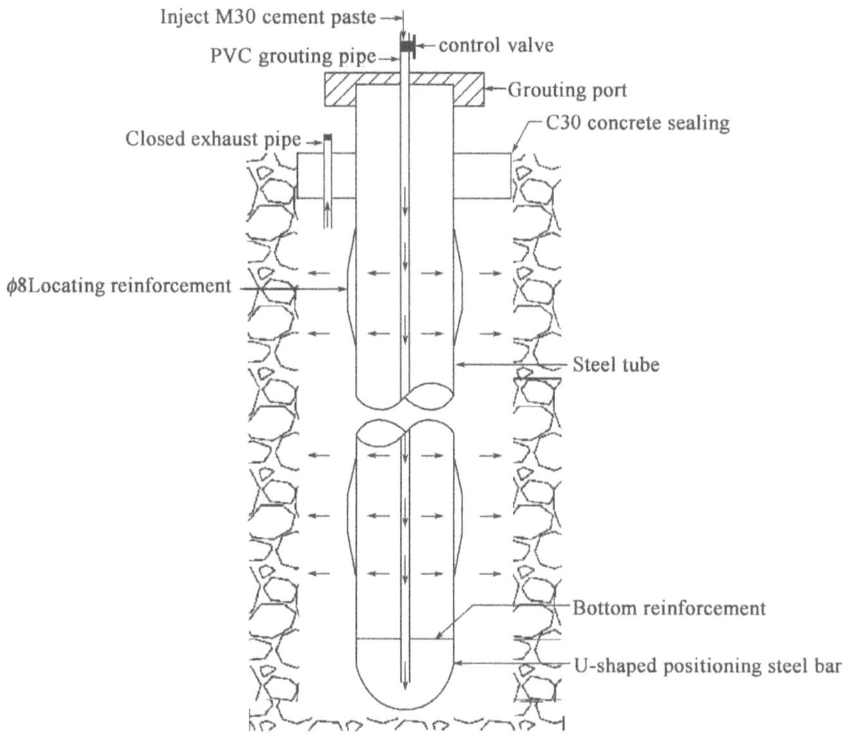

Fig. 5.10. The second grouting diagram.

else there is the risk of the pulp mouth being plugged with cement mortar, or reentry of slurry grouting after initial setting. After repeating the process several times, the pressure reaches the design size and is stable for a period of time, after which the control valve of the nozzle is closed and the grouting tube is removed (Fig. 5.10).

(7) When the grouting pressure reaches 2.5 MPa, one must close the pressure control valve and stop the feed, then the pressure will slowly decrease. The pressure gauge is basically immobile and can last 4–5 min when it is reduced to 2.0 MPa. The pressure is then stable.

5.4.3.3 *Verify the reinforcement effect*

Engineering examples: Verification of reinforcement effect of steel flower pipe grouting on surface steel inclined shaft in the Jialing Section of Qingdao Metro Line 13

5.4.3.3.1 Project overview

For the Qingdao Metro Line 13 Lingshan station, the tunnel vault was 10–16 m deep, and it was overlying or through deteriorated soil or coarse sand gravel of the quaternary period, with an overburdened and broken strongly weathered tuff layer, with relatively serious early tunnel water seepage, posing constraints to the stable excavation at the site.

5.4.3.3.2 Scheme design

According to the geological data of the tunnel area and the grouting reinforcement requirements, the surface grouting construction range is 6 m for the large mileage of the right line and the small mileage of the left line of the main line of 1 # inclined shaft. The upper boundary of the grouting range is 3 m above the vault, the lower boundary of the grouting range is the contour line of the tunnel vault, and the lower boundary of the grouting range of the two sides of the tunnel is 1.5 m below the vault. If the lower boundary of the strongly weathered rock is below the vault, the lower boundary of the grouting reinforcement range is adjusted to the lower boundary of the strongly weathered rock.

The surface vertical borehole drilling mode uses the two-sequence borehole layout: the II sequence for normal drilling grouting hole and the H order to check and supplement the grouting hole drill hole. The spacing between the holes in I sequence is 2 m, the drill hole is drilled in the H sequence drilling hole, and the grouting fan is enclosed in the area below the boundary of the fixed range. The construction process of the first order includes slurry drilling, drill hole injection of slurry in the sequence, the grouting of thin, weak zones, check points selection of drilling grouting effects of H sequence, and supplementing the grouting effect in poorer areas.

The drilling method adopts the vertical borehole of the surface, and adopts a two-sequence borehole layout. The first drill hole is a normal grouting hole, and the H sequence drilling is to check and supplement the grouting hole. The hole spacing of I sequence is 2 m, the hole in the H sequence is drilled, and the grouting range is the area below the boundary of the reinforcement range. In the construction process of the first order, after the drill hole grouting, the grouting area is weak due to part H series drilling grouting effect, and the grouting effect in poorer areas is complementary according to the grouting borehole layout as shown in Fig. 5.11.

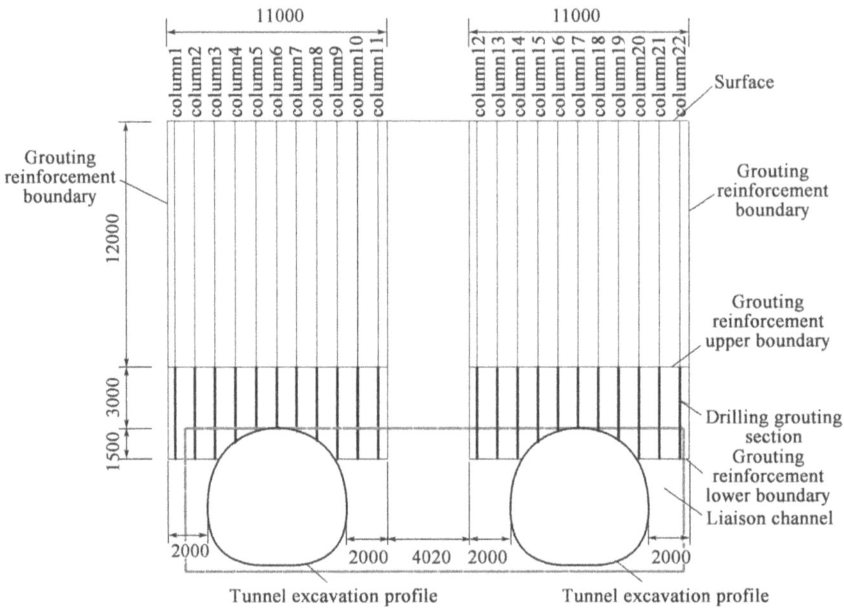

Fig. 5.11. Cross-sectional view of drilling layout (size unit: mm).

5.4.3.3.3 Implementation effect

To ensure the grouting effect, the surface grouting effect should be checked. If the grouting is reinforced, it should also be supplemented. The grouting effect is checked in the following four ways:

(1) By means of an inspection hole probe to the advance grouting reinforcement area before the detection of the tunnel excavation hole, and through the inspection hole water yield, the types of drill cuttings, hole collapse if any, and judging the front surrounding rock grouting reinforcement effect (Fig. 5.12).
(2) The surface inspection hole is used to detect the weak areas that may exist in the grouting process, and the grouting effect inspection is conducted through the II sequence inspection and a supplementary grouting hole to the grouting reinforcement area (Fig. 5.13).

Fig. 5.12. Excavation section.

Fig. 5.13. H sequence hole core sampling.

(3) For the water-rich section, a geological radar is used to detect the water-bearing state of the grouting area in the front. If the water content is more, the grouting should be supplemented.
(4) Using a borehole TV characteristic inspection holes are selected and searched for internal drilling lithology changes along with the drilling depth, water position, and geological information, assessing the grouting reinforcement effect.

5.5 Interior curtain grouting technology

The research and application of curtain grouting technology started in China in the early 1950s. In the 1970s, the curtain grouting method was widely used to prevent the dam foundation from being infiltrated when a high dam was built in a karst development area. After years of development, curtain grouting has been widely applied in many fields, such as water conservancy, construction, railways, and mining.

5.5.1 *Process principle*

Curtain grouting is a kind of grouting technique using hydraulic or pneumatic will sets according to the design of the slurry concentration through an *ad hoc* grouting hole, taking into consideration the pressure to the rock and soil layer, and filling gaps in the pore by geotechnical engineering, aiming to improve the physical and mechanical properties of the grouting object, so as to meet the needs of all kinds of engineering. According to its function, it can be divided into seepage prevention grouting, grouting reinforcement, and seepage control slurry injection which is to strengthen all kinds of foundations with seepage resistance and is widely adopted a method. In this technique, a reasonable drill hole is made, for injection of grout, making each hole's grouting body overlap each other to form a structure similar to the concrete diaphragm wall of the curtain, to truncate the flow of water, so as to fulfil the purpose of plugging, known as the curtain grouting.

5.5.2 *The construction method*

Curtain grouting is through drilling, geological exploration of holes and grouting holes in the ground to release hole pressure using sodium silicate cement or cement slurry, in an excavated section fissure, to ensure that the surrounding rock fracture is filled with a mix of a certain intensity and density of slurry, with rock consolidation as a body, forming a water curtain.

The quality assurance measures are as follows:

(1) **End standard of single-hole single-stage grouting:** The grouting final pressure, grouting volume and wall sealing condition are used

as the end standard of single-hole single-stage grouting. When the grouting final pressure reaches the design requirements, the grouting should be terminated. When the grouting volume meets the design requirements, the grouting should be continued until the grouting seal is unusable.

The method of intermittent reinjection: According to the diffusion conditions of slurry, the single injection quantity is determined, which is generally considered to be 30 m^3. A reinjection can be made for the flow sand layer and the straw soil layer. The injection hole can increase the injection amount, and then the injection volume can be reduced properly.

(2) **Control of grouting pressure:** Reasonable grouting pressure is determined according to the strength of borehole wall and different grouting depths. A pressure gauge and pressure relief dual-control valve are used.

5.5.3 *Verification of reinforcement effect*

Engineering examples: Verification of reinforcement effect of curtain grouting in small mileage tunnel of Linghuang shaft in the fourth working area of No. 13 Line of Qingdao Metro

5.5.3.1 *Project overview*

The risk assessment level is grade II in the main traversing middle, with coarse sand, a silty clay layer, and a middle weathering tuff layer within 140 m. The right side small mileage constraints on the steps of the upper part are due to sand, silt, and a silty clay layer, which is not a good geological condition. In this area as such the water content is higher, a stream of water flows out, forming into a hollow in places, there could be landslides at any time, and mud and sand are abundant. The final part of differentiation into the brecciated tuff, from the soft surrounding rock, under blasting excavation is still needed. Small mileage constraints of the left side of the wall rock are bigger, such as steps of sand, silt, and silty clay, which are poor geological conditions. In this area the water content is higher, a stream of water flows out, forming into a hollow, with the constant potential for landslides, and abundant mud and sand, on the right

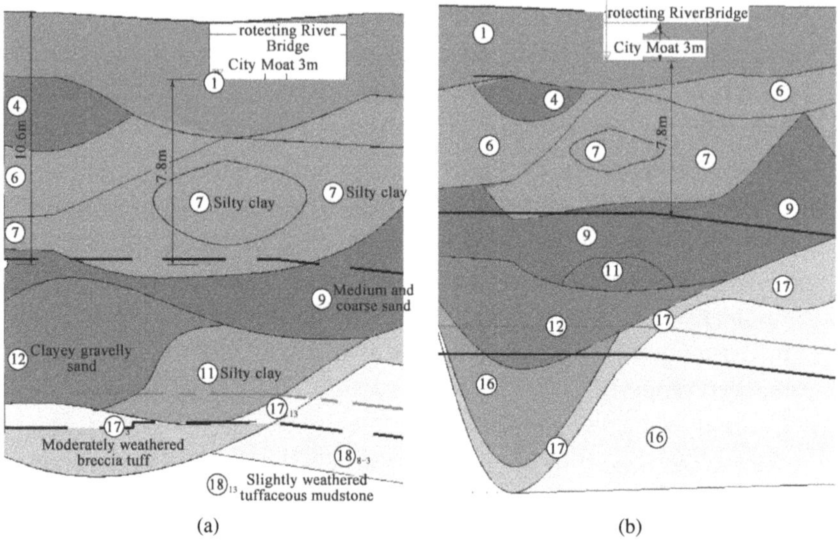

Fig. 5.14. Schematic diagram of the geological longitudinal section of the left and right sides.

side of the arch foot for weathering brecciated tuff, which still needs blasting excavation (Fig. 5.14).

5.5.3.2 *Scheme design*

5.5.3.2.1 The difficulty of grouting treatment.

The arched roof of this section directly reveals the sand layer of the fourth system, which has a poor self-stabilization ability, and the sand–rock interface has linear flowing water, which is at risk of surging any time. At the same time, the underground pipeline is intensive, and the surface uplift should be strictly controlled under the premise of a guaranteed effect.

1. The sand layer is located at the top of the arch and the sand has a low quantity of mud, and simple squeezing and splitting grouting is difficult.
2. It is very easy to cause the ground to collapse.
3. The stability of the surrounding rock is poor, and the excavation process can easily lead to the collapse of the landslide.

4. The underground pipeline is dense, and the surface elevation control is strict.
5. In the second half of the area, it is the weathered rock, which needs blasting and excavation, which can easily destroy the original grouting and solid structure, which affects the grouting.

5.5.3.2.2 Grouting reinforcement management design

(1) **Dynamic control of grouting**

Grouting is done above the tunnel through the gas, sewage and rainwater pipelines. All the tunnel constraints reveal sand holes, as does the left half section. The reinforcement and seepage control design should be aimed at the changes in the geological condition, surface uplift of the grouting hole, and grouting parameters such as dynamic adjustment (Fig. 5.15).

(2) **The grouting scheme**

To strengthen the grouting reinforcement effect, the safety of tunnel excavation is improved, with a grouting reinforcement ring of 3 m thickness, a grouting reinforcement angle of 150°, and a tentative grouting diffusion range of 1 to 1.5 m.

(3) **Design of grouting drill**

The semi-section pre-grouting treatment is designed with 12 m as a control loop, and 9 m is excavated, and the slurry disk is reserved for 3 m. The

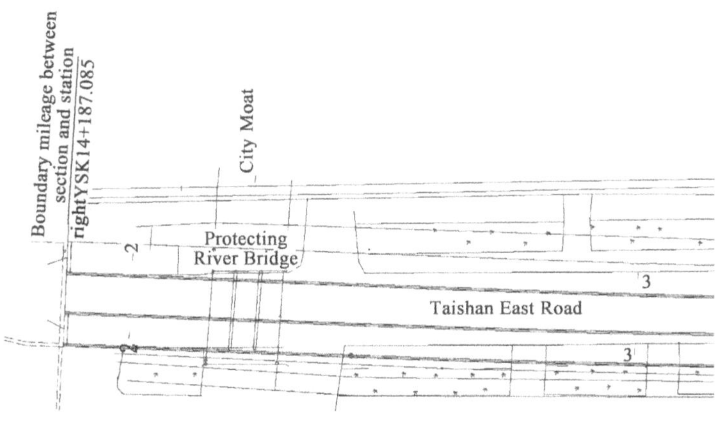

Fig. 5.15. Schematic diagram of pipeline layout of grouting reinforcement area.

four-order borehole construction reinforces the single circular tunnel within the scope of a half section quicksand layer of 3.5 m, sealing and plugging the leakage channels at the same time. The first sequence drilling constraint is for 0–5 m in the front section, the second-order drilling constraints for 5–8.5 m, and the third-order drilling constraints for 8.5–12 m. The four-order hole is reinforced with a sand–rock interface and tunnel arch shoulder. After the completion of the four-order drilling and grouting treatment, the inspection and filling grouting shall be carried out according to the engineering geological conditions. The first sequence grouting hole is set for 8, the second-order grouting hole for 7, the third-order grouting hole for 6, and the fourth-order grouting hole for 5, and should be checked to make sure drilling is according to the actual situation, which is temporarily set at 9 (Fig. 5.16).

In order to avoid grouting, the drilling in each sequence is carried out on two sides to the center.

(4) Design of construction parameters for grouting

According to similar engineering experience, in the production of injection grouting material with ordinary cement slurry and cement 425, a

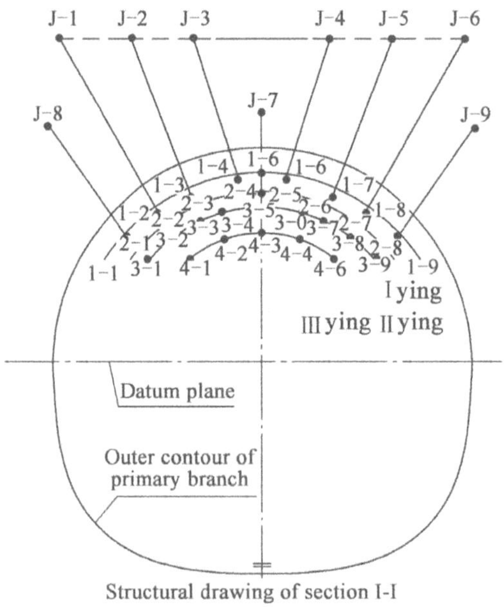

Fig. 5.16. Drilling and inspection of supplementary hole layout plan.

sodium silicate double liquid is used with a water–cement ratio of 1:1, with cement–monohydrate glass in a 1:1 ratio, and the initial setting time of about 40–50 s.

The grouting pressure is tentatively fixed at 1–1.5 MPa, but close attention should be paid to the surface uplift in the process of grouting.

One, two, three, and four sequences using phi 108 are used in the dye tube grouting hole method; the first order is used as a one-time process for hole grouting, while the second, third, and fourth sequences are used for advanced hole grouting, which is based on the analysis of the seepage effect of the flow in the sand bed in a study conducted by Shandong University which utilized a pressure of 1.5–2 MPa. However, close attention should be paid to the surface during the grouting process (Fig. 5.17).

In the process of grouting, it is necessary to make full use of borehole water inflow, grouting data records, and water inflow records of grouting, so as to facilitate the adjustment of the follow-up plan. Meanwhile, real-time monitoring of surrounding rock deformation during grouting can ensure the structural safety and provide the basis for determining reasonable grouting pressure.

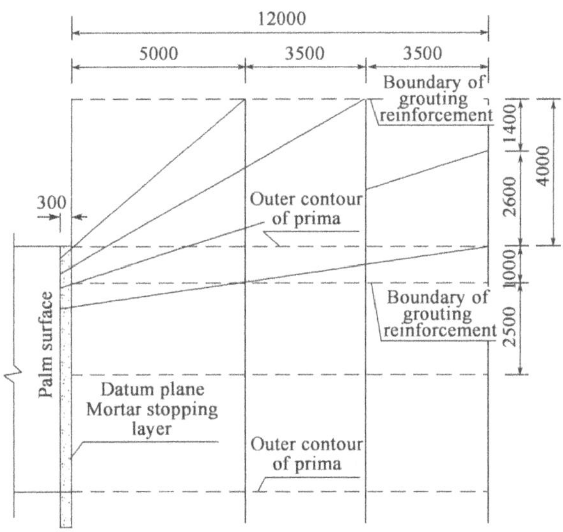

Fig. 5.17. Drilling and inspection hole grouting section diagram (size unit: mm).

(5) **Control of the grouting process**

In the process of grouting treatment, the structure and development of the flow sand layer and the bedrock intersection interface are analyzed by observing each grouting hole and the seepage grouting.

According to the situation of stratum injection ability and seepage grouting, dynamic adjustment of cement water glass proportioning is carried out to achieve the dynamic regulation of the reinforcement of the tunnel arch and the seepage sealing. During grouting, the process is stopped when the pressure of grouting reaches 1.5 MPa or when the surface uplift exceeds the danger limit.

5.5.3.3 *Implementation effect*

During the excavation of the grouting body, the slurry filling the grouting concretion body is in good condition, the palms are rich in blood vessels. After grouting treatment, the original loose sand layer with large water content in the tunnel has good integrity and high strength, no water in the face and low water content in the sand layer. There is no water on the surface of the tunnel, and the moisture content of the sand layer is very low. It is indicated that the grouting design is suitable for grouting in such strata (Fig. 5.18).

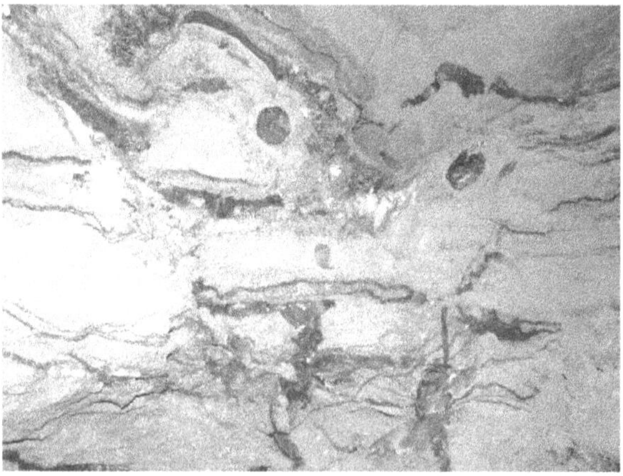

Fig. 5.18. Grouting governance effect.

5.6 Typical engineering project cases

5.6.1 *Water leakage treatment project of the first branch of Qingdao Metro Line 13*

5.6.1.1 *Project overview*

The section from Jinggangshan Road station to Huanghai East Road station starts at the intersection of Jinggangshan Road and Changjiang Road. The subway is built to Binhai Avenue along Jinggangshan Road in the south, and then to the intersection of Emeishan Road and Binhai Avenue in the West. The design mileage of the section is YCK7 +809.3 ~ YCK11 + 565.3 and the length is about 3756.0m (Fig. 5.19).

At a distance of 18.0 m out of Jinggangshan road station, there is a line spacing between mileage YCK9+525 on the left side of the districts and the wind hole and shaft (2); there is a connecting passage in mileage YCK11+419 set on the right side temporary construction shaft and connecting passage shaft (3), and at the right side of mileage YCK10+600,

Fig. 5.19. Schematic diagram of traffic position of proposed site.

establishing a deviated well and contact channel, 13.5 m into the carnival station. The line spacing interval mining method is used for the proposed construction, the Ming excavation method is used for the shaft construction, and a combination of the excavation method + mine construction is proposed for the slope. The bottom plate of the tunnel is at an elevation of –31.0 to –10.5 m, and the floor is buried at a depth of 16.0–36.9 m.

The body through the surrounding rock tunnel is mainly for the quaternary muddy opaque leaching soil, silty soil, medium breeze coarse sand, and granite porphyry, diorite, brecciated tuff, with local joint development of dense zone and cataclastic rock construction.

Overlaying for quaternary artificial filled soil, silty sand and silty soil, gravel, coarse sand, and leaching earth is done at a thickness of 6.6–18.6 m, under V late Yanshanian granite porphyry, diorite, and white main brecciated tuff. The weathering zone revealed in this interval is fragmented, and structural alteration is observed in multiple structures.

The tunnel hole is mainly located in the bedrock and the breeze belt, and the surrounding rock is ranked at the H–N2 level. Groundwater is found mainly in the bedrock fissure of the strong breeze rock, but the natural water content is not high. A pumping test reflects that its penetration is relatively poor, but the practical engineering example shows the formation of the bedrock fissure water channel; once formed after the excavation of the underground water channel, the water quantity increased obviously, especially during the cavern excavation process, often forming a dot or linear pattern, with oozing water forming inside the underground running tunnel, and the water being discharged upon pumping. The carnival station is located at Y(Z)CK11+245 in the silty soil, and the surrounding rocks are graded as VI. Influenced by the regional structure, the jointed zone and sandy soil, the massive cataclastic rocks are more fertile, and the surrounding rocks rank at the N2–VI level. The groundwater type is mainly the quaternary pore water and bedrock weathering fissure water. Joints are dense in the local development belt and cataclastic rock area in the structure, depending on the type of groundwater in the rock structure crevice; the degree of rock mass structural fissure and catchment condition influence the water, and the water permeability of surrounding rocks is different. In the process of tunnel excavation, the water loss of the rock layer will not cause surface subsidence and the construction of the ground will continue.

5.6.1.2 *Scheme design*

5.6.1.2.1 Difficulty in governance

Governance is given priority with four classes of surrounding rock and a small level 3 section of surrounding rock, according to water seepage exposing the surrounding rock in the coarse grain granite, and mainly in the breeze of the joint fissure development; at the same time, through the water quality analysis, the water supply of this rich groundwater is determined. According to the on-site investigation and preliminary analysis, the main bedrock fissure water seepage, part of the crack and the upper water unicom, and seepage of the inverted arch are mainly caused by rock fracture at the beginning of the flow behind the construction (Figs. 5.20–5.23).

Fig. 5.20. Leakage in the hole.

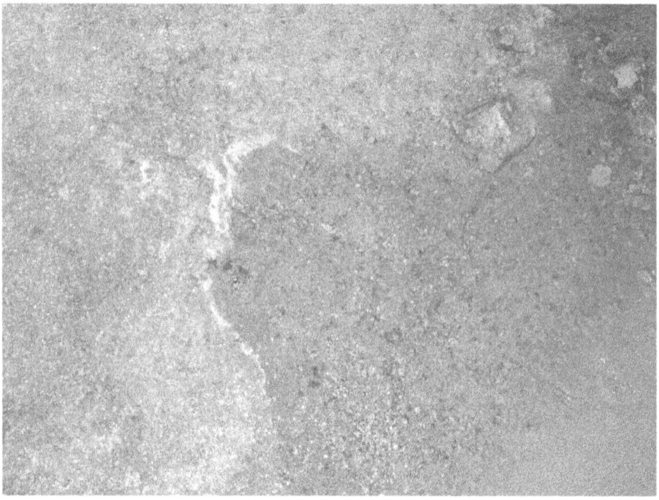

Fig. 5.21. Leakage in the inverted arch.

Fig. 5.22. Leakage diagram of the side wall.

Fig. 5.23. Leakage flow in the hole.

5.6.1.2.2 Cause analysis

According to geological data analysis and field investigation, there are two main reasons for the leakage:

(1) The groundwater in this area is rich, the fissure of the surrounding rock is relatively developed, and it is connected with seawater.
(2) Regarding the system of the water tunnel on the whole, before grouting construction, although it was a basic anhydrous vault, the early collection of groundwater through cracks in the surrounding rock to the area behind the inverted arch caused serious inflow of water behind the inverted arch.

The scheme is aimed at the current situation of leakage, and the overall scheme design is as follows:

(1) First of all, the geological radar of a 900-m antenna is used to comprehensively explore the tunnel, the rich water area is clearly defined,

the recharge range and supply mode of groundwater are clarified and targeted grouting is carried out.

(2) The water flows through the pipe from upstream to downstream of the governance area, and the water distribution point and the size of the water yield management area, and end ring grouting in the governance area, are close to the upstream position of the water.
(3) From the position of relative elevation to the lower position is a step forward.
(4) A preliminary analysis of the surrounding rock fracture determines the main water channel. In view of the sealing of the surrounding rock fissure, the radial hole grouting that is on the same ring as the six grouting holes effectively prevents pulp from forming, improving the efficiency of grouting.
(5) After the injection, if the water cannot be stopped completely, it is necessary to fill grouting in the backside and shallow fissure of the first branch, cut off the hydraulic connection behind the initial branch and strengthen the anti-seepage capacity of the initial branch.
(6) Finally, upon aiming at the formation of water droplets which is difficult to cure, epoxy resin is utilized for targeted treatment.

5.6.1.2.3 Scheme implementation

(1) **The deep wall rock fracture grouting**

As shown in Fig. 5.24, the boreholes will be evenly distributed within the same section of the tunnel grouting hole, with the end hole position at the beginning of a vertical distance of 3.5 m, according to field occurrence of the surrounding rock grouting hole angle. To achieve deep surrounding rock fracture grouting, a rubber check pulp plug must be installed at the depth of the 0.8-m position, preventing pulp from forming at the orifice, ensuring the slurry injection into the deep surrounding rock fracture.

According to the parameters shown in Table 5.7, the profile and top view of the grouting hole are shown in Figs. 5.25 and 5.26.

To prevent slurry from collecting near the drilling string without grouting, and at the same time improve the efficiency of drilling, as shown in Fig. 5.27, the following sequence is followed for sections 1, 2, 3 and 4, set in the governance area 4 working section:

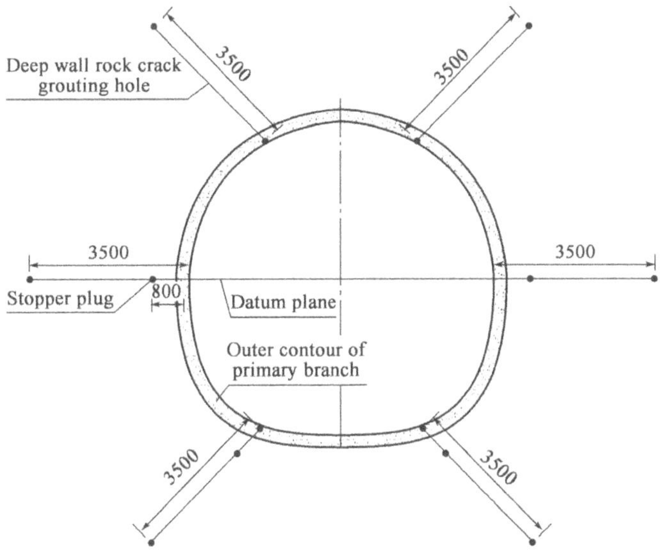

Fig. 5.24. Cross-section diagram of grouting hole of deep surrounding rock fractures (unit: mm).

Table 5.7. Parameter table of the grouting borehole in deep surrounding rock fissure.

Grouting hole diameter	Number of boreholes per ring	Ring spacing	Grouting pipe length	Grouting pressure	Hole depth
42 mm	6	3 m	1.5–2 m	0.5–1.5 MPa	Vertical distance from the position of the final hole to the initial branch 3.5 m

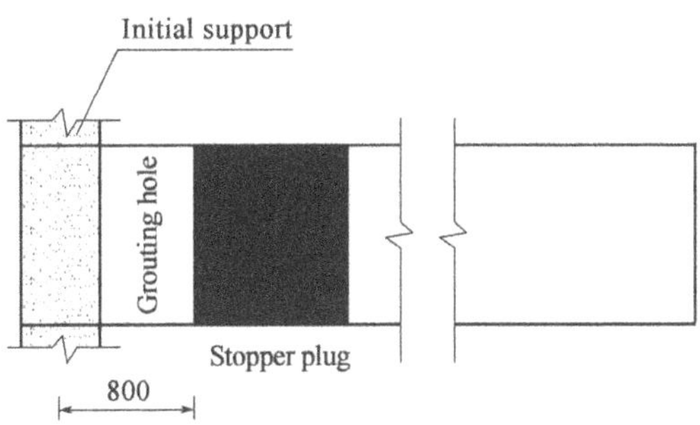

Fig. 5.25. Diagram of grouting hole profile (unit: mm).

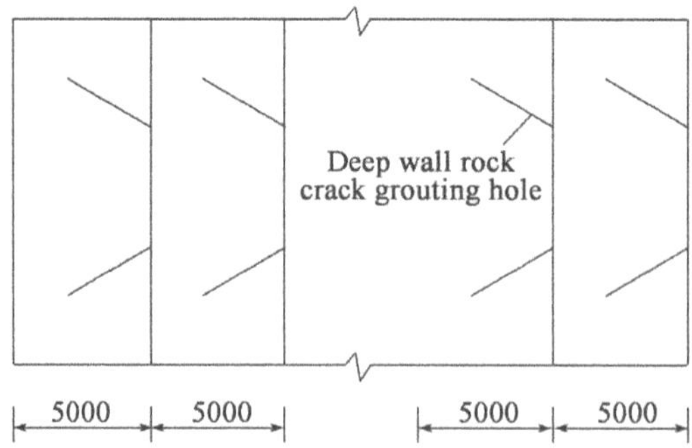

Fig. 5.26. Elevation view of grouting section (unit: mm).

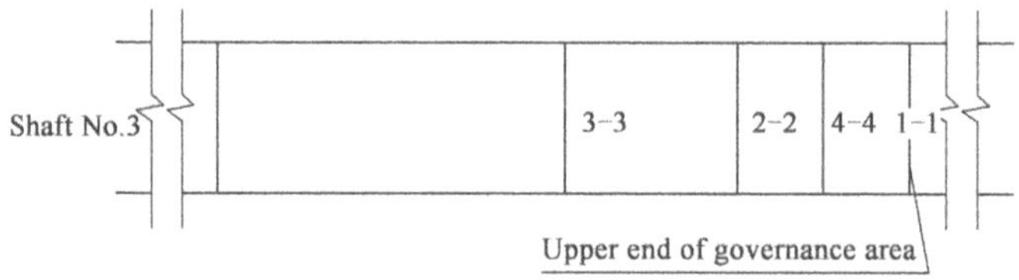

Fig. 5.27. Profile of grouting working face.

a. After the hole is formed, the borehole is used to take a high-definition video to clarify the position of the water outlet, adjust the length of the grouting pipe and to seal the outlet water point.
b. The water pressure test of the grouting hole is carried out to determine the development degree of the cracks around the grouting hole, and the effect of flushing the hole, improving the permeability and increasing the size of the slurry injection.
c. Between rings and ring grouting holes, to prevent slurry, grouting in the deep surrounding rock fracture can be divided into two parts during the construction, namely, the first ring spacing of 6-m grouting hole after which the construction to the governance unit grouting is completed, and the 3-m hole in the middle location, interspersed with the remaining grouting holes, complete with deep surrounding rock fracture grouting holes.

d. Grouting holes prevent the pulp from running, improving efficiency and shortening the construction period. On adopting a group of grouting holes, namely, once for the entire section 6 of the grouting, the grouting pressure gauge and flow meter monitor each orifice, ensuring the even spread of slurry.
e. If there is a slight run of pulp within the slurry, mixing a certain amount of sawdust, the grouting is continued, and the concentrated run of pulp is targeted using epoxy resin plugging in time, reducing the waste of slurry, improving the grout injection rate and reducing the pollution of the environment of the tunnel. If the above two methods cannot stop the running pulp, then the single liquid should be changed to double grout, and the ratio of cement and sodium silicate should be dynamically adjusted, thereby realizing the dynamic control of the grouting process.
f. The grouting material adopted is mainly cement slurry and c-s double-liquid slurry with water–cement ratio of 1:1. By observing the surrounding running slurry, a single- and double-liquid switch is carried out in time. At the same time, the structural development and spatial distribution characteristics of water-bearing structures are analyzed. Prior to the end of grouting, a certain amount of sulfoaluminate saline mud is injected to improve the seawater corrosion resistance of the calculi.
g. According to the stratigraphic situation, the grouting pressure is tentatively fixed at 0.5–1.5 MPa.

(2) **The initial support and shallow fissure grouting**

If deep fissure grouting in some regions has blocked all the seepage, the area is no longer in the backside and shallow fissure grouting can be done.

After grouting of the deep surrounding rock fracture, grouting on the back of the initial support and the shallow fissure, the residual seepage is treated, and the parameters are shown in Table 5.8. As shown in Fig. 5.28, the grouting holes shall be evenly inserted between the grouting holes of the deep surrounding rock, the grouting hole angle shall be determined according to the occurrence of the surrounding rock and the bottom of the hole is at a depth of 1 m.

Table 5.8. Parameter list of grouting holes at the back of initial branches and shallow fissures.

Grouting hole diameter	Number of boreholes per ring	Ring spacing	Grouting pipe length	Grouting pressure	Hole depth
42 mm	6	3 m	0.8 m	0.5–1.5 MPa	Vertical distance from the position of the final hole to the initial branch 1 m

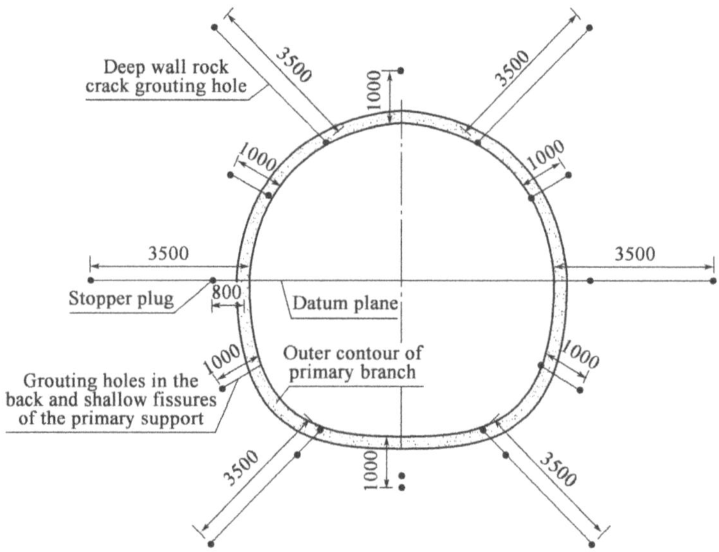

Fig. 5.28. The transverse profile of the grouting hole at the back of the initial branch and the shallow fissure (size unit: mm).

As shown in Figs. 5.29–5.31, the grouting pipe is made of a 42-mm seamless steel pipe, the grouting pipe is closed with a rubber stop plug and the rubber stopper is installed on the initial support to prevent slurry.

a. After the hole is formed, a borehole is used to make a high-definition video to clarify the position of the water outlet, adjust the length of the grouting pipe and to seal the outlet water point.

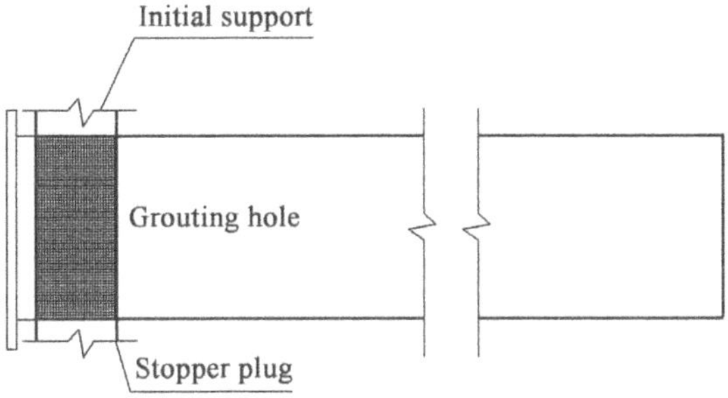

Fig. 5.29. Diagram of the grouting hole profile (size unit: mm).

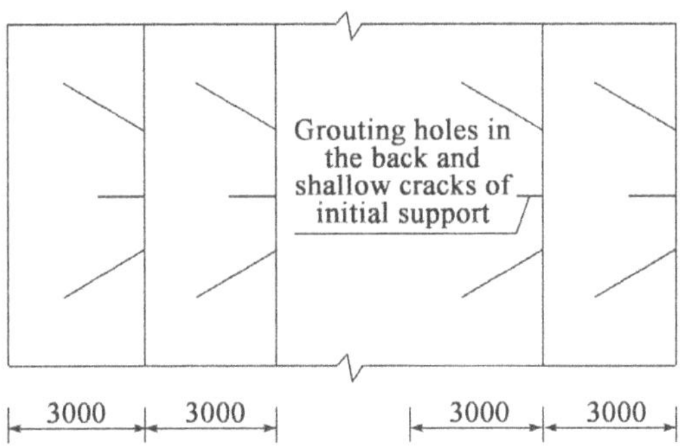

Fig. 5.30. Top view of the grouting section (size unit: mm).

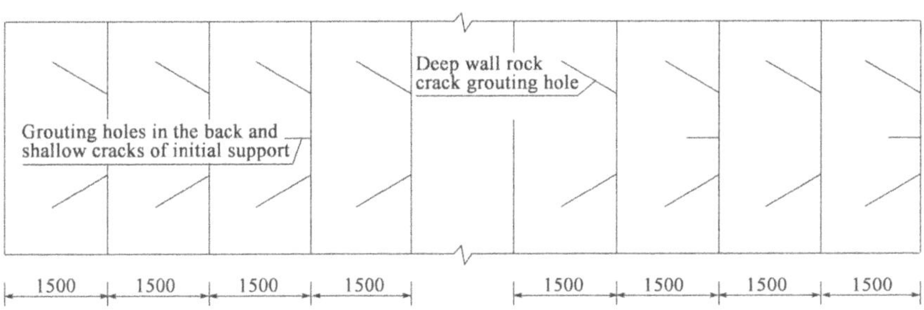

Fig. 5.31. Drilling position diagram (size unit: mm).

b. The water pressure test of the grouting hole is carried out to determine the development degree of the cracks around the grouting hole, and to wash the borehole, improving the permeability and increasing the amount of slurry injection.

c. Because the grouting hole is small and the slurry is considerable, grouting is used to ensure the grouting effect.
d. If there is a mild run of slurry, the cement is mixed with a certain amount of sawdust to continue the grouting, and the running pulp point is targeted using epoxy resin plugging in time, reducing the waste of slurry, increasing of grout injection rate and reducing the pollution of the environment of the tunnel. If the above two methods cannot stop the running pulp, then the single liquid should be changed to double grout and the ratio of cement and sodium silicate should be dynamically adjusted, thereby realizing the dynamic control of the grouting process.
e. In order to improve the seawater corrosion resistance of the calculi, the grouting materials mainly adopt the water–cement ratio of 1:1. Sulphoaluminate cement has the micro expansibility of 5, which can effectively control the water seepage of the rock microfissure.
f. According to the stratigraphic situation, the grouting pressure is tentatively fixed at 0.5–1.5 MPa.

After the grouting is finished, the grouting pump is stopped, the orifice valve is closed and then the grouting pump and the slurry pipeline are washed until the water is clear. The cylinder body of the grouting pump is removed, checked and rinsed again. No residue is allowed. After grouting is over, the equipment and waste products are cleaned according to their respective positions.

After grouting of the deep surrounding rock fracture and initial support and shallow fissure grouting, the water droplets that have not been managed can be blocked by epoxy resin.

5.6.1.3 *Implementation effect*

Through the field construction and the digital borehole television camera detection, the comprehensive analysis of the grouting effect is conducted.

After each ring of grouting is finished, the grouting effect is tested. The inspection method is mainly used to check the hole. Through the digital

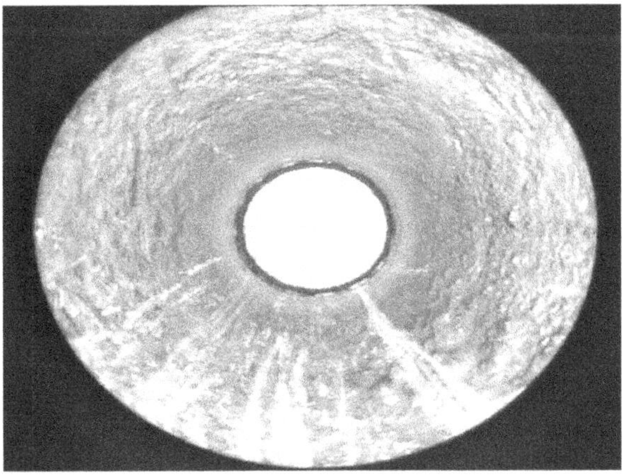

Fig. 5.32. Borehole TV camera.

Fig. 5.33. After grouting.

borehole television camera, as shown in Fig. 5.32, it can be seen that there is no water in the hole, and the slurry can fill the cracks effectively. As shown in Fig. 5.33, after grouting, the tunnel cavity has no water, and the effect of good grout sealing is achieved, ensuring the construction quality and safety.

5.6.2 *Qingdao Metro Line 13 Lingshanwei car running grave excavation grouting treatment project*

5.6.2.1 *Project overview*

Lingshanwei station is located at the east side of Mount Taishan road and along the east road of Mount Tai. The design center of the station is YCK14+119.085, and the length is about 190 m (Fig. 5.34). This station is an underground island, excavated using the Ming excavation method. The station structure floor is at an elevation of –10.21 m, buried at a depth of 14.9–15.8 m. The landform of Lingshanwei station is the coastal accumulation area, which is composed of coastal sediments and river sediments. From top to bottom, the station strata are as follows: 3 m in silt, 3 m in silty clay, 6 m in medium coarse sand, 2.2 m with leaching soil gravel sand and 2.6 m of strong weathering tuff. The base is located at the junction of the gravel sand with the strong weathering layer. During the excavation of the station, there have been many accidents in the side wall of the foundation pit.

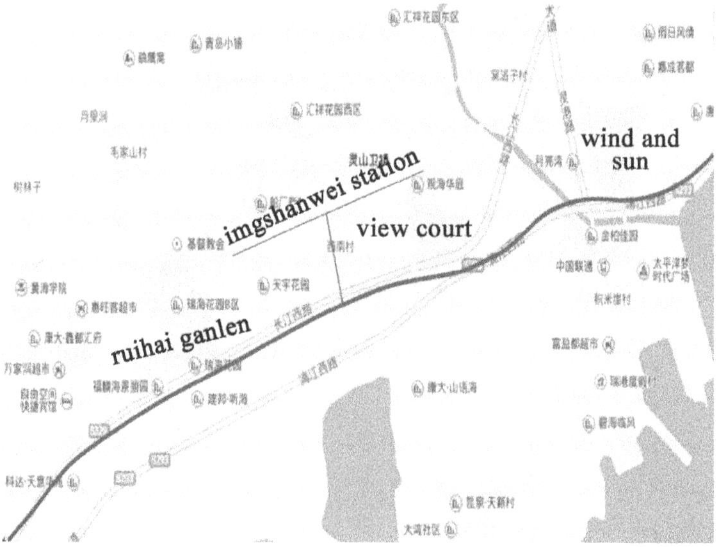

Fig. 5.34. The traffic position diagram of Lingshanwei station.

5.6.2.2 *Scheme design*

5.6.2.2.1 Difficulty in governance

(1) There is no bearing capacity of the sand layer, the flow of the groundwater is unhindered and it is easy to cause landslides under load.
(2) The grouting reinforcement requirements of the sand layer are high, and the grouting blind area cannot exist, otherwise it is easy to cause the collapse of a large area.
(3) When the curtain grouting around the foundation pit is carried out, the slurry diffusion radius is not easy to maintain, otherwise it is very easy to cause damage to the side wall of the foundation pit and cause engineering accidents.

5.6.2.2.2 Cause analysis

The excavation of the foundation pit in the sand layer shows a surge of water and sand, indicating that the high-pressure spinning pile failed to achieve the water-retaining effect in the middle coarse sand and gravel sand, and the water and sand poured into the foundation pit. There are several reasons for this:

(1) **Thick sand layer with gravel**

The thickness of coarse sand layer in this area is 6 m, and the thickness of gravel sand layer is 2 m. In the ninth layer, the coarse sand content is only 8%, the mud content is low and the cement slurry will solidify slowly after entering the sand layer. The mean value of 2–5-mm coarse grain in this layer is 24%, the local clip has gravel in a diameter of 1–5 cm and the cement slurry is lost to the periphery. In the 12th layer, the average content of the coarse grain of the leaching soil gravel is 26%, and the local clip has a block diameter of 1–5 cm. The high-pressure rotary chute piles in coarse sand and gravel in the sand pile, and slurry in the surrounding strata; the pile body contains coarse sand, gravel, sand, cement and gravel sand and gravel cemented pores, leading to water penetration.

(2) **Micro-confined groundwater**

The groundwater type at the proposed site is mainly diving and confined water. Diving: It is mainly in the 1st and 4th floors, which have a small amount of water and a stable underground water level. Confined water: It mainly occurs on the 9th and 12th floors. The pressure water of this layer is connected with the surface layer, and it is characterized by micro pressure. After the construction of the high-pressure rotary jet pile, under pressure, the micro-confined groundwater enters the unsolidified pile and dilutes the cement slurry.

(3) **Groundwater lateral to runoff**

The pressure water in the 9th and 12th floors is connected with surface diving, mainly receiving the lateral runoff and the atmospheric precipitation recharge, and the drainage is dominated by the lateral runoff, and the artificial mining is supplemented.

There is an old channel on each side of the 50-m line at both ends of the foundation pit, which is about 400 m away from the sea. It is connected to sea water and is affected by the fluctuation of seawater, and the groundwater has natural lateral runoff. After the construction of the high-pressure rotary jet pile, the water slurry flows to the periphery of the sand layer, which affects the pile quality because of the lateral flow of the groundwater.

(4) **A strong permeable layer**

The permeability coefficient of the 9th layer is 25, and the permeability coefficient of the 12th layer is 12, which is a strong permeable layer. The cement slurry infiltrates the surrounding sand layer.

(5) **Artificial exploitation of groundwater**

After investigation, the south station residents access tap water, both for production and living, from the well drilling extraction of groundwater, and the north side residents mostly use tap water, and industrial enterprises use groundwater; the south and north have a total of 15 Wells, at 15–30 m depth, a diameter of 40–80 cm, with water well spacing foundation pit edge of 10–20 m, and total water output about 20 m^3/d. Due to continuous water intake, lateral runoff is generated in the inland water of the sand layer, and the water loss in the high-pressure jet grouting pile with the flow of water leads to the poor quality of the pile (Table 5.9).

Table 5.9. Groundwater use around foundation pit.

Bit position	Well diameter (cm)	Use path	Degree of depth (m)	Standby note
Xinying steel	50	Domestic	18	
Building Materials Monopoly (cement)	50	Domestic	15	
Lathe processing	80	Produce	28	Production water supply
Vegetable distribution	50	Domestic	15	
Huibaichuan Building Materials General meeting	55	Domestic	20	
Everybody happy KTV	60	Domestic	18	
Anshunda tire supermarket	50	Domestic	15	
Venus repair plant (car wash)	60	External business	20	High water consumption
Huiyou beauty products	50	Domestic	22	
Hua Cheng advertisement	40	Domestic	30	
Shandong Longtai cement	70	Domestic	9	
Haokang pipe industry	40	Domestic	11	
Kimboda Machinery (North side)	55	Domestic	18	
Yu Xiaoming Store (North side)	50	Domestic	20	
Maolan Oil Seal (North side)	80	Site area	30	Production water supply

5.6.2.2.3 Scheme design

(1) **The drilling is arranged**

The surface vertical borehole drilling mode uses the two-sequence drill hole arrangement, for normal grouting hole drilling, I first, then II sequence, and then supplement the grouting hole drilling for inspection. The spacing between the holes in I sequence is 2.8 m, and the grouting range is in the area below the boundary. In the process of construction, grouting shall be carried out for I section of hole first, and after grouting for I section of hole, grouting effect shall be checked for II section of hole in the weak grouting area and supplementary grouting shall be carried out for the area with poor grouting effect.

(2) **Design of drilling structure**

Around the curtain grouting construction process control domain, the key to maintaining the effect of grouting reinforcement is to ensure that the injected slurry is retained in the target area, and the effective grouting quality within the target area is closely related with the grouting reinforcement effect. To prevent flow of slurry in shallow drilling into the formation caused by invalid grout diffusion, slurry injection formation in the area of the borehole grouting should be ensured, using a pressure expansion mold bag above the diffusion channel.

The working principle of the pressure expansion mold bag is as follows: the pressure expansion mold bag length is 3 m. Before grouting, the expansion slurry is injected into the pressure isolating expansion membrane bag

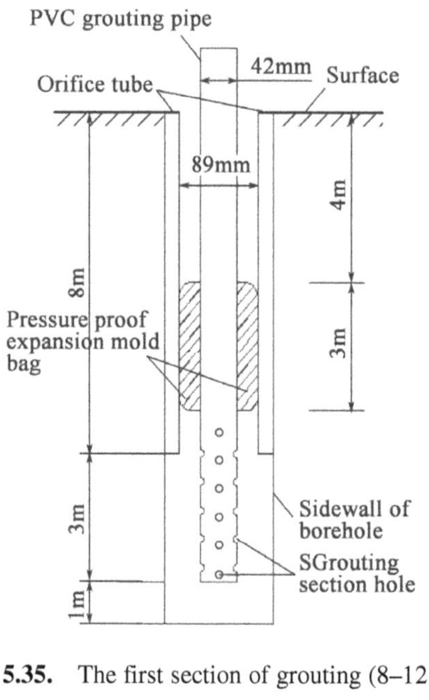

Fig. 5.35. The first section of grouting (8–12 m).

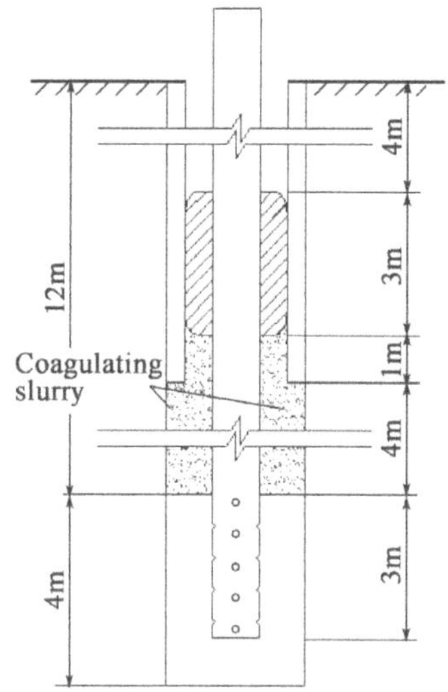

Fig. 5.36. The second section of grouting (12–16 m).

to make the membrane bag expand and compress the surrounding strata of the borehole. The soil above the target grouting layer is compacted and consolidated by the membrane bag. The expansion membrane bag and the soil compaction and reinforcement area form a grout stopping zone, so that the slurry cannot return from the grout outlet area upward. Only the slurry is allowed to enter the stratum by the hole grouting section, so as to realize the effective retention of the slurry in the target area.

5.6.2.3 *Implementation effect*

(1) Excavation of foundation pit reveals side wall: Excavation of the Lingshan guard station foundation pit revealed the side wall of the foundation pit, and one could observe the sidewall plasma pulse distribution, the density of the sand layer, the seepage situation, etc., and use the information to review the grouting effect, as shown in Fig. 5.37.

(2) The surface inspection hole probe, to examine the weak region that may exist in the process of grouting reinforcement, may be conducted through H sequence checking, supplementing the grouting

Fig. 5.37. Excavation of foundation pit reveals pulp veins.

Fig. 5.38. II Sequence hole.

hole effect and inspecting the grouting reinforcement area, as shown in Fig. 5.38.

To sum up, the grouting effect is better, and the evaluation result is that the station can continue to excavate. In the middle of November 2016, the excavation of the foundation pit was completed, and no large-scale leakage occurred during excavation (Fig. 5.39).

Fig. 5.39. Excavation drawing of station foundation pit.

5.7 Summary

In view of the poor geological conditions, this chapter introduces grouting reinforcement techniques, grouting materials, common grouting reinforcement works, and applications, through the specific engineering example of Qingdao Metro Line 13, detailing the specific operation of grouting reinforcement method and operation process. Based on his own experience, the author draws the following conclusions for reference:

(1) The conventional grouting processes, such as surface cyclone reinforcement, surface grouting reinforcement, and in-hole curtain grouting, can basically meet the requirements of poor geological grouting reinforcement in hard rock formations.
(2) For different geological features, only appropriate construction techniques and methods can be selected to reduce the probability of quality problems and ensure the normal construction of metro projects.
(3) The engineering experience shows that surface cyclone reinforcement does not adapt to the rich sand stratum.
(4) Each project has its own characteristics, and the geological characteristics for a specific project should be analyzed to accurately grasp the requirements for improvement, comprehensive use of conventional

grouting technology, safety and rapid construction, which have immense significance in engineering.

Bibliography

[1] Kuang Jianzheng. *Theory and Engineering Practice of Geotechnical Grouting*. Beijing: Science Press, 2001.

[2] Wang Mengshu *et al. Technology of China Tunnel and Underground Engineering*. Beijing: People's Transportation Press, 2010.

[3] Wang Mengshu. *Technical General Theory of Shallow Underground Excavation*. Anhui: Education Publishing House, 2004.

[4] Zhang Qingsong. *Mechanism and Management of Water Flow In Tunnel and Underground Engineering*. Beijing: People's Transportation Press, 2014.

[5] Zhang Zhongmiao. *Pile Foundation Engineering*. Beijing: China Construction Industry Press, 2007.

[6] Cui Zhengjiang. *Underwater Tunnel Grouting*. Beijing: People's Railway Publishing House, 1978.

[7] Xu Shu-Gen. Pre-grouting method for underwater tunnels. *Underground Engineering and Tunnel*, 2007 (a01): 55–57.

[8] Zhang Fengxiang. *Construction and Effect Detection of Earthwork Grouting*. Shanghai: Tongji University Press, 1998.

[9] Zhou Shunhua, Dong Xinping. *Calculation Principles and Application of Pipe Shed Construction*. Shanghai: Tongji University Press, 2007.

[10] Wu Shunchuan, Jin Aibing, Gao yongtao. The research and evaluation of the effect of the sleeve valve on the change of the soil body and the evaluation of the effect on the fruit. *Rock Soil Mechanics*, 2007 28 (7): 1353–1358.

[11] F. G. Bell. *Engineering Geology and Geotechnical Engineering*. Beijing: China Construction Industry Press, 1990.

[12] Z. Y. Fu. *The Latest Soft Ground Treatment Method.* Beijing: China Railway Publishing House, 1988.

[13] H. F. Winter Cohen, Fang Xiao Yang. *Basic Engineering Handbook*. Beijing: China Construction Industry Press, 1983.

[14] Bai Shijun. *Construction Method of Civil Engineering*. Japan: Shanhai Tang, 1974.

[15] Pandian NS, Nagaraj TS, Raju PSRN. Permeability and compressibility behavior of bentonite-sand/soil mixtures. *Geotechnical Testing Journal*, 1995 18 (1): 86–93.

[16] Fransson CFTJ. Estimation of deformation and stiffness of fractures close to tunnels using data from single-hole hydraulic testing and grouting. *International Journal of Rock Mechanics and Mining Sciences*. 2010 6 (47): 887–893.

[17] SWEDENBORGS, DAHLSTR0 m L0. Rock mechanics effects of cement grouting in hard rock masses. In *Proceedings of the 2003 Specialty Conference on Grouting at the Third International Conference on Grouting and Ground Treatment*. New Orleans, Grouting and Ground Treatment, ASCE, 2003, GSP120: 1089–1102.

[18] Nilsen AHVH. Rock mass grouting in the Loren tunnel: Case study with the main focus on the grout ability and feasibility of drill parameter interpretation. *Rock Mechanics and Rock Engineering*, 2014.

[19] Du Jiahong. *Concise Manual of Underground Construction Grouting Engineering*. Beijing: Science Press, 1992.

Chapter 6

Green Construction and Standardized Site Construction

6.1 Overview of green construction

6.1.1 *Overview of green construction abroad*

Green construction is the embodiment of sustainable development in the construction industry. The concept of green construction in the world proposes different ideas, such as sustainable building, green building, clean production, environmental construction and so on, all of which have a very high correlation.

Beginning in the 1930s, the United States, the European community, along with many other countries, the United Nations Conference on Environment and Development, the international standards committee, the ASTM International, the Association of American Architecture (AIA), ASHRAE, the United States Green Building Council (USGBC), American Association of Lighting Engineering (IES), and other international organizations and institutions started to carry out research on green construction, sustainable buildings and clean production, and applications in environmental protection construction were carried out.

Developed countries are particularly concerned with what kind of material to use in the construction to avoid wastage and save energy, reduce the environmental pollution caused by construction, reduce the construction cost, reduce labor intensity and extend the service life of buildings, and work in this field has made great progress. Some building research institutes and big companies worked together and published "the green building techniques manual" (for design and construction), "green building design and construction reference guide" and other books, which provide a certain amount of guidance and promote the development of green construction.

6.1.2 *General situation of green construction in China*

The construction industry in our country, first of all, started from the ideas of civilized construction and environmental protection, extending gradually to the "energy conservation and emissions reduction" and "economical construction site" concepts, and formed in the course of development of the green construction system

6.1.2.1 *Green construction strengthened at the national macro level*

Since the 1970s, our government has paid increasing attention to energy conservation, emission reduction and environmental protection, and issued policy documents for the same.

In 1978, China and other countries began to implement methods of environmental protection, control and prevention of pollution, energy saving, etc. They began to enact laws and regulations, such as in the constitution of the People's Republic of China, the environmental protection act, "the standards for environmental protection management method", "law on the prevention and control of water pollution", the section about energy management provisional regulations, "the water law", "law on the prophylaxis and control of water pollution in the detailed rules for the implementation", "implementing rules for the prevention of air pollution", electric power law, the law on the prevention and control of environmental noise pollution, "the outline of energy-saving technology policy", etc.

In July 2005, the state council issued a notice regarding the construction of a conservation-oriented society. The main idea of the notice was to emphasize resource utilization efficiency as the core, with energy saving, water saving, material saving, land protection, comprehensive utilization of resources and the development of circular economy as the key points, and the need to speed up the construction of economical society.

In 2007, the ministry of construction issued guidelines for construction of green buildings, then issued national standards titled "building green construction assessment standard", "green construction of construction engineering" and so on successively, effectively promoting the development of the concept of green construction. The green construction branch of China Construction Industry Institute was founded in 2013, and has carried out the national green construction demonstration projects, further promoting the popularization and application of thought and knowledge innovation, methods, experience and research results of green construction in the country. In 2015, at the fifth plenary session of the 18th Communist Party of China central committee, it was put forward that a period of "much starker choices and graver consequences" was emerging, and we must firmly establish and earnestly implement the concept of green development at the national level, coordinating, enforcing and sharing innovations, developments and ideas, to further green development, thereby encouraging the development of the construction industry in the right direction.

6.1.2.2 *Definition of green construction*

The definition of green construction is the same in China as elsewhere. In October 2011, the national standard for green construction of construction engineering (gbit50640-2010) was defined: under the premise of guaranteeing quality and safety requirements, through scientific management and technological progress, to save resources as far as possible, reduce the negative impact to the environment, and to achieve green "four sections" (energy saving, material saving, water saving, land saving) of construction project in construction activities.

6.2 Green construction management and engineering practice of metro engineering

6.2.1 *The environmental protection*

6.2.1.1 *Basic environmental protection requirements*

(1) Environmental protection work should follow the principle of prevention and control, with clear ideas on who pollutes and who governs, while strengthening process control, and implementing a top-down and side-to-side management system.

(2) The construction site must implement the requirements of environmental protection. In Qingdao city, the appoint project construction management put forward "seven one hundred percent" inspection points: the bare earth sandstone covering, construction road hardening, vehicle washing, producing homework dust suppression, temporary non-greening of land, complete protection against strong noises, and treatment of "three wastes" up to standard.
(3) Construction site environmental protection should follow the principle of "six must, four is not allowed, one guarantee":
    (a) *Six Obligations*: Must be a fence-closed operation, must harden field roads, must set up washing facilities, must be an industry, must be equipped with wet cleaning personnel, must regularly clean the construction site;
    (b) *Four Prohibitions*: No vehicles should be allowed to carry mud on the road, no high-altitude throwing of dust, no water at the site, no on-site incineration and waste disposal;
    (c) *One Guarantee*: Ensure the effective implementation of the prevention and control measures.
(4) The construction site should set up an environmental control supervision notice board. The supervision board shall indicate the name of the project, the construction unit, the name and contact number of the on-site supervisor, the project duration, the environmental protection measures and the report telephone number.
(5) Environmental investigation should be carried out strictly before construction. Before construction, a detailed survey of the construction area, surrounding constructions and pipelines shall be carried out, and an investigation report shall be prepared. Communication shall be established with the relevant authorities and their protection requirements shall be understood. When necessary, the security of the structure should be evaluated comprehensively through special house structure inspection and identification. An environmental investigation may be conducted by a professional entity, and the relevant evidence shall be retained in the event of a dispute.
(6) Geological and hydrological work should be done before construction so as to understand the geological and hydrological conditions in detail.

(7) An environmental management system should be implemented to identify and evaluate environmental factors, determine environmental objectives and indicators, develop environmental management plans, implementation and operation, process inspection and effect verification.
(8) Environmental risk management shall be conducted according to risk identification, risk assessment, risk control (risk early warning and preplanning) and risk monitoring.
(9) Environmental management plans shall be prepared, along with environmental emergency plans and monitoring plans, with environmental monitoring tools, reserve emergency supplies and equipment kept ready and emergency drills conducted regularly.
(10) Construction in densely populated and busy streets, near schools, hospitals and government agencies, should be managed, such that the field of shaft construction, application of yards, material processing plants, and the outdoor spray machine, mixer and air compressor that generate dust, noise and light pollution should be fully controlled. When it is located in a sparsely populated area, simple preventive measures can be taken according to the actual situation.

6.2.1.2 *Comprehensive environment monitoring*

The site shall establish an environmental monitoring system, with all-weather automatic continuous monitoring, collecting environment data such as dust, noise, temperature, humidity and wind speed, and monitoring the same and prominently displaying it in public, ensuring that real-time data are available to the dispatch center at the same time (Fig. 6.1).

6.2.1.3 *Dust and exhaust control*

6.2.1.3.1 General requirements

1. The emission concentration of waste gas and pollutants produced by the construction shall not exceed the relevant provisions of the state and local areas.
2. The facilities and workplaces that produce air pollution on site must take effective preventive measures.

(a) (b)

Fig. 6.1. Environment monitoring display screen.

3. It is strictly prohibited to discharge gas and dust containing toxic substances into the atmosphere. It is strictly prohibited to burn asphalt, linoleum, rubber, plastic, leather, garbage and other substances that produce smoke and odor.
4. Construction dust pollution prevention and control measures should be strictly carried out, and the construction site to do closed management, factory road hardening, canopy covering, water cleaning, material waste residue airtight transportation and vehicle cleaning.
5. In the stage of constructing the foundation pit and earthwork, spraying, covering and other measures of dust removal technology are adopted to spray concrete. It is necessary to adopt strong exhaust ventilation and dust filtration purification measures in the tunnel and other confined space, and the spraying concrete should be sprayed with wet shotcrete or cement, and the dust prevention measures should be taken.
6. When transporting soil, garbage, equipment and materials that are easily scattered and leaked, measures must be taken to ensure that the vehicle is clean and not defiled. The construction site exit should set up car washing facilities.
7. The materials that are liable to produce dust such as cement, fly ash, bentonite and grouting granule materials should be covered using dustproof measures, and materials that collect dust easily should be closed and stacked, stored and transported.

8. Dust control measures such as mechanical demolition and sprinkler should be prepared before dismantling. Wind hour should be select wisely for blasting operations. Before blasting and removing, wet, pre-wet substances, water storage and spray mist can be collected, and an anti-dust grid can be set up.
9. In the tunnel and underground space, the use of power transportation equipment can replace the internal combustion machinery, and one can change the construction process, reduce the welding operation and implement other control measures to reduce the harmful gas emission.

6.2.1.3.2 Wet spraying technology

1. The use of wet shotcrete construction technology, low homework dust, can largely improve the working environment, reduce resource waste and meet the requirement of occupational health, thereby ensuring concrete of stable quality and good construction quality (Fig. 6.2).
2. The mixing equipment should be equipped with a strong mixer and a raw material automatic metering device. The transportation equipment should be equipped with automatic mixing function.
3. It is recommended to use the JS750 and the above model mixer group, and the data upload function should be used.
4. It is advisable to use the non-alkali accelerator in the confined space inside a tunnel, so as to avoid chemical corrosion to the human respiratory tract.
5. The wet spraying machine must be placed in a safe place, with stable placement, with all supporting wheels and legs firmly on the ground.
6. When entering the spraying site, the jet hand must wear a safety helmet, dust respirator and other labor protection products.

6.2.1.3.3 Hydraulic blasting technology

Water pressure blasting was conducted in the 1990s by our country's expert the famous Professor He Guangyi. The blasting design and the design of the traditional tunnel smooth blasting scheme are basically the

(a) (b) (c) (d)

Fig. 6.2. Commonly used type of wet shotcrete.

same, with appropriate adjustments just on the charging structure and hole plugging. The principle is to "inject a certain amount of water to a certain position in the hole, and using special equipment and work stemming of back filling wall plug into the perforation", using the propagation of shock wave in the incompressibility of water, after the explosion energy was lost almost intact in the water nearby the surrounding rock; at the same time, the effect of explosion of gas expansion of the "water" is beneficial to secondary crushing rock and the water in the hole with atomized dust effect, reducing dust pollution in the environment. The tunneling water pressure blasting and conventional blasting (no backfill blockage of the perforation) have the same difference between 6:00 and 2:00 (Fig. 6.3).

In hole, hole quantity, hole distribution, cut form depth degree, detonating sequence and initiation time interval are six aspects that are similar. The water pressure blasting is the same as conventional blasting and

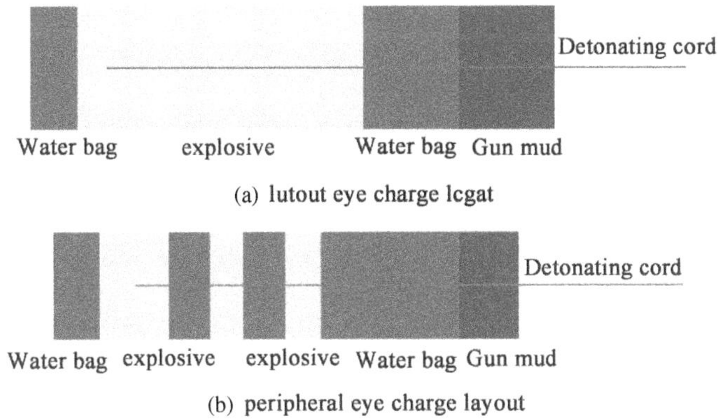

Fig. 6.3. Loading structure diagram.

compared to the broken hydraulic blasting, conventional blasting does not add any workload. Two points are different: the water pressure blasting is compared with the conventional blasting, which only adds the water bag and mortar in the hydraulic blasting.

This process can improve the working environment of the construction site, improve the effective energy efficiency of the explosives, reduce the blasting vibration speed, reduce the disturbance to the surrounding buildings and residents, and thereby play a significant role. Compared with the actual use effect at a site, the amount of dust produced by blasting is greatly reduced and the dust removal effect is remarkable.

6.2.1.3.4 Pulse bag filter

The pulse bag filter working principle is as follows: the dust is collected by dust collector inlet gas into the middle and lower housing, through the process of filter bag into the box body, due to various effects of the filter bag dust, gas separation, dust adsorbed on the filter bag, and gas filter bag by a venturi tube into the box body and out of the outlet line (Figs. 6.4 and 6.5).

As for the body through the filter bag dust gas purification process, with the increase of time, the area of the filter bag dust increases, therefore increasing resistance force of the filter bag and reducing the amount of gas through the filter bag gradually, to make the dust collector work normally,

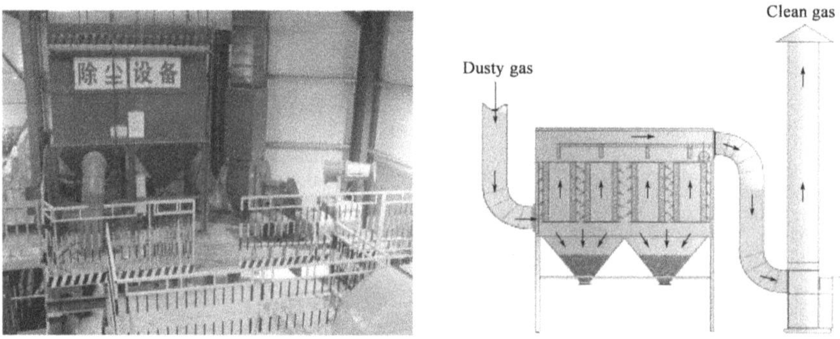

(a) (b)

Fig. 6.4. Pulse bag filter.

Fig. 6.5. The dust removal system around the shaft.

so the resistance is controlled within the set limit range (a kind of 12–150-mm water column). When resistance is maintained within a limited range, control instrument gives instructions to trigger each pulse valve or control valve open, and air bags of compressed air spray torch each hole scripture's tube into the corresponding filter bag, under the effect of instant air reverse. A sharp expansion makes the product on the surface of the filter bag, dust, fall off, and the filter bag is fresh again. The removed dust falls into the ash bucket and is discharged from the machine body, which is removed by the powder dust fixed on the filter bag, and the cleaned gas body is passed normally, so as to ensure the normal work of the dust collector.

In order to control and collect the dust produced in the process of construction, the shaft wellhead used to collect and expel the dust, combined with the location of the site, should be in the mouth of the well chamber set on both sides, equipped with an air suction hood and gas collecting dust hood, which adopt an effective wind technology multi-layer capture mode to achieve the greatest degree of capture of dust that escapes the wellhead.

6.2.1.3.5 Underground tunnel dust removal spray

The tunnel construction should reduce the dust concentration in the tunnel by utilizing dust control spraying. Spray should be utilized on two roads, the first one 50 m from the construction work surface, and the second road 50 m from the tunnel opening (Fig. 6.6).

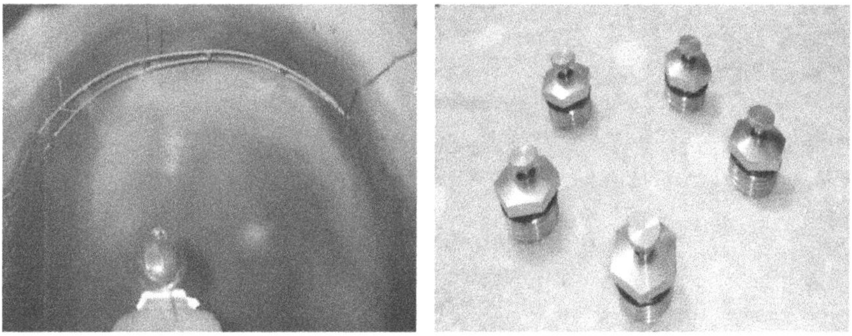

(a) Full section tunnel spray (b) Sprinkler style

Fig. 6.6. Tunnel dusting spray.

6.2.1.3.6 Other dust control measures

Dust control in the field can also use other kinds of dust removal measures. The unhardened road and exposed soil in the site are covered by the network in a timely manner. The remote spraying machine, sprinkler, mechanical sweeper and slag spraying system are widely used at the site (Figs. 6.7 and 6.8).

(a) (b)

Fig. 6.7. Dust removal by the sprinkler.

Fig. 6.8. Slag spraying system.

6.2.1.4 *Noise and vibration control*

6.2.1.4.1 General requirements

1. In addition, it is necessary to reduce the working time especially during the night. In the night, the application of the work and the procedures for continuation are in accordance with the national and local regulations.
2. In order to solve the problem of noise control, sound insulation should be adopted.
3. For monitoring and control of the noise, the noise emission should not be more than the accepted standard of environmental noise in the

construction field and the regulations of the local authorities. The noise during the day is no more than 70dB, and the construction noise at night must not exceed 55dB.

4. Low-noise, low-vibration construction equipment should be used, such as silent vibrating equipment; sound insulation and vibration isolation measures should be adopted for construction machinery such as damping muffler, micro-pore muffler, etc., and construction noise and vibration should be avoided or reduced. Mechanical equipment maintenance should be conducted on a regular basis, and machine shut down when not in use.
5. Preparation and review of blasting scheme should be done before blasting of earthwork; noise reduction, dust control and stone control measures should be adopted. In urban areas or in densely populated areas, it is advisable to adopt a blasting scheme with low-noise and low-environmental impact. Blasting construction should adopt sound insulation, sound insulation boards, sound insulation doors and noise reduction measures.
6. Site carpentry, electric saw, planer, air compressor, mixer and stationary concrete pump mechanical equipment should be set against the tent, used within a confined space, and when it is necessary used in a protective shed which has the function of noise reduction. When the protective shed cannot be set up, it should be set on the side of the construction area far away from the residential area, and noise pollution can be reduced by means of a sound insulation cloth and low-noise equipment.

6.2.1.5 *Blasting noise control*

Figure 6.9 shows some sample images for blasting noise control measures.

6.2.1.6 *Equipment noise control*

Figure 6.10 shows some sample images for equipment noise control measures.

6.2.1.7 *Water pollution control*

6.2.1.7.1 General requirements

1. The construction site should set up drainage and sewage facilities, and rain and sewage diversion discharge. The construction site sewage

(a) elfiat drawling of wellhead cable stayed soundproof (b) Physical drawing of wellhead cable-stayed acoustic cover plate

(c) shaft mouth steel plate closed (d) shaft mouth insulation quilt

(e) Acoustic door at the entrance (f) Acoustic door at the entrance

(g) sheaf side panels are soundproof panels (h) water pressure blasting cannon is covered

Fig. 6.9. Blasting noise control measures.

(a) Borehole sound proof shed

(b) Effect drawing of noise elimination box added to fan

(c) Silencer

(d) Fully enclosed maintenance shed

Fig. 6.10. Equipment noise control measures.

discharge should meet the requirements of the national standard or "comprehensive sewage discharge standard". Qualified units are entrusted to carry out the water quality inspection and provide the sewage test report. Water quality should be tested according to the actual situation when using non-traditional water sources and circulating water.

2. Corresponding sewage treatment facilities should be set up at the construction site, including drainage, sewage pipeline setting filtration and precipitation measures. The canteen shall have a grease trap. Toilets should be fixed with septic tanks. Grease traps and septic tanks should be treated with anti-seepage materials, and timely cleaning and disinfection.
3. Measures should be taken to avoid groundwater pollution in geotechnical investigation and basic engineering construction.

4. During the use and maintenance of construction machinery equipment, oil pollution should be controlled; the waste water and waste oil of the cleaning machine shall not be discharged directly. Environmentally degradable grease and foam can be used in a shield to reduce pollution to the environment.
5. The oil and chemical solvents and other items should be stored in a special storage room. The ground is hardened and leakproof. Waste oil and chemical solvents should be collected and disposed.
6. Construction site area sewage and construction and washing wastewater drainage should be collected, and a centralized precipitation treatment system should be used, with uniform discharge after standardization. It is forbidden to discharge sewage into the urban sewer or storm sewer. It is forbidden to dump waste, waste residue, mud and other materials into the sewer. The construction unit shall contact the municipal departments regularly to inspect the pipes and ensure smooth flow through the pipes.
7. Wastewater treatment equipment should be treated with sewage treatment equipment.

6.2.1.7.2 Emission standards

1. The discharge of sewage and waste water must be clear with no distinct smell.
2. If the concentration of chloride ion does not meet the requirements, sewage and waste water should not be discharged into the sewage pipe.
3. Construction of the sewage water and waste water discharge line should fulfill the requirements of "water quality standard for water discharge in town" (gbit31962-2015). The content of easy to sink solid cannot be greater than 10 mLI (L·15 min), and the detection method is the volume method according to the urban sewage water quality testing standard (cjit51-2004). The content of suspended matter cannot be greater than 250 mgIL, and the detection method is the weight method according to the urban sewage water quality standard (cjit51-2004).
4. The discharge of sewage and waste water into the rainwater pipe shall meet the requirements of the comprehensive discharge standards for

water pollutants in the Shandong peninsula (DB 37 i676-2007). The content of the solid matter cannot be greater than 10 mLI (L·15 min); the content of suspended matter cannot be greater than 50 mgIL, and the detection method is same as article (3).

6.2.1.7.3 Mud treatment technology for pile foundation construction

The disposal of mud in construction of pile foundation has always been a difficult problem, especially as the construction of metro pile foundation is mostly in urban areas, and the construction requirements of are high and the transportation is difficult. The normal drying is only to dry the surface mud so that the transport vehicles can be loaded. The surrounding environment of the foundation pit is complex, and the foundation pit in flood season is prone to instability and the risk of collapse. If the mud is treated with mud, the cost of transportation is greatly increased.

The people's square station of Qingdao Metro Line 1 adopts a mud purification device numbered zx-250 (black cyclone, Fig. 6.11), and its working principle is as follows: The slurry pumped from the bottom of the back circulation gravel pump is sent to the coarse sieve of the mud purification device through the total feed pipe, and the particle size is separated from the slag of over 3 mm through vibration screening. Coarse sieve

Fig. 6.11. Zx-250 mud purification plant (black cyclone) workflow.

screening of mud into the slurry storage tank of the mud purification device occurs in the mud purification plant from the slot suction mud slurry pump, and the pump outlet has certain energy storage of mud along the slurry soft tube from the hydrocyclone tangential to the pulp inlet, through hydrocyclone separation, and the mud that settles at the bottom of cyclone mouth discharges into a fine screen. After sifting by a fine sieve, it is separated from the dry fine ballast. After sifting through the fine sieve, the slurry is returned to the slurry tank. After treatment, the clean mud flows from the cyclone overflow pipe into the middle storage tank, and then the slurry pipeline is sent back.

Drum and spiral high-speed rotation is performed at a certain differential speed, whereby mud is fed by the feeding tube into the spiral feeding chamber in a row; after accelerating in the rotary drum, under the action of centrifugal force, the heavier solid deposition in drum walls forms a sediment layer. The spiral pusher will deposit the solid contents continuously in the drum cone end, after zone drying from the slag discharge closes the mouth, and lighter liquid forms an inner liquid ring on the sides by a continuous overflow drum that interrupts the overflow, and the liquid discharges outside.

6.2.2 *Energy saving*

The overall energy-saving requirements are as follows:

(1) Make reasonable use of energy consumption indicators. We will improve energy structure, improve energy efficiency, and choose efficient and environmentally friendly energy sources such as electricity and gas to reduce coal use (Fig. 6.12).
(2) Use energy-saving, efficient, environmentally protective construction equipment, and use compensation technology to save electricity. Use frequency conversion technology, energy construction equipment and energy-saving lamps.
(3) As for zoning electricity, each region and each large equipment set the electricity control index separately, and the partition control sets the meter to measure. Carry out regular measurement, accounting and comparative analysis of temporary electricity, and take preventive and corrective measures.

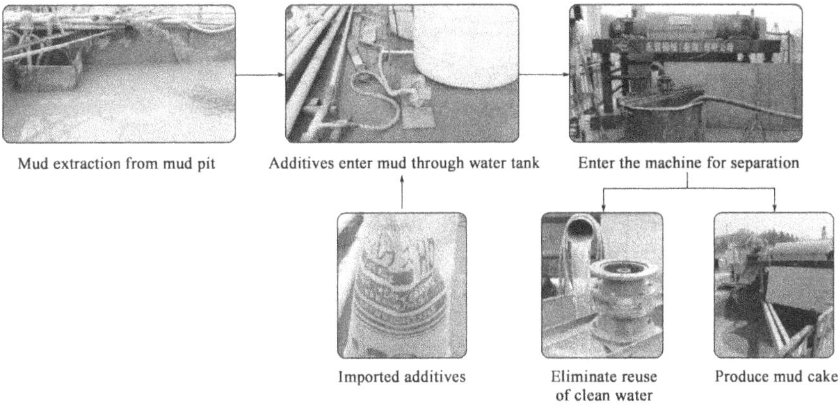

Fig. 6.12. LW360 × 1580Y horizontal spiral discharge settling centrifuge working process.

(4) Arrange the construction sequence and working face reasonably, reduce the number of machine tools and try to share resources in different areas. The priority is to choose the construction technology that consumes power or other energy the least.
(5) Avoid overloading of equipment.
(6) Optimize line selection and circuit layout, reduce the distance between the distribution box and electric equipment, and reduce unnecessary loss.

6.2.2.1 *Engineering examples: Qingdao Metro Line 13 ancient town mouth vehicle base, Dongjiakou car park photovoltaic power generation system*

(1) *Basic information* (Fig. 6.13): Qingdao Metro Line 13 designs a large number of solar photovoltaic power generation systems, especially for the use of the parking lot, car depot united maintenance, parking column library screening, by setting the large capacity of photovoltaic power generation systems, in addition to the direct supply field, within the period of lighting power load; the battery can also be incorporated into the medium voltage network, to provide power to the nearby station, which has obvious economic comparative advantage, see Table 6.1.

(a) (b)

Fig. 6.13. Poly-silicon panels and installation renderings.

Table 6.1. Program investment returns statistics.

S/N	Monomer name	Roof area (m^2)	Installation capacity (kw)	First year electricity generation (10,000 degrees)	Initial outlay (10,000 dollars)	Dynamic investment recovery period (year)	Load ratio
1	Application library	25,200	1008	139.81	987.8	12	
2	Overhaul Library	32,000	1280	177.54	1254.4	12	
Subtotal		57,200	2288	317.35	2242.2		90.67%

Notes: All the above values are based on the theoretical calculation of the poly-silicon solar photovoltaic panel. The initial investment includes solar panels, energy storage batteries, installation accessories and network matching costs. The actual usable area of the solar photovoltaic panel is considered to be 50% of the total area of the roof. It is estimated that the annual power consumption of the power lighting load of the Guzhenkou vehicle base is 3.5 million degrees, slightly higher than the annual output of the photovoltaic system. The photovoltaic power generation can be absorbed and utilized by the low-voltage power lighting load. The dynamic payback period includes operating management, photovoltaic panel cleaning and other costs.

(2) The ancient vehicle base in the south of the Guzhenkou covers an area of 34.7 hectares, and within the set period use, maintenance, and comprehensive building of a library, adjustable machine garage, wheel dynamic detection of the garage, tent, material mixing substation, GongJian tent, maintenance and air compressor, sewage, station, boiler room, test line and doorman building (Fig. 6.14) have been planned. The design can install a 2.3-mW photovoltaic power generation equipment.
(3) In the Dongjiakou parking lot in Dongjiakou town, the site is equipped with a traction step-down substation of mixed-use combination, car washing, special garage and complex building, library building and a waste water treatment station with a total construction area of 31,000 m^2 (Fig. 6.15). The design can install a 1-mW photovoltaic power generation equipment.

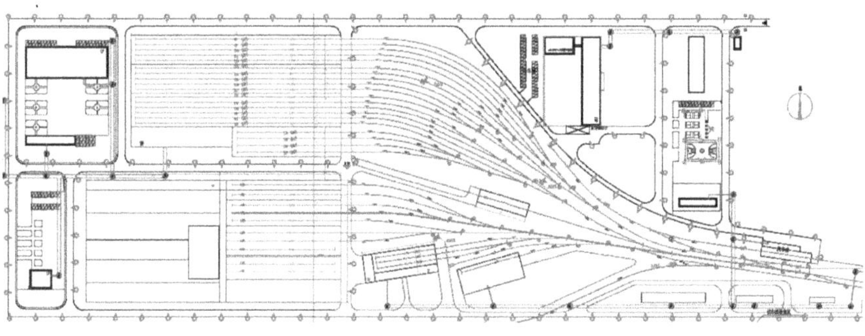

Fig. 6.14. Schematic diagram of plane layout of the vehicle base in Guzhen Town.

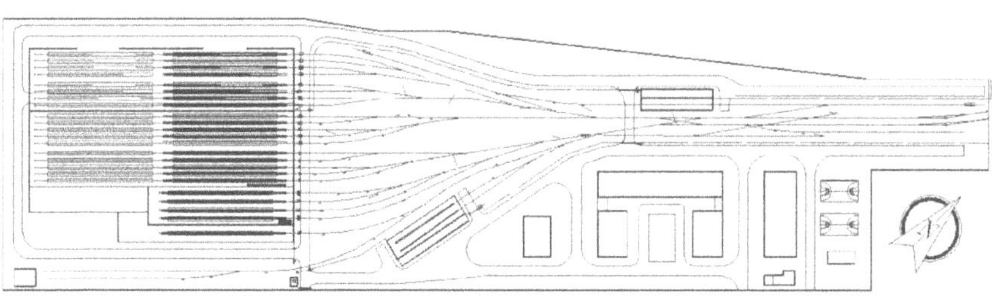

Fig. 6.15. Layout of Dongjiakou parking lot.

(4) According to the national industrial policy, the country was sent [2013] 24 files of the state council on promoting the healthy development of the photovoltaic industry according to the dry opinions of He GuoFa on change of price [2013 J 1638], "about play a role of price lever to promote the healthy development of photovoltaic industry general knowledge", perfecting the photovoltaic power price policy; for the distributed photovoltaic power generation project, carried out in accordance with the generation of electricity price subsidies policy, the feed-in tariff standard is 0.42 yuan/kWh. The people's government of Shandong province was sent [2013] 24 files to promote healthy development of photovoltaic industry, regarding the implementation in the country, stating the following: "mentioned in the 2013–2015, into the state's annual guide distributed photovoltaic power generation projects, by all power in the state subsidies standards on the basis of 0.42 yuan/kWh, provincial again to 0.05 yuan/kWh of electricity price subsidies." The above policy is comprehensive, and the project set up a photovoltaic power generation system in Shandong province, which will get national finance and in turn finance certain subsidies; the improved efficiency of photovoltaic power generation system, the method and investment returns are shown in Table 6.2.

Table 6.2. Program investment returns statistics.

Monomer name	**Roof area (m^2)**	**Installation capacity (kwp)**	**First year electricity generation (10,000 degrees)**	**Initial outlay (10,000 dollars)**	**Dynamic investment recovery period (year)**	**Load ratio**
Application combination library	25,600	1024	142	1003.5	12	71%

Notes: All the above values are based on the theoretical calculation of the poly-silicon solar photovoltaic panel. The initial investment includes solar panels, energy storage batteries, installation accessories and network matching costs. The actual usable area of the solar photovoltaic panel is considered to be 50% of the total area of the roof. It is estimated that the power lighting load of the Dongjiakou parking lot is 2 million degrees annually, which is higher than the annual output of the system. Photovoltaic power generation can be absorbed and utilized by low-voltage power lighting load. The dynamic payback period includes operating management, photovoltaic panel cleaning and other costs.

6.2.3 *In rhythm*

Land planning and general layout are as follows:

- The construction process and equipment of the section are selected to reduce the occupation of land. In combination with the site conditions and construction requirements, the temporary facilities are reasonably determined and the temporary facilities are as light as possible and can be moved. The floor area of the temporary facilities is designed according to the minimum area required for land use.
- Under the premise of safe and civilized construction and environmental protection requirements, plan a reasonable construction site and make sure to use it in a compact and non-wasteful way. The general layout management of the construction is scientific, reasonable and orderly.
- Local conditions can be used to construct buildings, roads and vacant sites. In combination with engineering situations, use the existing buildings or factories that have been built in a safe manner to reduce and avoid a large number of temporary building demolitions and relocations.
- Temporary offices and living rooms should be economic and beautiful, covering a small area, causing less impact on the surrounding landscape environment, with an easy to tear down and assemble prefabricated structure, and adopting a multi-layer buildup.
- The living area and the production area should be separately arranged, and set up with an enclosure, wall and other separation facilities. The construction site should be equipped with effective isolation measures, such as continuous closed, light steel structure with a prefabricated enclosure. Traffic and pedestrian safety should be ensured during temporary occupation.
- When necessary, take the foundation pit to set up the site and road method to solve any site problems.
- The construction site mixing station, warehouse, processing plant, work shed, material yard and other facilities should be close to the road, to reduce the distance of transportation.
- Steel reinforcement and steel frame processing use centralized processing and the centralized distribution mode on a priority basis.

- The road to the construction site is arranged according to the principle of permanent road and temporary road.
- Reasonable storage of building materials and a JIT (Just in time) management technique are used for material storage. According to the chronological order of the use time of materials, it is easy to take out and avoid clutter, inconvenience and waste of space.

6.2.3.1 *Engineering examples: Qingdao Metro Line 13, no. 3, beam field, double-deck storage beam pedestal*

On the Qingdao Metro Line 13, no. 3, the beam deposit area adopts a double beam design (Figs. 6.16–6.18), with the lower U-shaped beams stored on a concrete pedestal, the upper beams stored on a door-type steel structure, according to the peak of storage calculation of 400 U-shaped beams, rather than a single deposit economical use of land of about 85 mu.

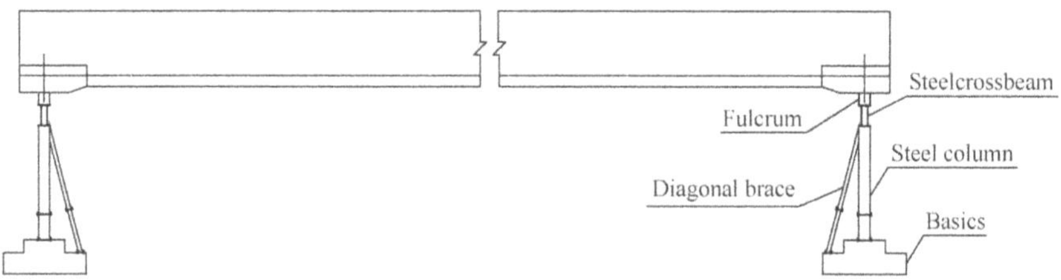

Fig. 6.16. Vertical elevation of the beam pedestal.

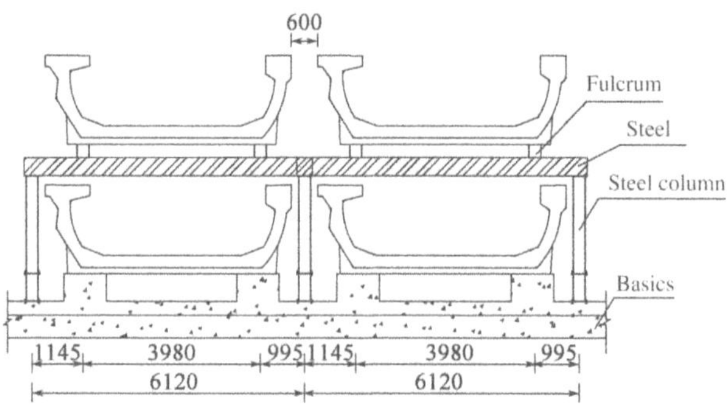

Fig. 6.17. Horizontal elevation of the pedestal platform (unit: mm).

Fig. 6.18. Double-layer storage beam.

6.2.4 *Water saving*

The general requirements for water conservation are as follows:

- The first step is to choose the first section of the water supply, such as rotary drilling rig construction, long auger drill, full barrel drill, dry masonry and so on. In construction, pile grouting requires a large amount of water, and hence a reasonable selection of mix proportion shall be adopted, and the recycling and waste pulp control and treatment process shall be adopted to save water.
- The construction site road will be sprayed, and greening, watering, flushing toilets and washing cars will be done without municipal running water, so as to reduce the non-traditional water sources such as rain water.
- Effective water conservation measures will be taken. To ensure strong section water management, the construction personnel should carry out section water training. Effective measures should be taken to reduce leakage of pipe networks and water equipment, and prevent water running, dripping and leaking.

- The concrete should be made of plastic film and heat preservation material to protect from moisture and preserve heat. When performing sprinkling or fog curing, the water used is collected rainwater.
- The construction site establishes recycling for the water collection and treatment system. It set up a circulating water system for dirty water obtained after washing equipment and vehicles.
- According to the different working areas, it is necessary to determine the index of fixed amount of water in the production area, the living area and the public area, and the management of the amount of the meter.
- Special metrology assessment of the area and work point of the concrete mixing station and other water concentration spaces is carried out.

6.2.4.1 *Engineering examples: The water circulation system of girder field on line 3 of Qingdao Metro Line 13*

6.2.4.1.1 U-shaped beam spray regimen

In traditional artificial water conservation, water is not uniform, and curing water waste is a serious problem. By drench curing the device, according to the U-shaped beam characteristics of the homemade chute, one can achieve full maintenance and save water resources.

Chute spray curing of the device is done by the longitudinal main line and branch pipe and by a number of chute heads of longitudinal main line on top of the wing. The longitudinal head is set every 1.5 m away from the connectors, to stay inside and outside the connecting pipe. The pipe end connection chute head cures the water that enters through the main line, which is picked out by the chute ends and sprinkled on the beam body.

6.2.4.1.2 Water circulation system of the factory

According to the functional division at the jobsite, the selection will help beam area maintenance and the stone washing machine in the use of water, living water and rain water, respectively, by three levels of precipitation, precipitation through pipes into the reservoir for further processing and to field area after water is recycled (Fig. 6.19).

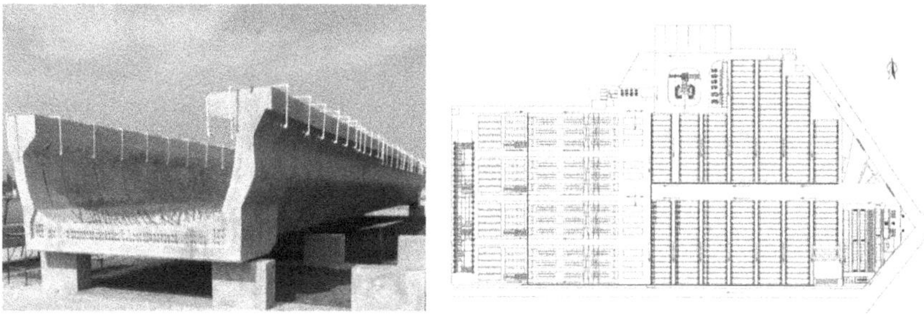

(a) (b)

Fig. 6.19. Schematic diagram of the water circulation utilization of the U-shaped beam spray.

6.2.5 *Saving material*

6.2.5.1 *Material management measures*

1. When the drawing is reviewed, the relevant contents of materials and resources are put forward.
2. Use green building materials, make full use of local resources and local materials, and reduce transportation costs and energy consumption.
3. Promote the use of the factory prefabricated assembly process, carry out station structuring, decoration and electromechanical construction.
4. Choose green building materials, use high-performance concrete, high-strength steel and high-strength steel materials. Select efficient templates and support systems. Strengthen the repair and maintenance work, and increase the turnover number of the template.
5. Arrange material purchase and approach, and reduce inventory according to schedule. Establish a material inventory system, material quota system and material use reward and punishment system, and regularly analyze the material consumption situation and formulate specific material-saving measures. Establish and improve the material management and warehousing system, implement material management responsibility and strengthen the management at all stages of collection, issuance, storage and use. Strengthen security to avoid material theft.
6. In order to avoid damage and loss of materials, it is necessary to select suitable means of transport and loading and unloading methods before

performing those activities. When loading and unloading, try to perform those activities as near as possible, make sure that it is done in a clean and orderly manner, and avoid or reduce secondary transport and accidental loss.

7. Maximize the construction efficiency, and achieve “the completion of the finished materials, construction operation level and construction quality, to avoid rework phenomenon”.

6.2.5.2 *Engineering examples: Qingdao Metro Line 13 two-river station, water concrete construction*

6.2.5.2.1 Basic information

The two-river station is located at the east side of the intersection of Xueguan road and the two-river road. It is a two-story island station with a total length of 203 m, a standard length of 19.7 m, an effective platform of 80 m and a platform width of 11 m. Water concrete is used in the top side wall, pillar and roof of the station.

6.2.5.2.2 Process practice

Mixing sequence of as cast finish concrete in Lianghe station: binding of frame column reinforcement in station hall floor → formwork support and removal of frame column in station hall floor → binding of reinforcement for side wall of station hall floor → formwork support and removal of side wall of station hall floor → formwork installation of roof support → top plate reinforcement binding → concrete pouring of roof (Fig. 6.20).

6.2.5.2.3 Implementation effect

The water concrete station directly uses the natural texture of concrete after forming as the finishing effect, making the concrete more expressive and fully emphasizing the artistic quality of the modeling and the specificity of the material. At the same time, the lack of use of various decorative coatings greatly saves any decoration cost (Fig. 6.21).

(a) Clean water column

(b) Side wall formwork

(c) Use high frequency vibration to vibrate on site

(d) Field vibration

(e) loverevl qoteotile

Fig. 6.20. Concrete pouring process using clear water.

(a) (b)

Fig. 6.21. The finished concrete products of the two-river station.

6.3 Selection and layout of construction site

6.3.1 *Site selection principle of construction site*

- Adjust measures according to local conditions and economical use of land, environmental protection, safety and reliability, convenient transportation, smooth communications, a reasonable principle of layout and proper site selection; priority should be given to the clearing of land and as far as possible municipal public land should be avoided.

  Investigate the surrounding roads, buildings and pipelines, and indicate the corresponding position in the layout plan.
- It is strictly prohibited to select a site in dangerous areas prone to debris flow, landslide, rockfall or flooding, or an area near steep cliffs and dangerous rocks. Avoid high-voltage lines and tall trees, and maintain a safe distance from underground pipelines such as communications and natural gas.

6.3.2 *Site construction of the tomb requirements*

- The construction site shall be managed in a closed manner, surrounded by enclosures, and the envelope structure must be stable and the external compress shall be neat and beautiful. The containment structure has a foundation. The ground height of the basic structure is 30–50 cm, and the local potential is higher than that of the water stagnation level.

The basic structure can be made of brick or concrete, and the structure must be firmly fixed.

- The construction site shall be divided into an office area, living area and construction work area according to the general layout plan of construction. The office area and living area should be separated from the construction work area and a safe distance between them ensured. The isolation material can be a brick wall, grid column, etc., of a height not less than 1.3 m. All districts shall have clear indication signs.
- The construction site should be built such that it improves the drainage system and keeps it open. The capacity of water displacement should meet the requirements of the most unfavorable environment, and prevent damage caused by water in the rainy seasons and snow in winter.

6.3.3 *Typical site layout*

6.3.3.1 *Overall requirements of site layout*

The construction site of the urban rail transit is influenced by the surrounding environment, and it is necessary to determine the method of construction by taking into consideration the influence of the working field on the environmental. The construction site is divided into several parts, such as the vertical shaft, inter-zone inclined well, dark open-cut station air well and mining station foundation pit. The field cloth should be filled with the production base of the current field, which should be taken as follows:

(1) **Analysis of construction management needs:** Generally, only a small amount of office space should be set up at the construction site, which should be used for housing and necessary warehouses, and facilities such as the office, accommodation and canteen should be resettled in the field.

(2) **Construction material stacked layout:** Steel processing in principle was to be focused in a work area or line setting, i.e., a steel processing zone, with vertical shafts at intervals. Without reinforcement in the processing zone, the station construction was to be according to the site situation with an appropriate design and steel processing

area. The location and size of the site production should be reasonably determine according to the needs of the sand silo, steel reinforcement, construction area and material stacking (warehouse).

(3) **The layout of temporary water and electricity pipeline:** The water supply and the power supply requirements shall be determined, and then the pipelines shall be disposed to the terminals for various needs.

(4) The layout should be constructed in a safe and civilized manner, taking into consideration fire hazard, drainage, sewage treatment, identification of placard placement and gate position.

(5) Proper placement should be done of construction machinery and equipment, such as the hoisting machine and mixing station.

(6) The layout of the construction road should ensure the smooth operation of the construction vehicles.

6.3.3.2 *Area of production and service facility calculation*

6.3.3.2.1 Floor space calculation

The production area does not contain the management and construction personnel's accommodation room, but should contain a room for the layout production facilities. The room should be furnished in a double-deck arrangement, with a guard room, standard equipment, toilet, etc. The single room can occupy an area of about 25 m^2, and the construction vertical incline takes into consideration 10 occupied rooms, and the light and dark open-cut station considers 15 floor areas.

The construction area of the vertical slant well is about 250 m^2, and the floor area of the building is about 375 m^2.

6.3.3.2.2 Mixing station and sand silo

1. The mixing station and sand silo include the cement tank, mixing machine, control room, loading ramp, sand silo, etc.
2. The diameter of the cement tank is 2.1 m, with a base of 3 m × 3 m, covering an area of 9 m^2.
3. The size of JS500 is 4.7 m × 3 m, the size of JS750 is 5.4 m × 3.3 m, the batching machine is 2.2 m × 6.6 m, and considering the height of material receiving, the floor area is considered as 10 m × 7 m.

4. The sand and gravel bin of the vertical inclined shaft in the mine law interval is 160 m^2: The average footage of each of the four faces of the II, III grade surrounding rock is 2 m, spray quantity per day: 2 m/d × 4 face × 5 m^3/liner meter = 40 m^3/d, plus over excavation and loss of shotcrete, 80 m^3/d in total. The weight of cubic stone and stand is 832 kg respectively, it is required to store the materials for shotcrete for three days on site: 60 m^3 × 0.832 t × 2 ÷ 1.3 × 3 day = 307.2 m^3, the average height of the pile is 2 m, and the required height of the pile is 2 m, and the required aggregate warehouse is 160 m^2.
5. Underground wind well sand bin is 130 m^2 (wind mixing station in one of the well site settings): Air duct early shotcrete material 21.8/liner meter, footage 1.5 m/d, plus over excavation and loss of shotcrete, 65.4 m^3/d in total. The weight of cubic stone and sand is 832 kg respectively it is required to store the materials for shotcrete for three days on site: 65.4 m^3 × 0.832 t × 2 ÷ 1.3 × 3 day = 251m^3,the average height of the pile is 2 m, and the required aggregate warehouse is 125 m^2. The layout area is 130 m^2.
6. The station sand bin is 60 m^2: Excavation 1000 m^3/d, need to spray the surface about 100 m^2, plus over excavation and loss of shotcrete, 20 m^3/d in total. The weight of cubic stone and sand is 832 kg respectively, it is required to store the materials for shotcrete for three days on site: 20 m^3 × 0.832 t × 2 ÷ 1.3 × 3 day = 76.8 m^3, the average height of the pile is 2 m, and the required aggregate warehouse is 38.4 m^2. Considering the volume of feed trucks, emergency material preparation, the layout area is 60 m^2.

6.3.3.3 *Refuse storage site*

(1) **Interval vertical slope is 300 m^2:** The II and III grade surrounding rock is 2 m, 8 m by 30, and the loose coefficient is equal to 360, which can be stored in 2d, with a height of 2.5 m, and 360 × 2 is 288, which requires an area of 300 square meters.
(2) **Underground mining air shaft is 310 m^2:** The main section of the station is 171 m^3/liner meter, with a daily footage of 1.5 m, 171 m^3/liner meter × 1.5 m × 1.5 loose coefficient = 385 m^3, two days storage, height 2.5 m, 385 × 2 ÷ 2.5 = 308 m^2, the layout area is 310 m^2.

6.3.3.4 *Rebar site*

When conditions are favorable for off-site central processing at the working surface in a timely manner:

(1) **The original material is stacked: 52 m^2:** The site is 13 m × 4 m, stacked with 4–5 kinds of steel bars, and storage of the steel bar is 12 m high.
(2) **Steel reinforcement processing: 52 m^2:** The site is 13 m × 4 m, and the main functions are cutting, spinning, bending, rod straightening and other steel processing, as well as the temporary stacking of semi-finished reinforcing bars.
(3) **Semi-finished products stacked: 32 m^2:** The site is 4 m × 8 m, mainly storing steel arches, mesh, anchors, lead, lock feet and so on. Steel arch shelf: If each section requires 2 pieces a day and each piece has 6 units, a total of four working faces need 48 units. Based on one half inserted into the hole in time and another half temporarily on the ground, each unit has a chord length of 3 m, a height of 0.7 m, and the ground stacking not more than 5.6 m, a total of 6 piles are needed to cover an area of 3 m × 6 m. Screen: for each phase, 104 pieces are required, and 1 pile is required, and 1.2 m × 2.4 m is required to cover the area: 1.5 m × 3 m. Anchor: 1 m × 4 m. Advanced ducts: 1 m × 4 m.

6.3.3.5 *Turnover*

Steel pipes, square wood, steel formwork, construction steel, bamboo plywood, top brackets and fasteners are required. Interval of the vertical incline: 30 m^2; underground mining air shaft: 100 m^2; and mining station: 200 m^2.

The following is the calculation of the width of the turning road, taking Dong Feng EQ1141 as an example (Fig. 6.22):

(1) The minimum turning radius of the car is $r1 = 10.5$ m. The inner radius of the car is $r = \sqrt{}(r12 - 12) - (b + n)/2 = 7.03$ m; the outer radius of the car: $R = \sqrt{}(L + d)\ 2 + (R + b)\ 2) = 11.33$ m; outer radius $R0 = R + x = 11.\ 73$ m; inner radius: $r2 = r - y = 6.63$ m; minimum width of ring road: $W = 11.\ 73 - 6.63 = 5.1$ m; x, y is the distance from the outermost point to the outer edge of the ring road and the distance from the inner edge to the inner edge of the ring road, respectively, greater than or equal to 0.25 m, taking 0.4 m Lane

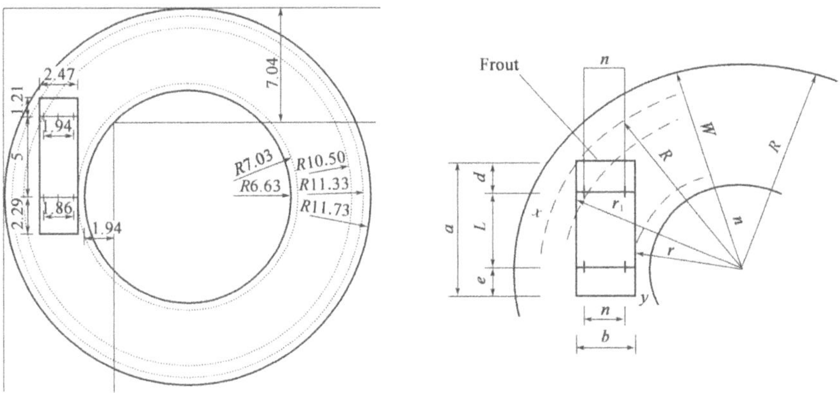

(a) (b)

Fig. 6.22. Schematic diagram of width of turning road (unit: m).

Setting width and add $r^2 - r^2 \div \sqrt{2} = 1.94$ m on the basis of ring road, lane set width $5.1 + 1.94 = 7.04$ m.

(2) The length of the vehicle is 8.49 m, width $b = 2.47$ m, the wheelbase $L = 5$ m, the front chord $d = 1.205$ m, the rear chord $e = 2.285$ m, the rear wheel distance $m = 1.86$ m, the front wheel distance $n = 1.94$ m and the minimum turning radius $R = 10.5$ m.

(3) To sum up, the road setting of 90 curves should be more than 7.1 m.

(4) The vehicle's turns need to cover a large area and there is a safety hazard; the layout should take into account the location of the material yard and the gate, and avoid setting the turn road close to the same.

The open-cut station road setting of the mine dig around the foundation pit should be at least at a safe distance on one side so to decorate the vehicle channel, especially when the concrete pump truck and the crane operation are considered, and on the other side according to the 2-m lateral distance of the foundation pit.

(1) The 37-m concrete pump truck is selected for the two-story open-cut station (Fig. 6.23).

Section 1 arm length: 8.7 m; the second arm length: 7.86 m; the third arm length: 8.05 m; and the fourth arm length: 8.39 m. Vertical height $H = 37$ m; arm frame horizontal length $R = 33$ m; and arm frame vertical depth $D = 21.3$ m.

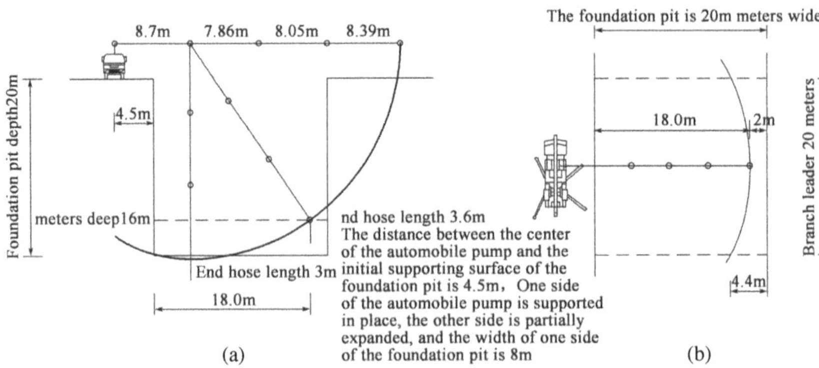

Fig. 6.23. 37-m concrete pump truck schematic diagram.

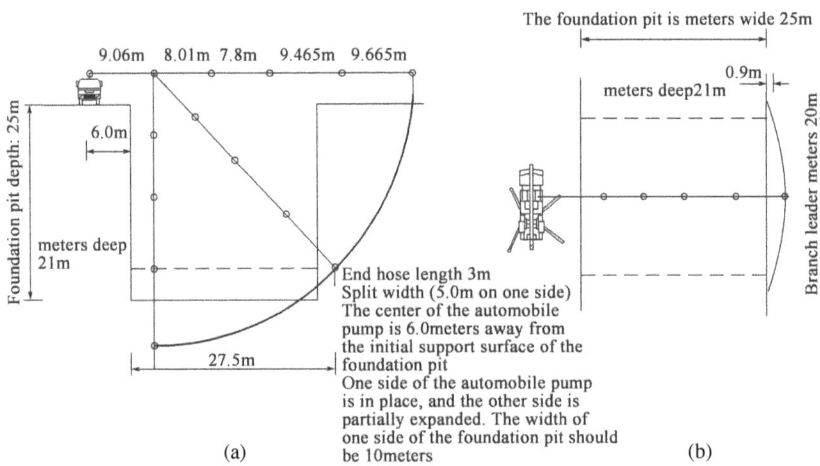

Fig. 6.24. 48-m concrete pump schematic diagram.

(2) The 48-m concrete pump truck (Fig. 6.24) is used for the three-story open-cut station.
   i. The first arm length: 9.06 m; the second arm length: 8.01 m; the third arm length: 7.8 m; the fourth arm length: 9.465 m; and the fifth arm length: 9.665 m.
   ii. The vertical height $H = 48.0$ m; the horizontal length $R = 44.0$ m; and the vertical depth $D = 30.8$ m.

(3) The road width of the side of the foundation pit is calculated as follows (Fig. 6.25): According to the length of the extension of the supporting leg, it is appropriate that the outer distance of the underground

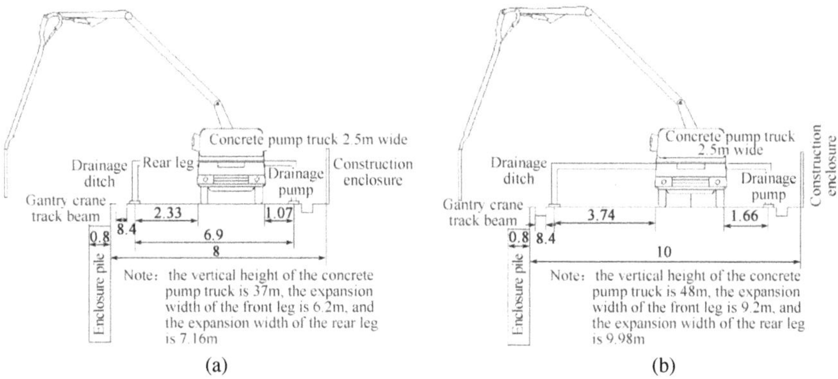

Fig. 6.25. Calculation of road width on the side of foundation pit.

floor is about 8 m outside the foundation pit, and the outside distance of the three-floor underground station is about 10 m.

The area of typical site production and service facilities is shown in Table 6.3.

Table 6.3. Summary of typical site production and service facilities.

		Area (m^2)				
S/n	**Usage**	**Pothole**	**Incline opening**	**Station air shaft**	**Open digging station**	**Specifications**
1	Structure and guardrail occupation	60	600	140	6000	The vertical shaft is 6 m × 10 m, the inclined shaft is 6 m × 100 m, the station air shaft is 7 m × 20 m, and the station is drilled by 30 m × 200 m.
2	Board house	250	250	375	375	
3	On-site road	180	180	270	270	The front channel of the board is calculated by 5 m
4	Reinforcing non-usable part of enclosure	160	300	170	160	

(*Continued*)

Table 6.3. (*Continued*)

S/n	Usage	Area (m^2) Pothole	Incline opening	Station air shaft	Open digging station	Specifications
5	Mixing plant	79	79	79	79	Mixing machine 70 m^2, cement tank 9 m^2
6	Gravel bin	160	160	130	60	
7	Reinforcement site	32	32	160	160	Only semi-finished product stacking area in vertical inclined well
8	Ballast temporary storage	300	300	310	0	
9	Car washing tank	0	0	0	0	Calculated in the common area of the site
10	Safety Quality Sample Room	26	26	26	26	
11	Cement warehouse	0	0	13	13	
12	Air compressor room	26	26	26	26	
13	Generator room	26	26	26	26	
14	Transformer room	26	26	26	26	
15	Temporary storage of explosive materials	8	8	8	8	
16	Oxygen quick	10	10	10	10	
17	Oil storage	6	6	6	6	
18	Turnover material	30	30	100	200	
19	Fire sand tank	10	10	10	10	
20	Amount	1389	2069	1885	7455	More than 10% increase in public areas in actual calculations

Note: Some data in the table are empirical data, and the actual construction process is adjusted according to the requirements.

6.3.3.6 *Layout plans of each typical site*

Figures 6.26–6.29 show the layout plans of each typical site.

6.4 Construction of standardized site

6.4.1 *External image of construction*

6.4.1.1 *Gate*

(1) The construction site of rail transit shall be managed in a closed manner, and the door shall be set at the entrance and exit of the site (Fig. 6.30).

The door is divided into the side gate, the drive gate and a small door for a person to walk through, so as to separate the people and the equipment. According to the position of the small door, it can be set up. Opaque material to be applied onto the small door and it can be positioned either on the left or right side according to the present situation. There should be a card reader affixed on the small door so

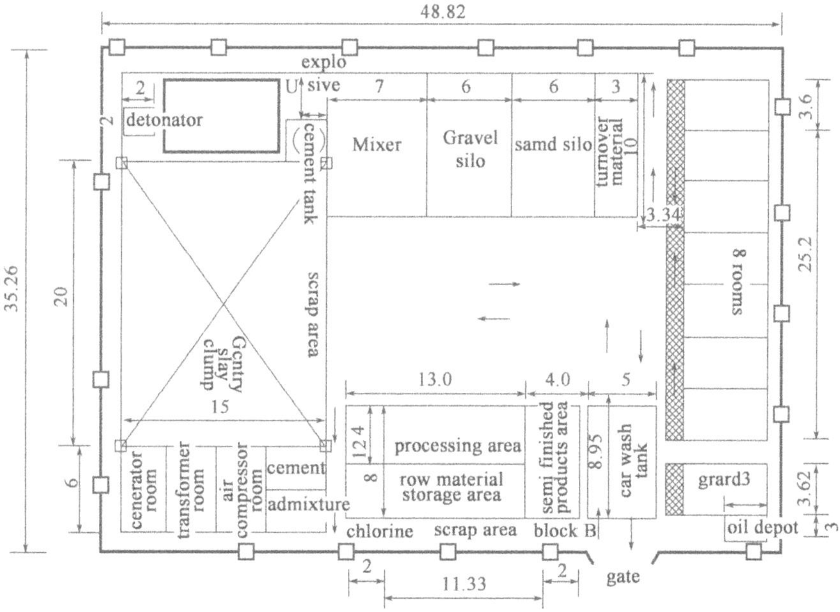

Fig. 6.26. Layout design drawing of section shaft (unit: m).

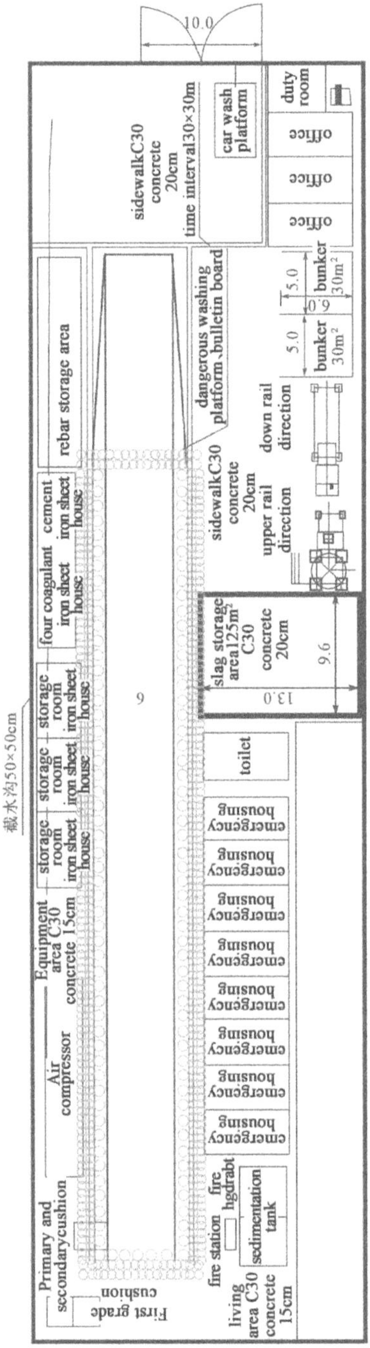

Fig. 6.27. Layout design drawing of inclined shaft in section.

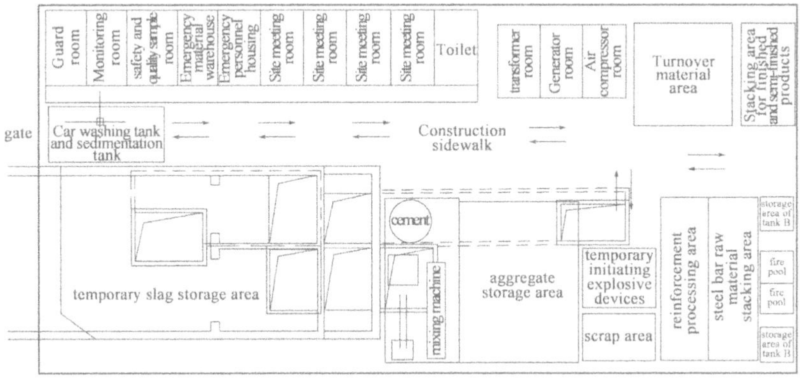

Fig. 6.28. Layout design drawing of air shaft of underground excavation station.

as to permit only authorized personnel to enter and strictly prohibit non-authorized people.

(2) The door leaf is not fully closed, and it is installed at the bottom of the wheel, making it flat like a push–pull doors. The door leaf box height is not less than 2200 mm. Skeleton steel pipes, angle steel or other production materials are available. The thickness of the tin is not less than 1 mm and column interval is not less than 6000 mm; between pillars lighting lamps and lanterns are installed.

The color of the gate is blue [C100 m^{3}0]. The size of the gate is measured and an appropriate piece is selected.

i. Station name font: Ye Genyou blade black grass, size 73 cm × 77 cm; the name of the construction unit: in Chinese text; and the size: 38.5 cm × 38.5 cm.
ii. Qingdao Metro Logo: 80.5 cm × 80.5 cm, in grass green color. The text size is 28 cm × 24 cm and it is located on the left side of the gate.
iii. Construction enterprise Logo: size: 108 cm × 72 cm, white or white with red, with an appropriate text size.
iv. The typeface and logo are painted on the spot, and all fonts should be white of 28 cm × 24 cm, located on the right side of the gate.

(3) The security room can be set up on the side of the gate according to the actual situation.

(4) The gate should be set up with a video monitor, and the road surface should be covered by the adjacent road. The surface of the steel plate

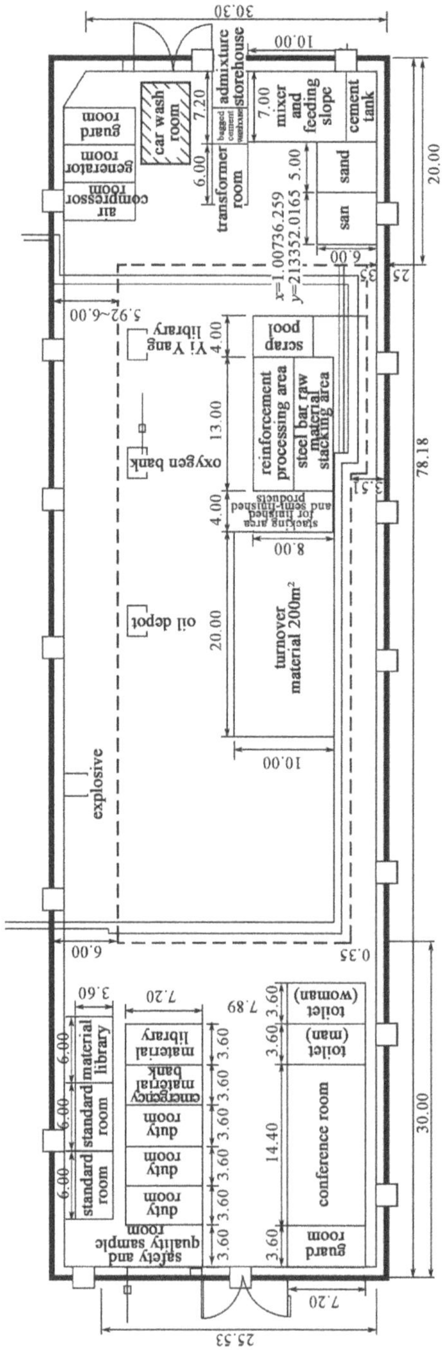

Fig. 6.29. Open excavation foundation pit layout design drawing (unit: m).

Fig. 6.30. Enterprise logo and enterprise name chart.

should be covered with yellow and black striped warning paint, and the shock belt should be set and the communication line should be set.

6.4.1.2 *Enclosure*

(1) Qingdao city rail transit construction uses Wei Dang print plates, and baffles, square tube columns and wall base welding, which are collected into three parts around the retaining wall in a row, with line migration, basins, such as in the construction of the metro engineering fence (Figs. 6.31–6.34).

Technical parameters:
Damper:
Size: 2000 mm × 5000 mm, 4000 mm × 9000 mm;
Steel frame: L40 × 3 mm;
Welding panel: 0.5-mm plate;
Picture: light box cloth spraying;
With the whole combination, the baffle is welded on the square pipe column.

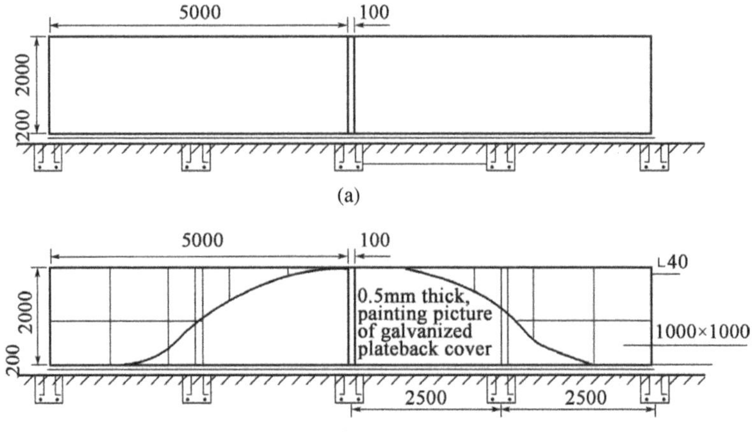

Fig. 6.31. Long-term enclosure elevation (2000 mm × 5000 mm) (size unit: mm)

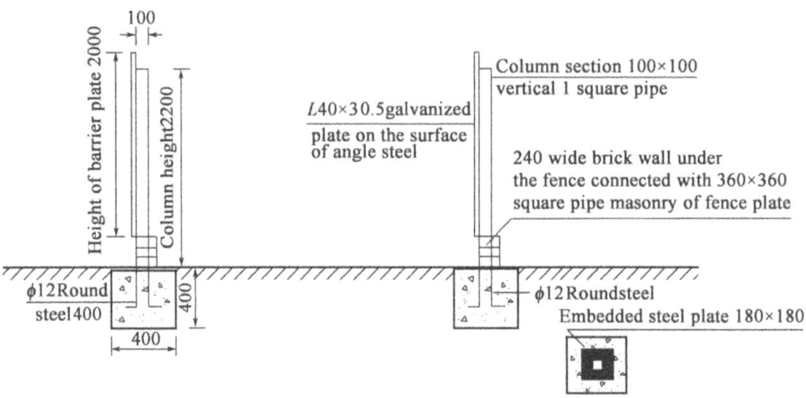

Note: cement, asphalt road and other hard ground shall be reinforced with 12mm×150 mm expansion bolts; 360×360 bricks shall be laid at the bottom of column for reinforcement, and 400×400×400 concrete embedded parts shall be fixed on the soft ground; reinforcement

Fig. 6.32. Long-term enclosure support section plan (2000 mm × 5000 mm) (size unit: mm).

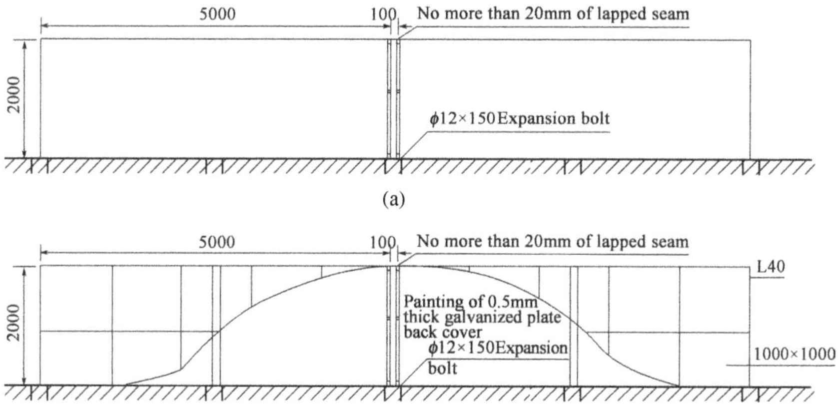

Fig. 6.33. Temporary enclosure elevation drawing (2000 mm × 5000 mm) (size unit: mm).

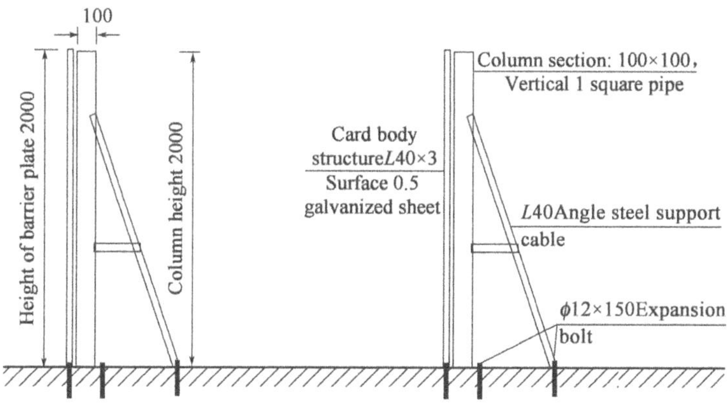

Fig. 6.34. Temporary block support section plan (2000 mm × 5000 mm) (size unit: mm).

Square pipe column:
Size: 100 mm × 100 mm × 2.6 mm.
Height: 2200 mm, 3800 mm.
Square steel column base plate thickness: 200 mm × 200 mm × 5 mm.
Reinforcement on the inside of square steel column: L40 × 3 mm.
Wall foundation (long-term enclosure).
The height of cement plaster after wall brickwork is not less than 200 mm.
Soft ground foundation is 400 mm × 400 mm × 400 mm with concrete pouring.

6.4.1.3 *Enclosure protection network*

(1) It is suitable for protection and as anti-stop painting for a large number of vehicles and heavy flow of traffic.
(2) The frame of the enclosure canvas is 2 mm × 5 m by 40 mm × 40 mm × 2 mm of square steel.
(3) The size of the mesh screen is 3 cm × 3 cm and the diameter of the mesh is 3 mm.
(4) A square steel with a length of 40 mm × 40 mm × 2 mm is added to the frame of the ring-fence, and the distance between the protective screen and the canvas is 8 cm. In order to maintain the image of the city and reduce the influence on the surrounding environment, the metro construction should be enclosed in urban areas and scenic spots (Figs. 6.35 and 6.36).

6.4.1.4 *Shed*

When adopting the enclosure, a special design should be used. The design should consider the lighting, ventilation, dust removal and other factors to meet the requirements of the production and working environment. A case shall be submitted for examination and approval (Fig. 6.37).

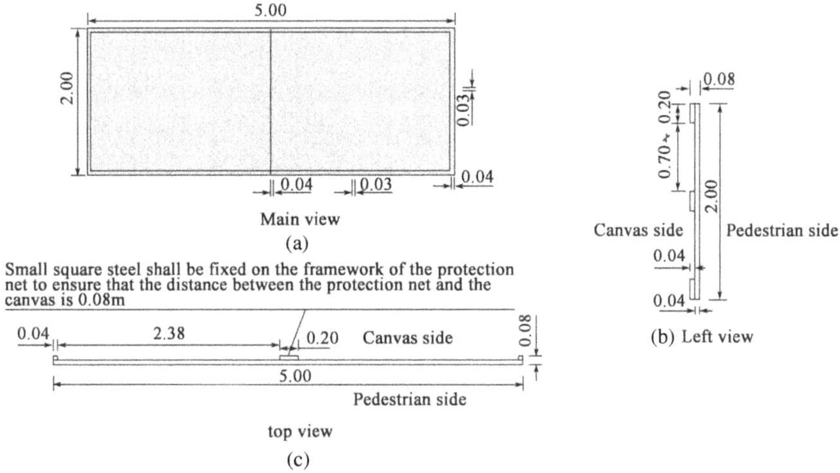

Fig. 6.35. Schematic diagram of enclosure protection network (unit: m).

Fig. 6.36. Effect picture of enclosure protection net (steel enclosure).

The frame is based on the independent foundation of reinforced concrete and adopts the anti-wind pressure h-shaped steel column. The core is C120 × 50 × 20 × 2.0. The wall and roofing materials are 75 mm polyurethane sealing glass fiber-reinforced fireproof panels (polyvinylidene fluoride, PVDF, spraying, internal punching).

(a) Effect drawing 1 of fully enclosed steel structure shed

(b) Effect drawing 2 of fully enclosed steel structure shed

(c) Wind resistant H-shaped steel column+C-shaped steel purlin frame

(d) Wall sound insulation material and C-type steel purlin frame

Fig. 6.37. Greenhouses.

6.4.1.5 *Flow control facilities*

(1) Hazard warning lights shall be set at the starting and stopping points of the construction site, so as to have a positive influence on the traffic safety of vehicles and pedestrians. The construction work on the roadway is required to set up warning signs and traffic guidance cards in the direction of the construction to guide the vehicles in an orderly and safe manner (Fig. 6.38).

(2) Night, fog and sudden bad weather might suddenly occur around the boundary of the working area, and thus flashing warning lights or red lights above 40W shall be placed at not more than a 4-m distance.

Fig. 6.38. Construction buffer setting diagram.

(3) Pipeline defects caused by construction, expanding or rebuilding projects should be covered in a timely manner and the construction units shall set up a guardrail, enclose any pothole and hang warning lights. In the area used by vehicles and pedestrians, there shall be a hole cover and obvious construction signs.
(4) Engineering construction should be avoided during rush hour. The necessity of limiting vehicles or implementing traffic control measures shall be approved by the municipal public security traffic management department. Any construction should set up notices in advance and temporary traffic control signs in appropriate locations and thus traffic guide.

6.4.2 *On-site production and service facilities*

6.4.2.1 *Site and road hardening*

Temporary roads to the construction site should be given appropriate hardening treatment using material of thickness not less than 200 mm and strength grade not lower than C20 concrete pavement base line for processing route drainage. The temporary road construction site should be ring shaped with a width of 4.5–6.0 m. The construction site material yard and material processing field must be equipped to deal with any material. A C20 concrete casting base of appropriate thickness and strength grade no less than 100 mm should be set up the for the processing field with a drainage ditch as required.

6.4.2.2 *Guard room*

The door of the site should be equipped with security features, and a security room nearby with a regular bright red warning lamp installed and an indoor relevant management system on the wall (Fig. 6.39).

(a) (b)

Fig. 6.39. Guard room.

6.4.2.3 *8 Cards*

A construction site of municipal work board or placard shall be set up at the gate of the construction or other reasonable places (Fig. 6.40):

(1) The construction site should have "eight CARDS and two pictures" at the gate or other reasonable locations.

The eight cards should have the following information: project summary list, management personnel and supervisor, relevant phone numbers, safety features of the production site, advanced construction, fire control measures, construction site management system information and environmental protection features.

As for the two pictures, the general plan of the construction site and the elevation (or effect) of the project should be set according to the situation.

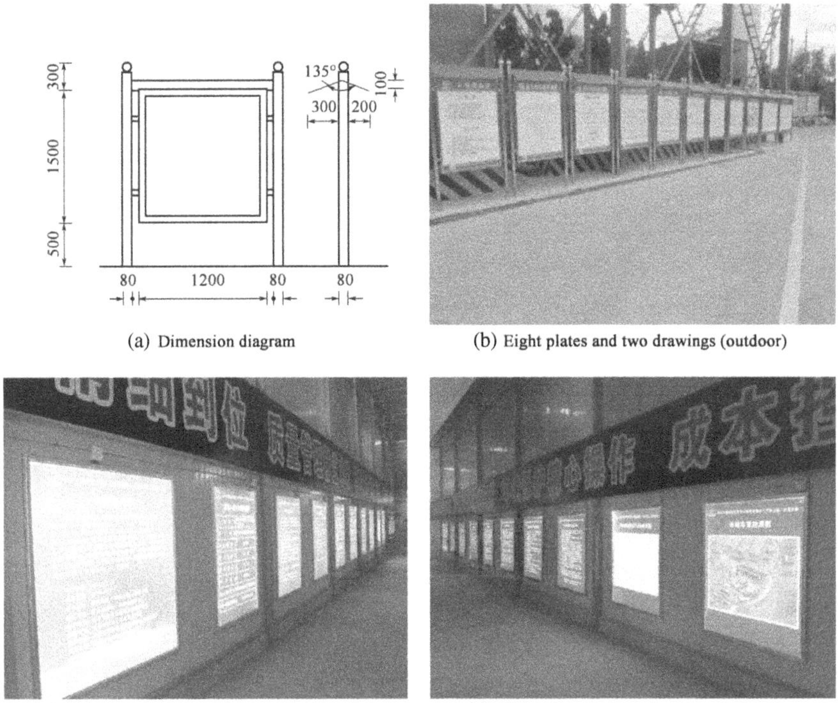

(a) Dimension diagram

(b) Eight plates and two drawings (outdoor)

(c) Eight plates and two drawings (indoor1)

(d) Eight plates and two drawings (indoor2)

Fig. 6.40. Eight boards, two pictures.

(2) Drawing standard: A card sized 1.5 m × 1.2 m (high × width) should be set 0.5 m from the ground, with a stainless steel flat plate and a stainless steel supporting frame, and set at the top of the sunshine board, unified for a mounting bracket, consisting of a base to ensure a solid foundation. The color of the card should be blue and white.

6.4.2.4 *Washing platform and sedimentation tank*

Set drainage ditches, car wash pools and sedimentation tanks in a high-traffic, well-populated place, and configure a high-pressure water gun. Clean the car body and wheel thoroughly before allowing vehicles to enter.

6.4.2.4.1 The car washing platform must be set up at the site of the construction

The motor vehicles driving out of the construction site must be washed and cleaned in the well-populated washing area at the construction site. An automatic car washing device is preferred for the car washing platform, but an ordinary car washing device can also be used (Fig. 6.41).

(a) Fully automatic car washing platform (b) Immersion car wash station

Fig. 6.41. Car wash platform.

6.4.2.4.2 Drainage channels and sedimentation tanks

1. A drainage ditch is arranged around the car wash tank, and the sludge in the gutter is cleared in time.
2. The drain panel is made of more than 32 steel bars to ensure that the vehicle is under heavy load.
3. The drainage ditch is connected with the sedimentation tank, and the sedimentation tank should have at least two levels for precipitation.
4. The sedimentation tank is usually made of bricks or concrete. The capacity shall be determined according to the discharge of waste water, and the measures to prevent mud and dirt from plugging the drainage pipes shall be set at the same time, and a special person shall be assigned to clean it regularly.
5. The tank will be painted yellow and black paint with a Baffle brush, with a spacing of 150 mm (Fig. 6.42).

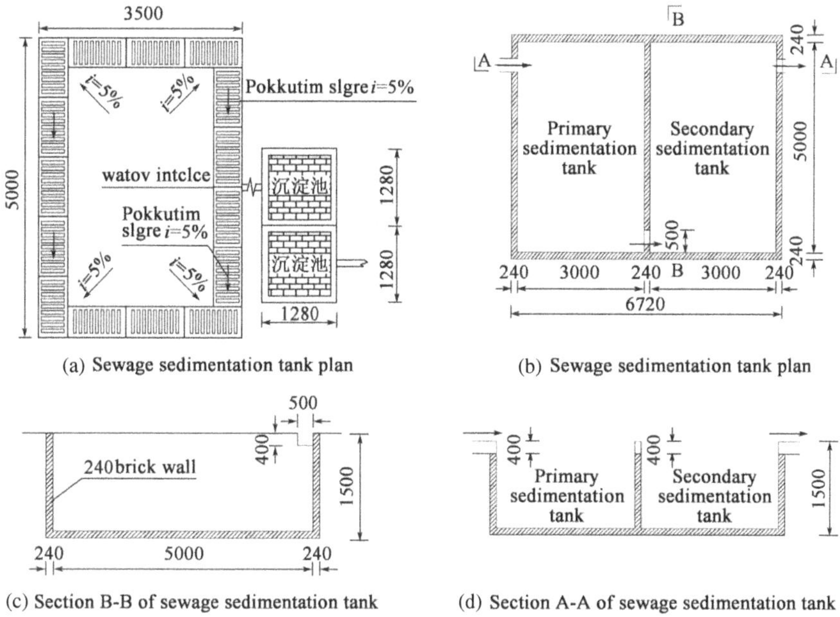

(a) Sewage sedimentation tank plan (b) Sewage sedimentation tank plan

(c) Section B-B of sewage sedimentation tank (d) Section A-A of sewage sedimentation tank

Fig. 6.42. Schematic diagram of drain and sedimentation tank (size unit: mm).

Notes: (1) The sedimentation tank adopts a 240-mm-thick brick wall, and the surface is plastered. (2) The sediment in the sedimentation tank should be cleared regularly, and the surrounding area should have protective features.

6.4.2.4.3 Sedimentation tank

1. The sewage and waste water tanks at the site must be equipped with tertiary precipitation facilities. The sedimentation tank cannot be less than the section size. The construction site can be appropriately increased according to the groundwater situation and site conditions.
2. The depositing tank uses a 240-mm-thick brick wall, the surface of which is plastered, or can be concrete casted, but the floor must use commercial concrete.
3. The wall and floor of the sedimentation tank shall be treated with waterproofing to prevent leakage during storage.
4. The sediment in the sedimentation tank should be cleared regularly and the tank can be covered with a reinforced concrete slab or steel plate.

5. The sedimentation tank should be surrounded by border protection, with warning signs set up and safety precautions in place to prevent accidents.

The sedimentation tank shall be connected with a site drainage system, and the location of the sedimentation tank shall be arranged according to the layout of the construction site.

The tertiary sedimentation tank for sewage is shown in Figs. 6.43 and 6.44.

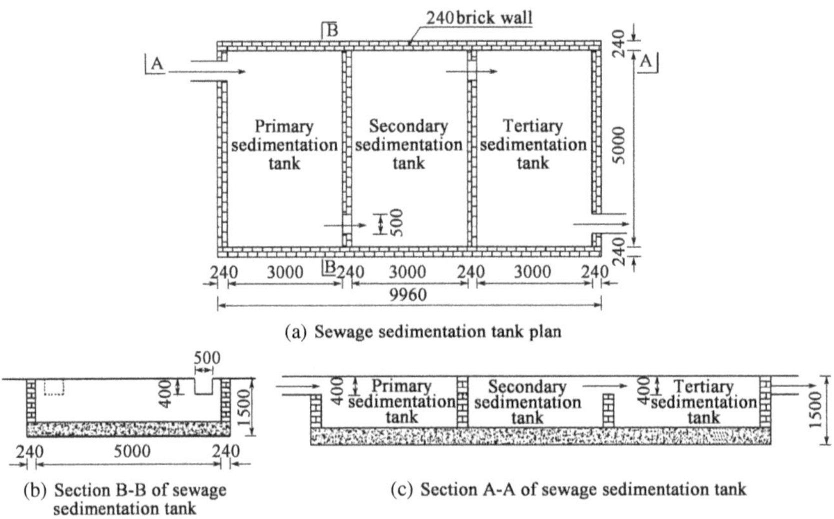

Fig. 6.43. Size of sedimentation tank (size unit: mm).

(a) (b)

Fig. 6.44. Tertiary sedimentation tank.

6.4.2.5. *Border protection*

6.4.2.5.1 The foundation pit and shaft lock port are adjacent to the border protection

1. The protective handrail painted with yellow warning paint, the facade of the balustrade is protected by steel mesh and the mesh is no more than 25 mm thick.
2. The protective railing has a concrete or masonry foundation not less than 300 mm high, with 200-mm-wide yellow and black color police paint (Figs. 6.45–6.47).

Protect the rail against the top of the bar.

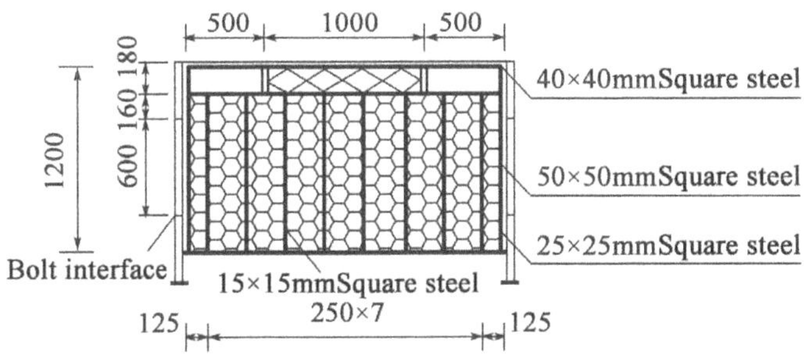

Fig. 6.45. Large sample of the safety guard railings (size unit: mm).

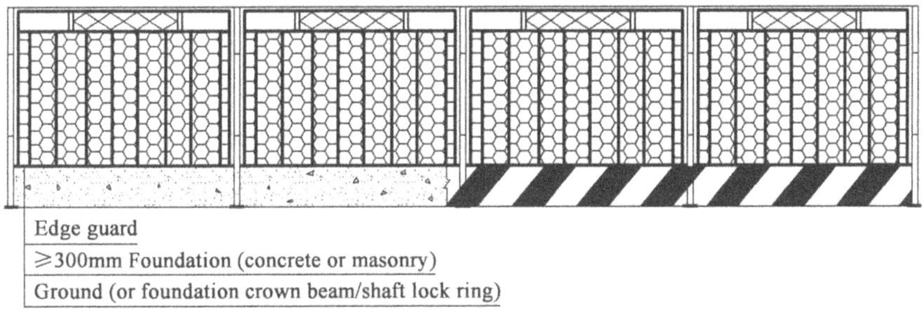

Fig. 6.46. Elevation of the safety guard railings.

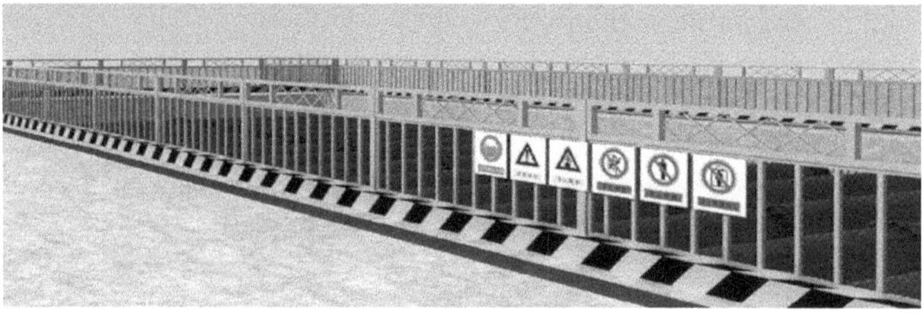

Fig. 6.47. Protective effect diagram.

6.4.2.5.2 Foundation pit strengthening protection

If the foundation pit is close to the perimeter of the enclosure, it is necessary to set up measures in the case of a crash in the foundation pit (Fig. 6.48).

Fig. 6.48. Foundation pit strengthening protection effect diagram.

Reference practice: Steel support in the inner side of the enclosure shall be provided. The section steel shall be more than 250-mm-thick type-I steel, with a height of about 1.5 m, and the spacing of the vertical column is less than 3 m, which is embedded in the underground concrete foundation piers, so as to ensure its firmness and reliability. Type-I steel can be used for reinforcing the surrounding enclosure.

6.4.2.5.3 Border protection practice

- Practice 1: Installation of steel pipe (Figs. 6.49 and 6.50)
  1. The protective railing is painted with red and white warning paint with brush spacing of 300 mm thickness, and the facade of the railing is protected by a thick mesh.
  2. The steel pipe is connected with a fastener, and the frame should be firmly anchored.
- Practice 2: Steel pipe or square pipe is set up (Figs. 6.51–6.53).

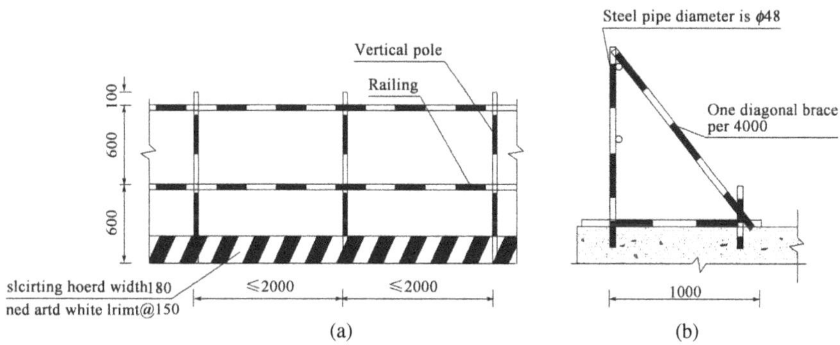

Fig. 6.49. Protective dimension figure (size unit: mm).

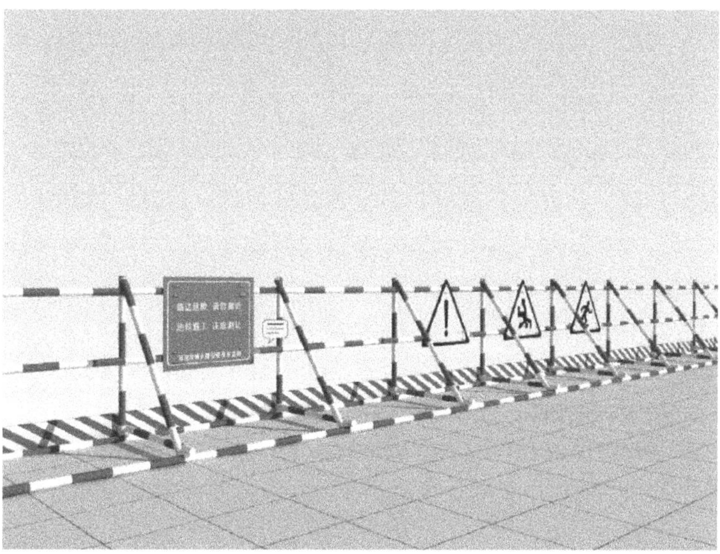

Fig. 6.50. Protective effect diagram.

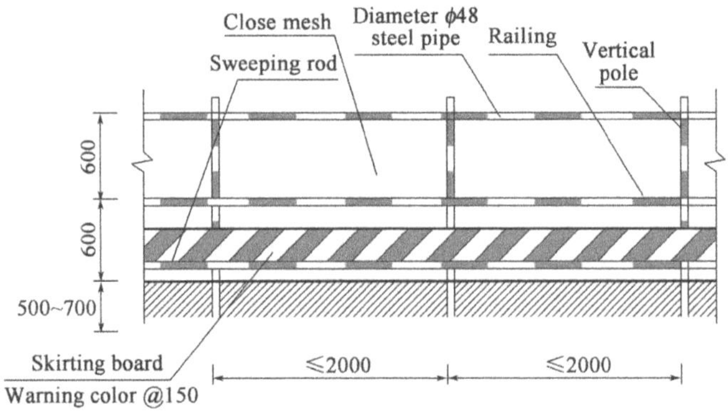

Fig. 6.51. Dimensions of the fence (size unit: mm).

Fig. 6.52. Protective effect of steel tube.

1. The protective material is made of steel pipe or square pipe, and the anchor is fixed in the underground or concrete structure, or is firmly connected with the underground embedded parts.
2. The fence is set not less than 180 mm from the skirting board and warning signs. Guard bars and kickboards are painted red and white as part of the inter-phase security guard ring. Lighting fixtures are provided during the night operation.

Fig. 6.53. Square pipe protection object map.

3. A safety net with protective barriers is provided. The mesh is not greater than 25 mm thick when it is provided for protection.

6.4.2.6 *Sand silo*

1. The materials that easily produce dust, such as sand and stone, should be enclosed in a closed storage mode. Electric rolling shutter doors should be used in the inlet and outlet, and a protective cover should be used so as to ease difficulty (Fig. 6.54).
2. The storehouse is arranged according to the warehouse, and the materials are stored according to different grain sizes and different varieties, and they are not allowed to mix or pile up. The floor shall be hardened with concrete of thickness not less than 20 cm. The floor of the inner warehouse shall be higher than the floor of the outer warehouse, with a drainage slope of no less than 4%, and a water hole in the wall.
3. The dividing wall should have enough anti-extrusion strength, for which concrete walls, steel plate walls, brick cement plaster walls, etc., can be used, and the high degree should meet the requirement of separation. A brick wall is used as a reference: wall thickness "37", wall height 1.5 m and cement mortar surface.

(a) (b)

Fig. 6.54. Sand silo.

4. The winter insulation measures are set according to the construction site climate.

6.4.2.7 *Mixing station*

1. A closed protective shed shall be set up for the mixing station shall set up, which shall be equipped with colored steel plates, water mists and dust removal devices. Where there are special requirements for noise emission, the shed will be set up with soundproof material.
2. The mixing and batching machine should be embedded in the ground, and enclosed, with the inlet and outlet covered with a canvas shade.
3. A dust removal device should be installed in the cement tank to prevent dust diffusion when adding cement (Figs. 6.55 and 6.56).

(a) (b)

Fig. 6.55. Concrete shelter for noise reduction and mixing station.

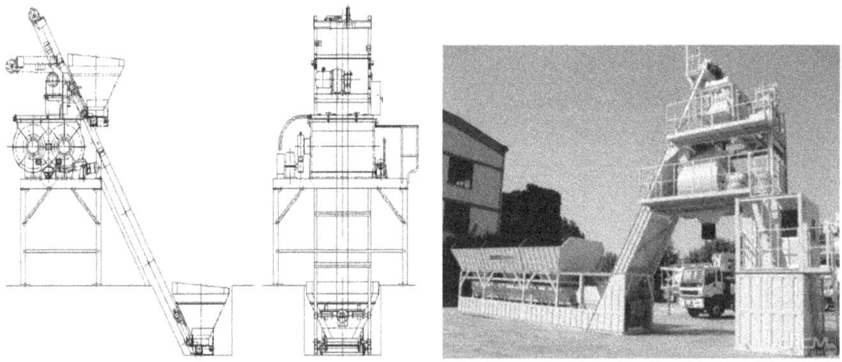

(a) (b)

Fig. 6.56. Sample diagram of the mixing station.

6.4.2.8 *Material stacking*

1. The building materials and components should be stacked in a neat and orderly manner according to the overall plane design. The material pile is marked with the name, variety, specification and so on.
2. Bulk cement shall be piled up separately according to different manufacturers, varieties, labels and production dates. The cement storage period is not more than three months. The cement warehouse must be kept dry. It should be set on high and well-drained ground. The ground of the warehouse should be hardened and the cement should be stored away from the wall.
3. The material identification card shall be set up in the storage area of the construction site.

The material identification card and safety protective equipment acceptance card, as shown in Fig. 6.57, are of 0.4 m width, 0.3 m height and made of aluminum plastic plate, and worded in black mortar. The material signs and safety protection facility acceptance cards can be set up with support, or can also be hung on the wall or the fence.

The material stacking area should be hardened, and the material should not be less than 100 mm away from the ground with a proper cover. The material shall be stacked according to different specifications and categories, ensuring that one end is flush against the wall and the material identification card shall be placed in the corresponding position.

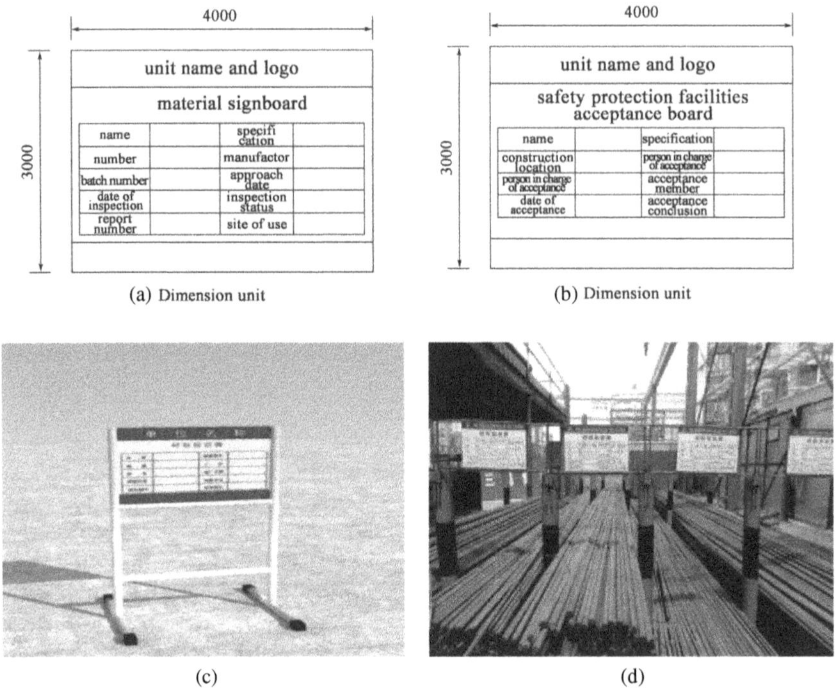

Fig. 6.57. Material identification card (size unit: mm).

6.4.2.9 *Processing protective shed*

1. All steel bars and woodworking processing fields in the construction site shall be equipped with processing protective booths (Fig. 6.58).
2. The height and width of the protective shed shall be determined according to the site conditions and in order to meet the work and operation requirements. The frame system of the protective shed shall be made of steel, which shall be firmly connected and calculated. The top should be covered with wood with a thickness of more than 5 cm and covered with rainproof material. Within the radius of the falling object or within the radius of the crane boom, a double layer of 5-cm-thick planks should be erected, and the upper and lower floors should be 0.6 m apart.
3. The processing shed shall be strengthened and tested to ensure safe use in typhoons and other hazardous weather.

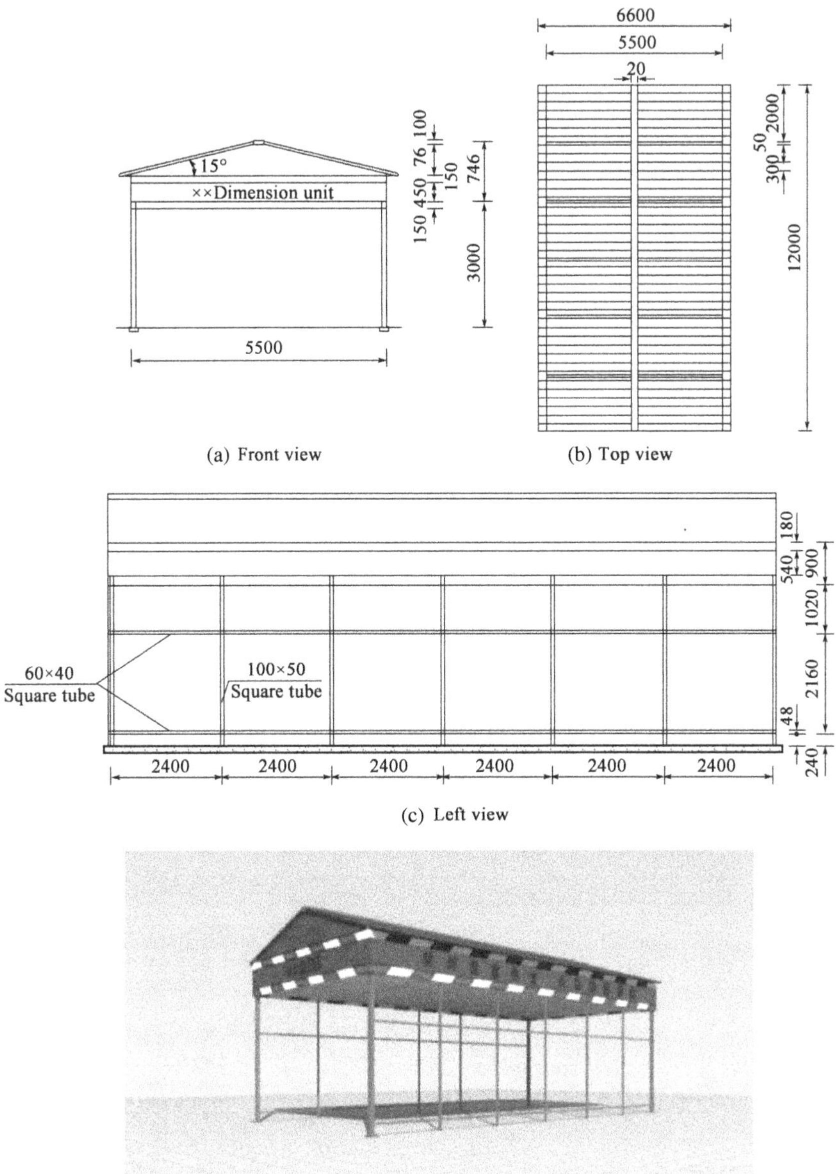

(a) Front view (b) Top view

(c) Left view

(d) Design sketch

Fig. 6.58. Processing protective shed (size unit: mm).

4. The top of the processing shed should be painted with red mortar and safety warning signs should be hung.
5. The processing equipment for wood and steel needs to be used in the processing shed, and the transmission parts shall be equipped with protective measures.

6.4.2.10 *Vertical shaft upper and lower channels*

(1) The height of the stair railings is 950 mm (starting from the front edge), and the height of the stair rail is 1050 mm (Fig. 6.59)
(2) Stair width, step height to width and the size of the rest platform should not be adjusted, but stair quantity can be adjusted according to the height of the shaft if it does not satisfy the integer times last run stair calculation as determined based on the scene.
(3) When the vertical shaft width is greater than 5 m, the number of steps should not be more than 18.
(4) A protective steel wire mesh should be set up on the open side, and a 180-mm-high ladder foot plate should be set up on the stairs, and red and white inter-phase warning paint of 150-mm-thick stripes should be applied.
(5) The stair steel plate adopts a decorative pattern plate, and the thickness of the steel plate is designed by the professional

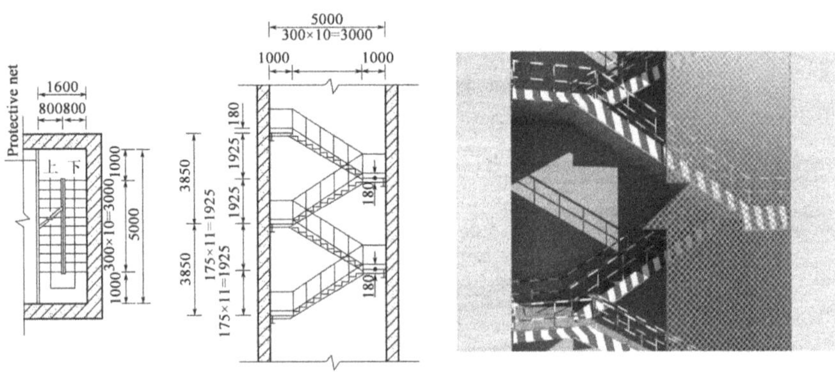

(a) Standard floor plan (b) I - I Profile (c) Shaft ladder

Fig. 6.59. Vertical shaft upper and lower channel (size unit: mm).

manufacturer who also installs it. Anti-falling measures at the bottom of the stairs can be modified according to the actual situation to make them more effective.

(6) The vertical ladder of the deep foundation pit can be made by referring to the relevant standards.

6.4.2.11 *Flammable and explosive goods storehouse*

(1) Oxygen b-type tanks must be set up in the temporary storage cabinet, the used bottles on the left and unused full bottles on the right, with a space of 1.5 m in between, which can be used by the department of products for publicity. This design of the left-side area being used to store the empty bottles, the right-side area containing the full bottles and the middle being used for publicity is adopted from the assembly system, and can be combined with real international ideas to disassemble and assemble. The lower part of the middle propaganda area is set aside as a shockproof circle storage area and fire extinguisher storage area, while the upper part is set aside for construction safety regulations and the relevant safety instructions for the different recipients and daily check patrol records (Fig. 6.60).

Fig. 6.60. Oxygen b-type express bottle bank.

(2) Temporary fast transport vehicles are equipped with oxygen b-type tanks following specifications, and colored in strict accordance with requirements, and are advantageous for on-site operation and the daily management of the project.

6.4.2.12 *Fire protection facilities*

(1) The living area, floor, warehouse, material yard and templates of the construction site should be equipped with corresponding types of firefighting equipment, which should be replaced periodically and utilized only during its period of validity (Fig. 6.61).
(2) Fire equipment should be placed in easy-to-access and conspicuous places.
(3) Temporary facility areas, each commonly 100 m^2 with at least two levels, should be equipped with firefighting equipment, not less than 3 fire extinguishers, and the overlay facilities shall be equipped with Taiping barrels for fire, fire shovels, fire axes, reservoirs, sand pools and other fire equipment.
(4) The distance between each group of fire extinguishers should not be greater than 30 m, and each group of fire extinguishers should not be less than an S size.

6.4.2.13 *Electrical equipment*

6.4.2.13.1 General provisions

The electrician does special type of work and must hold a special operation certificate to work.

The plan of the temporary electricity utilization project must be checked, examined and approved, by the relevant the department the construction unit, and can be put into use after being qualified.

The construction unit shall establish technical files for safety purposes, including electric scheme, technical clarifications, with information regarding electricity engineering tables, grounding resistance and insulation resistance, leakage protection, leakage motion parameter measurement records, electrical installation patrol maintenance and removal of hidden danger rectification work record.

(a) Fire Equipment

(b) Fire Equipment

(c) Fire Equipment

(d) Fire Hydrant

(e) Fire belt

(f) Fire blanket

(g) Fire box (pool)、pump

(h) Hand lift fire pump

Fig. 6.61. Fire control facilities.

For an exposed work point line at the construction site, an insulation fire protection sleeve should be added.

LED lighting should be used for the lighting of the construction site.

The construction unit shall establish an inspection system and conduct regular and irregular inspection of the on-site electricity consumption.

6.4.2.13.2 Laying of distribution lines

The overhead wires must be insulated and the wires must be installed on a special electric pole and shall not be erected on trees, the ground and other facilities. When laying the overhead wires, the insulator is fixed and the height is higher than 2.5 m. When the wires are to be laid across a road used by vehicles, the distance from the ground should be greater than 6 m.

In the process of circuit laying, it is necessary to ensure that the phase sequence and phase color are consistent. The N line and PE line are strictly prohibited.

The cable line shall be laid underground or overhead, and it is forbidden to be dragged along the ground.

The route of the buried cable shall have bearing marks. The laying mode and depth of the cable shall be in conformity with the requirements. The hard protective layer shall be covered directly.

When laying the cable, it shall be laid along the pole, bracket or wall and fixed with an insulator. When laying the cable along the wall, the maximum arc distance shall not be less than 2.0 m. Laying overhead cables on the ground, trees or other facilities is strictly prohibited.

The main trunk line should not be less than 2.5 m from the ground. The outdoor end of the overhead line shall be fixed with an insulator, and the over-wall line shall be protected by a tube and shall not be less than 2.5 m from the ground.

It is advisable to use armored cable for burying in the ground. In the case of unarmored cable, it should be waterproof and anti-corrosive. The cable should be protected from mechanical damage, and placed away from heat sources, buildings, other structures and traffic.

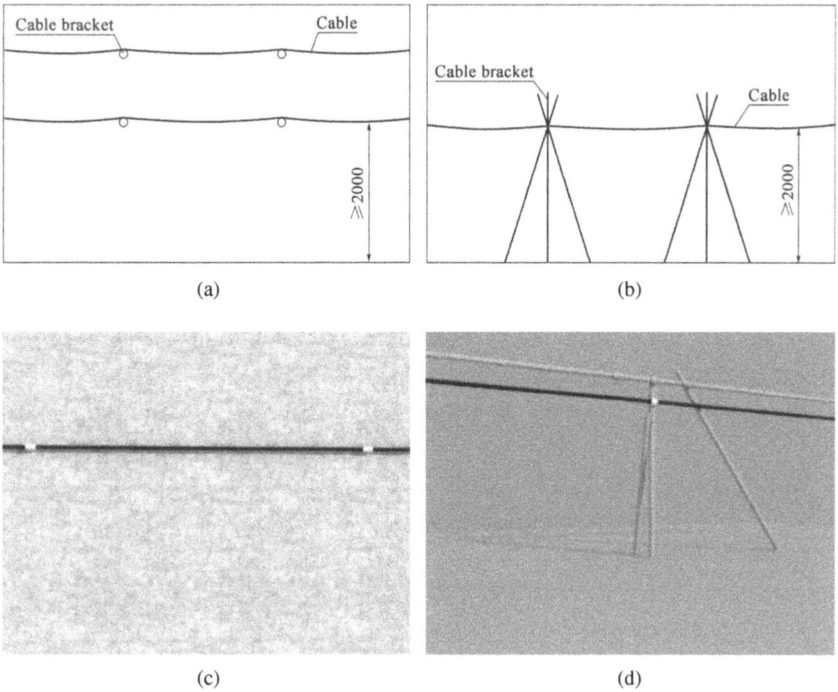

Fig. 6.62. Cable diagram.

The cable (Fig. 6.62) shall be buried in a specially excavated cable trough, and the depth of the groove shall not be less than 0.7 m. The sand should be laid evenly between 50 and 100 mm thick on the cable, and the hard protective layer covered on top. Application of a steel pipe or hard engineering plastic pipe cover should be made when crossing the driveway. The open cable trench must be cast with concrete, and the cable groove is made to be closed with thick planks or steel plates, and the cable trench should be protected from water. The cable shall be marked on the corners and in the line segment every 20 m.

The aerial cable should be along the electric pole or wall; the insulator is fixed and the tie line must use insulated wire. Laying the overhead cables along the wall when the maximum sag distance is less than 2.0 m at the mouth along the scaffold is forbidden, as with trees or other facilities (left: laying along the wall; right: laying port along the bracket).

6.4.2.13.3 Protection of fixed electrical box

1. There should be enough space and channels for two people to work together around the electric box. No items that interfere with operation or maintenance shall be piled up, and shrubs and weeds shall not be allowed to grow (Fig. 6.63).
2. The electric box shall have a label, indicating the name, number, marking, system wiring diagram, the name of the electrician and the contact number.
3. The daily inspection by the electrician shall be carried out for the management of electricity. The inspection form shall be posted on the door of the electric box, and the form shall be signed and confirmed

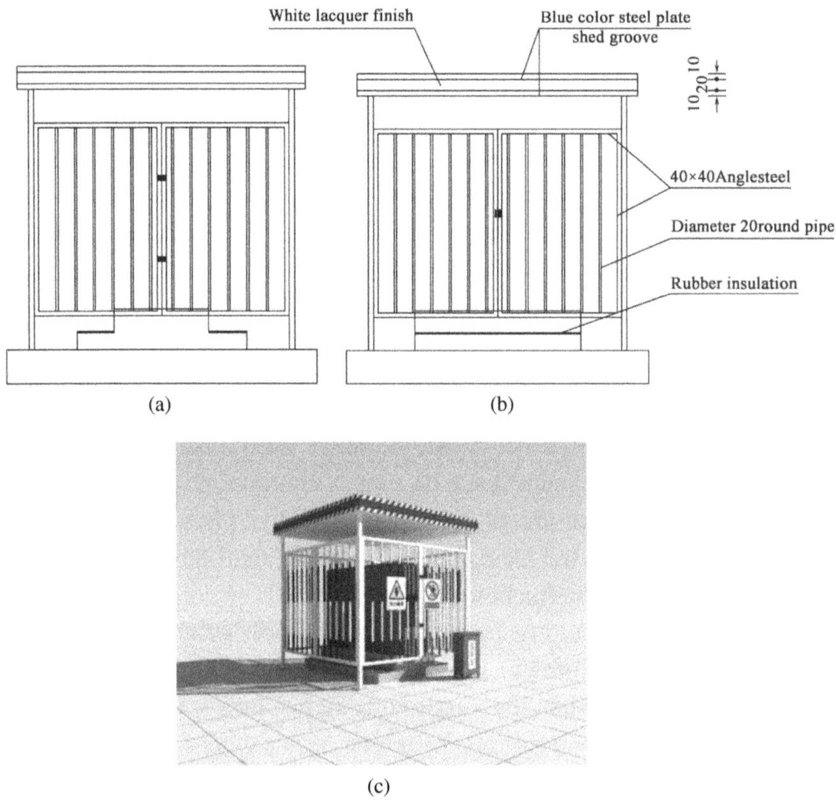

Fig. 6.63. Protection of fixed electrical box (size unit: mm).

after the inspection by the electrician on duty, and if a problem is found it shall be dealt with immediately.

6.4.2.13.4 Protection of mobile electric box

1. The electrical installation board of the distribution box must be divided into an N wire terminal board and a PE wire terminal board. The N wire terminal board must be insulated from the metal electric appliance mounting plate. The PE wire terminal board must be connected with the metal electric appliance mounting plate. The N line and PE line must be connected by the corresponding terminal boards (Fig. 6.64).
2. The door of the electric box should be locked, and a person shall be responsible for it. Metal case doors and metal cabinets must be electrically connected through a braided copper wire.

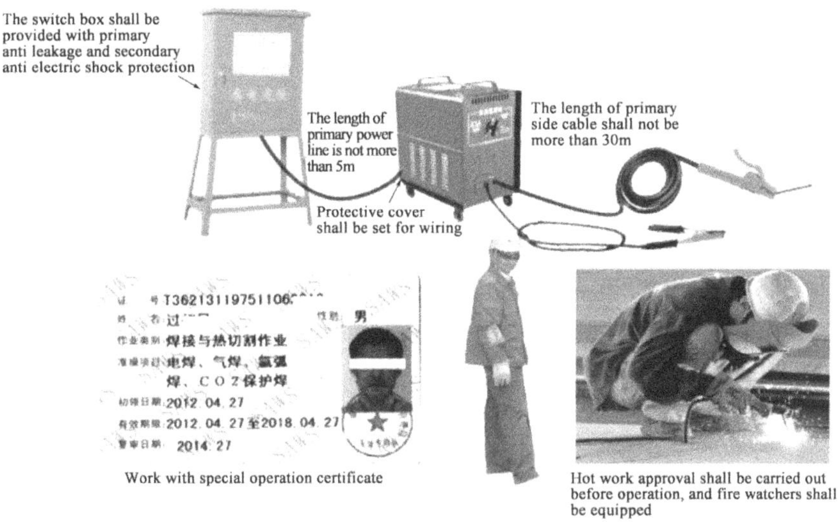

Fig. 6.64. Mobile electrical box and special operation ID card.

6.4.2.14 *Security monitoring*

The construction unit to form a security monitoring organization, and clear management responsibility. The unit should prepare an

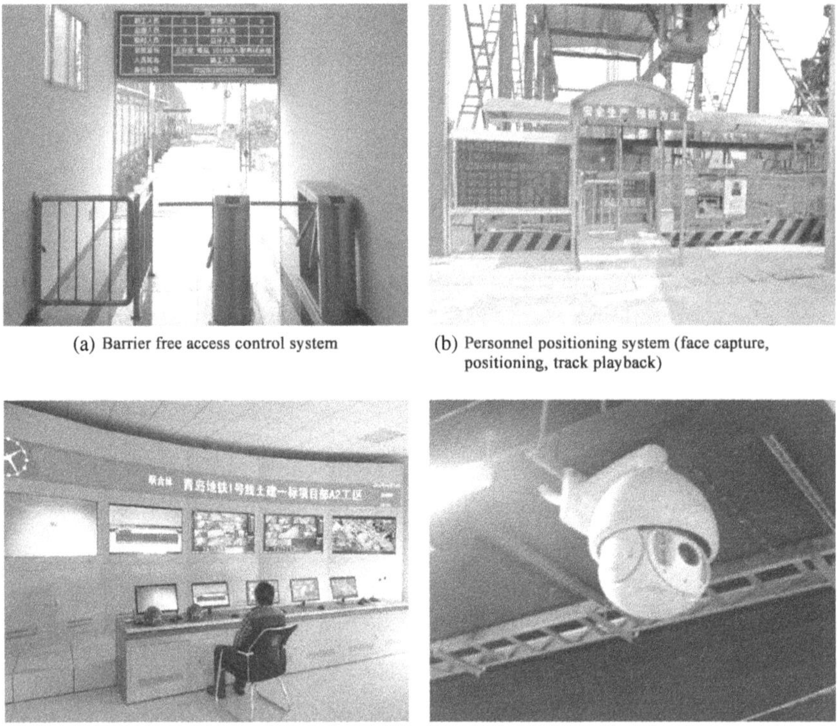

(a) Barrier free access control system

(b) Personnel positioning system (face capture, positioning, track playback)

(c) Video monitoring system 1

(d) Video monitoring system 2

Fig. 6.65. Safety monitoring facilities.

implementation plan, submit it to the supervisor for approval before implementation, according to the monitoring scheme of setting safety monitoring facilities in a timely manner (Fig. 6.65).

The construction unit shall purchase the formal products which meet their requirements of performance parameters, and it is strictly forbidden to purchase fake or substandard products that do not meet the requirements.

The construction unit shall conduct acceptance inspection and supervision of the supplier and other units before and after the installation of monitoring equipment. All parties shall sign and confirm the acceptance form.

The construction unit should actively promote educational activities and publicize the importance of video monitoring system. The unit should check the monitoring system equipment daily, find the problems, fill in the monitoring log and file it in a timely manner. All kinds of monitoring terminals or monitoring points should be set up and updated in time.

Video monitoring shall be set up in the living area, office area, construction work area, installation area and other places of supervision and control in each construction stage. The number and location of video monitoring shall be determined according to the situation of the site, ensuring that each work surface can be monitored in place.

Sealing management shall be carried out at the site of construction (including underground). Video and brake systems are set up in the population department, and the duty of professional staff is arranged. The personnel should be admitted with a valid card and strictly prohibited from the field of other personnel. The security monitoring system will be gradually improved such as the position of operators and the monitoring and information management. A new type of access control, such as a positioning chip and receiving device, is implemented to enable the personnel to take photos and track the work.

6.4.2.15 *Fire spacing*

(1) With the construction of an inflammable and dangerous explosive goods warehouse, a fire prevention project should be taken up with spacing not less than 15 m; with a combustible material yard and its processing field, the fire prevention spacing should not be less than 10 m. For the temporary housing, temporary facilities and the construction process, the fire prevention spacing should be not less than 6 m.

(2) The main construction site and temporary housing should have fire separation of not less than the provisions mentioned in Table 6.4; for office premises and group lodging arrangement, the fire separation can be appropriately reduced, but should meet the following requirements:

   i. The number of buildings in each group should not exceed 10, and the fire spacing between the group should not be less than 8 m.

   ii. The fire spacing between temporary rooms in the group should not be less than 3.5 m. When the combustion performance level of building components is grade A, the fire distance can be reduced to 3.0 m.

Table 6.4. Main temporary rooms and temporary facilities at the construction site E (unit: m).

Room/facilities	Office accommodation, dorm	Generator room, Transformer substation	Combustible material warehouse	Kitchen operating room, Boiler room	Combustible material yard and its processing plant	Fixed fire field	Flammable, explosive goods warehouse
Office accommod-ation, dorm	4	4	5	5	7	7	10
Generator room, transformer substation	4	4	5	5	5	5	15
Combustible material warehouse	5	5	5	5	7	7	10
Kitchen operating room, boiler room	5	5	5	5	7	7	10
Combustible material yard and its processing plant	7	7	7	7	7	10	10
Fixed fire field	7	7	7	7	10	10	12
Flammable, explosive goods storeroom	10	10	10	10	10	12	12

6.4.3 *Off-site service*

6.4.3.1 *Office area management*

6.4.3.1.1 Site selection and basic construction requirements

1. Construction should be in accordance with the principle of orderly, standardized occupancy, with complete function and convenience, and the hardware facilities must meet the requirements of the contract. Lightning protection devices should be set up according to the situation.

2. A room should be set on the first floor of the temporary housing, and the evacuation door should be easy to open in the event of an evacuation.
3. Dormitory and office space should not be built with a kitchen operation room, boiler room, transformer room and so on.

6.4.3.1.2 Civilized construction

The construction of the site should be carried out in a civilized manner in accordance with the local culture, with the health and well-being of the field workers in mind, so as to continuously improve their living conditions and enrich the local culture. The basic requirements are as follows:

1. The living area should contain a recreation room, library and so on.
2. According to the conditions, the quarters should be equipped with outdoor sports facilities such as basketball court, badminton court, outdoor fitness equipment, etc.
3. The construction site should formulate emergency medical treatment measures, and be equipped with a medicine box, with commonly used medicine and first aid equipment, and should have cooling and heat prevention measures in the high-temperature season.
4. Health issues in the construction site, office area and living area should be handled by a specially appointed person, with clear responsibilities.
5. The office area and the living area shall establish a management organization and formulate a health care and epidemic prevention system. Garbage should be stored in airtight containers, regularly sterilized, regularly disinfected and disposed in a timely manner, in order to keep the area clean.

6.4.3.1.3 Office area management

1. The office shall set up a conference room, library, offices, evening school for migrant workers and laboratory work area, which should be safe so as to avoid hidden trouble, complaints and reports of human rights violations from the staff in the suggestion box. All this should be according to the requirements of the enterprise.

2. The conference room shall be equipped with drawings of project overview, construction images, progress charts, and informational manuals regarding security, quality, ring protection, emergency rescue, occupational safety and health, and other management instructions. A projector and LED display should be installed.
3. Each office shall have a job description badge, which shall cover the content of safety responsibility.
4. The office area should be landscaped to create a beautiful work environment.

6.4.3.2 *Concentrated residential areas*

6.4.3.2.1 Dormitory management

The combustion performance level of the building components shall be class A. The use of new standardized box housing shall be promoted (Fig. 6.66).

The building shall not exceed the second floor, and only one floor will be allowed in the environmentally sensitive areas. The floor area should not be larger than 200 m^2, and no less than two evacuation staircases should be set.

The net width of the evacuation walkway should not be less than 1.0 m, and it should not be less than 1.5 m when furnishing a room. The passageway floor shall be equipped with firefighting equipment, installation of emergency lamps and evacuation signs. The net width of the evacuation staircase should not be less than the net width of the evacuation walkway.

The dormitory has a unified number, and on each dormitory door hangs a responsibility card indicating the class name, the public security, the health and the fire control persons, and the duty schedule.

The dormitory is not allowed to be very large, with the floor no less than 1.9 m $\times$ 0.9 m, and the beds should not be more than two bunks, with each bed having a name card. In the dormitory, there is a special cabinet for articles of daily life, which are to be arranged neatly.

The living area lighting power supply should adopt a safe voltage; the dormitory uses 36-V standard safe voltage and uses a USB interface to charge. Summer cooling and winter heating for the housing unit should be

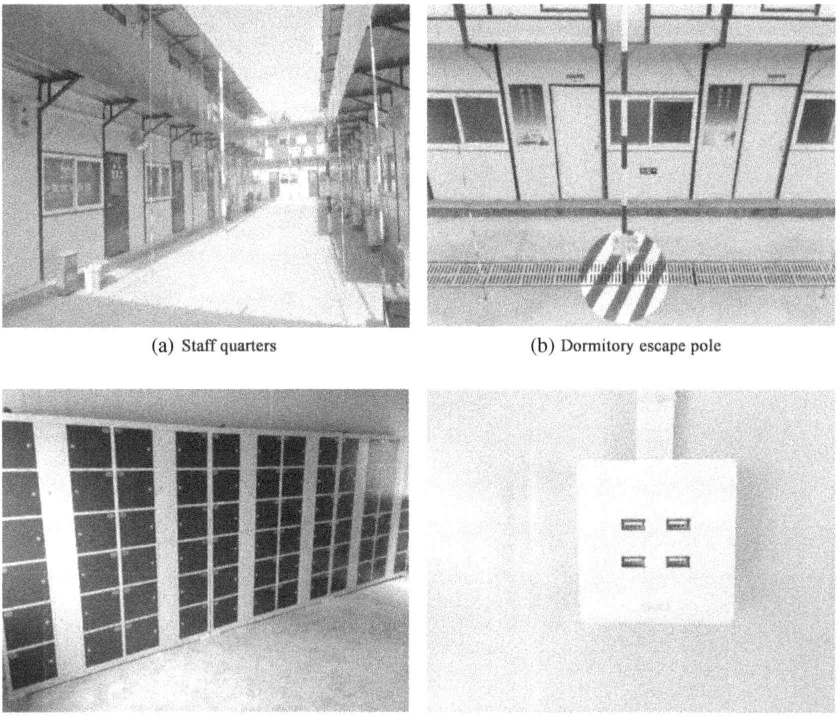

(a) Staff quarters

(b) Dormitory escape pole

(c) Central charging cabinet

(d) 36V USB charging interface

Fig. 6.66. Dormitory management.

by means of air-conditioning or central heating. The machine should be removed after charging. It is strictly forbidden to connect the charging power supply to the outdoor charging.

A video monitoring terminal should be installed in the living area to ensure full coverage and fire protection facilities according to the plan.

6.4.3.2.2 New standardized box-type room

The production and living area should be used to promote the use of standardized box-type rooms, which have the advantages of stable structure, good wind effect, reusability and energy saving (Figs. 6.67–6.69, Tables 6.5 and 6.6).

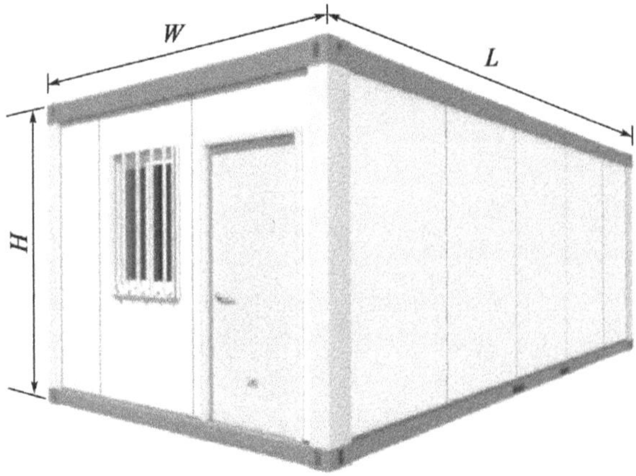

Fig. 6.67. Standard box schematic.

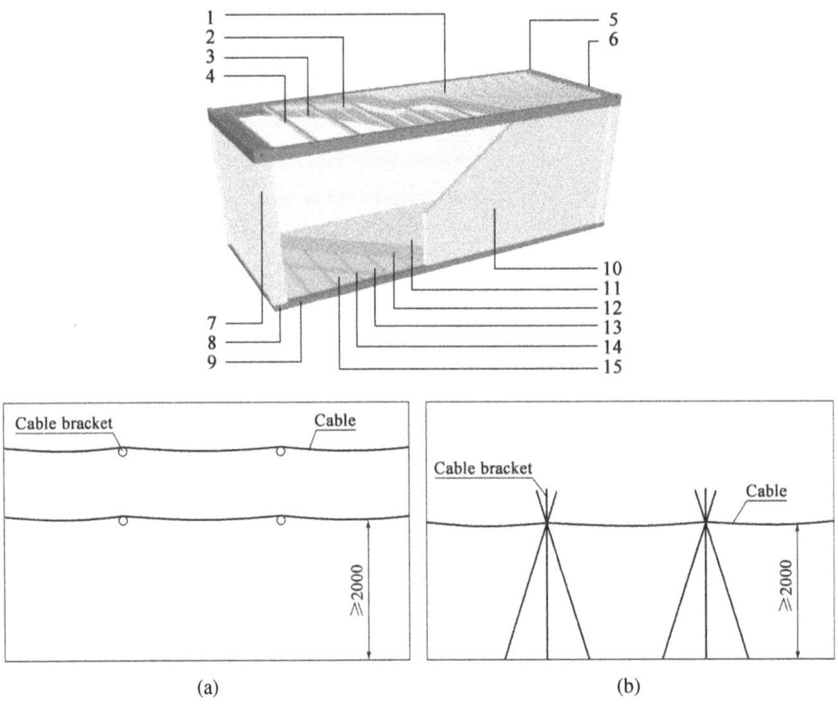

Fig. 6.68. Standard box construction.

Notes: 1- colored steel roofing tile; 2- glass filament insulation cotton; 3- colored copper ceiling board; 4- roof report; 5- roof corner pieces; 6- roof girder; 7- column; 8- ground corner piece; 9- bottom main beam; 10- angle; 11- rubber floor; 12- cement board; 13- glass filament insulation cotton; 14- height to the ground; 15- seal plate.

Fig. 6.69. Standard box renderings.

Table 6.5. Standard box size.

Category		Inch gauge grid (mm)	
		Type a	**Type b**
Enclosure	Inside	$L605 \times W2435 \times H2896$	$L6055 \times W2990 \times H2896$
	Outside	$L5840 \times W2225 \times H2540$	$L5840 \times W2780 \times H2540$
Window		$H \geq 1100$ $W650 \times H1100/W1500 \times H1100$	
Door		$H \geq 200$ $W \geq 850$	
Beam height top	Top	$H \geq 180$ (Thickness of the steel plate ≥ 4)	
	End	$H \geq 140$ (Thickness of the steel plate ≥ 4)	

Table 6.6. Standard container room material list.

S/N	Name	Detailed description	Inch gauge grid	Standby note
1	Surface	Standard box white	6055 mm × 3000 mm × 2890 mm	
2	Structure	Floor pounding frame beam	Thickness ≥ 4 mm, Plating amount ≥ 100 g	
		Floor pounding frame beam	Thickness ≥ 4 mm, Plating amount ≥ 100 g	
		Plating column	Thickness ≥ 3 mm, Plating amount ≥ 100 g	
3	Roof	Color plate	0.4 mm, white ash	
		Glass wool	100 mm thick, bulk 10 kg/m^3, single-side aluminum foil	
		Yx12-110-880 Colored steel ceiling	0.35 mm	
4	The earth's surface	Glass mirror fiberboard	18 mm	
		PVC floor board (blue)	1.6 mm	
5	Wall	Glass fiber cotton, colored steel sandwich panel	0.5-mm color plate, I75-mm glass wool, Thermostat, Bulk weight 60 kg/m^3	Double-sided plate
6	Window	U PVC Window frame, push–pull series	1100 mm × 1100 mm, White, with screens	Insulating glass
7	Door	Steel door	880 mm × 2020 mm	
8	Electrical system	Concealed distribution box		
		Industrial head trap	Voltage 220 V, 50 Hz, 3pole32A	
		Single tube fluorescent lamp	18W	
		Trap (air-conditioning)	16a three holes	
		Capture (Wall)	10a five holes	
		Lamp switch		

6.4.3.2.3 Fire protection facilities

1. Fire protection facilities mainly include fire hydrants, fire extinguishers, fire sand, fire hooks, axes and buckets, and the site should be equipped with a sufficient number of these.
2. The construction site and living area should be equipped with special fire hydrants and fire hoses, in sufficient quantity to cover the whole area (Fig. 6.70).
3. The fire hydrant should ensure that the pressure can meet the fire needs, and the pressure pump should be installed when the pressure is insufficient. When a fire hydrant is beyond its permissible use limit, it can be replaced by a fire water tank and fire pool, of capacities not less than 6 m^3. Fire hydrants and pipes should be protected against cold and heat. The site should be equipped with a lift fire pump according to the situation.
4. Key areas for fire protection shall be equipped with firefighting equipment suitable for any fire type.
5. The firefighting equipment shall be inspected, repaired and replaced regularly to ensure its effectiveness.
6. The firefighting equipment should be placed in easy to be easy to put and more prominent place.
7. For temporary facility areas, each 100 m^3 shall be equipped with at least two fire extinguishers with a fire extinguishing level not less than 3A. The spacing between each group of fire extinguishers should not

(a) Fire fighting rack (b) Fire fighting facilities in living area

Fig. 6.70. Fire protection facilities.

be greater than 30 m, and each group of fire extinguishers should not contain less than 2.

Bibliography

[1] Qingdao Urban Rail Transit Project Safety Civilization Construction Management Standard.

[2] Qingdao Urban Rail Transit Project Safety Civilization Construction Standardization atlas.

[3] Yao Kai, Zhang Siyuan. *Standard Atlas of Safety and Civilization Construction of Urban Rail Transit Engineering in Hefei City*. Beijing: People's Transportation Press Co., Ltd., 2017.

[4] Xiong Guirong. A brief discussion on the standardization management of construction materials of Beijing Subway Line 8. *Municipal Technology*, 2016 (S1): 11–14.

[5] Yang Zhaolong. Construction management technology of rail transit engineering. *Engineering Construction and Design*, 2017 (11): 165–167.

[6] Li Wei. *Research on Green Construction of Metro Engineering*. Guangzhou: South China University of Technology, 2014.

[7] Han Jianqiang, He Jian, Luo Zhenhua. Research on noise pollution management of metro green construction. *Science and Technology Information*, 2015.

[8] Jia Jinlong. Analysis on the process of rebar factory construction for metro construction. *Construction Technology Development*, 2016 (12): 3–4.

[9] Ding Hongming, Wang Jian, Yang Shuai. Discussion on construction safety control system of urban rail transit construction. *China Science and Technology Information*, 2014 (07): 108–111.

[10] Gao Zhanyong. Discussion on underground pipeline protection in rail transit engineering construction. *Technology and Innovation*, 2014 (04): 54–55.

[11] Wu Xianguo, Chen Yuqing, Zhang Liao *et al.* The study on the pre-warning management and evaluation standard of the work safety supervision and evaluation of the geotechnical workers' safety supervision and evaluation. *Iron Road Engineering Paper*, 2013 (05): 107–111.

[12] Su Haibin. Research on *Green Construction and Cleaner Production of Rail Transit Engineering*. Suzhou: Suzhou Institute of Technology, 2014.

[13] Liu Tienan. Standardized construction of metro extension shaft and horizontal channel. *Dalian Daily*, 2010-07-20 (AO2).
[14] Wang Min. Preliminary exploration of the green construction evaluation system of Qingdao Metro project. *Journal of China Ocean University*, 2015.
[15] Zhang Shaokuan. Tunnel, the shallow tunnel construction green production facilities form a complete set of standardized research. *2012 in China Railway Tunnel Group Guaranteed High Quality Low Carbon Ring Communication Workshop on Engineering Construction Technology*, 2012, p. 4.
[16] Uranus. Environmental protection methods and practices in China during the construction of rail transit engineering. *Northern Environment*, 2011 (09): 42–43.
[17] Chen Hong. *Application of Construction Settlement Monitoring System in Xi 'an Metro*. Xi 'an: Xi 'an University of Science and Technology, 2012.
[18] The study on safety evaluation criteria of metro engineering construction. *Journal of Civil Engineering*, 2011 (11): 121–127.
[19] Zhang Qiufu. *Research on Construction Safety Management of Zhengzhou Metro Project*. Nanjing: Nanjing University of Technology, 2011.
[20] The Ministry of Housing and Urban–Rural Development of the People's Republic of China. *Gb50720-2011 Construction Site Fire Safety Technical Specifications*. Beijing: China Planning Press, 2011.

Index

Lightning Source UK Ltd.
Milton Keynes UK
UKHW022145260820
368879UK00003B/176

9 789813 278080